AF902245

THEORETICAL FOUNDATIONS AND APPLICATION OF PHOTONIC CRYSTALS

Edited by **Alexander Vakhrushev**

Theoretical Foundations and Application of Photonic Crystals
http://dx.doi.org/10.5772/intechopen.69145
Edited by Alexander Vakhrushev

Contributors

Renat Gainutdinov, Marat Khamadeev, Albert Akhmadeev, Myakzyum Salakhov, K Senthilnathan, Renju Rajan, P Ramesh Babu, Ngoc Diep Lai, Guoyan Dong, Liangshun Han, Sunnam Kim, Seiji Kurihara, Han Ye, Yumin Liu, Zhongyuan Yu, Eugene N. Odarenko, Viktoria Mizernik, Viktor Lysytsya, Alexander Shmat'Ko, Jingxia Wang, Junchao Liu, Yun Lai, Jie Luo

Published by InTech
Janeza Trdine 9, 51000 Rijeka, Croatia

© The Editor(s) and the Author(s) 2018
The moral rights of the editor(s) and the author(s) have been asserted.
All rights to the book as a whole are reserved by InTech. The book as a whole (compilation) cannot be reproduced, distributed or used for commercial or non-commercial purposes without InTech's written permission. Enquiries concerning the use of the book should be directed to InTech's rights and permissions department (permissions@intechopen.com).
Violations are liable to prosecution under the governing Copyright Law.

(cc) BY

Individual chapters of this publication are distributed under the terms of the Creative Commons Attribution 3.0 Unported License which permits commercial use, distribution and reproduction of the individual chapters, provided the original author(s) and source publication are appropriately acknowledged. More details and guidelines concerning content reuse and adaptation can be found at http://www.intechopen.com/copyright-policy.html.

Notice

Statements and opinions expressed in the chapters are these of the individual contributors and not necessarily those of the editors or publisher. No responsibility is accepted for the accuracy of information contained in the published chapters. The publisher assumes no responsibility for any damage or injury to persons or property arising out of the use of any materials, instructions, methods or ideas contained in the book.

Publishing Process Manager Edi Lipovic
Technical Editor SPi Global
Cover InTech Design team

First published April, 2018
Printed in Croatia
Legal deposit, Croatia: National and University Library in Zagreb

Additional hard copies can be obtained from orders@intechopen.com

Theoretical Foundations and Application of Photonic Crystals, Edited by Alexander Vakhrushev
p. cm.
Print ISBN 978-953-51-3961-4
Online ISBN 978-953-51-3962-1

We are IntechOpen, the first native scientific publisher of Open Access books

3,350+
Open access books available

108,000+
International authors and editors

115M+
Downloads

151
Countries delivered to

Our authors are among the
Top 1%
most cited scientists

12.2%
Contributors from top 500 universities

WEB OF SCIENCE™

Selection of our books indexed in the Book Citation Index
in Web of Science™ Core Collection (BKCI)

Interested in publishing with us?
Contact book.department@intechopen.com

Numbers displayed above are based on latest data collected.
For more information visit www.intechopen.com

For EU product safety concerns:
IN TECH d.o.o., Prolaz Marije Krucifikse Kozulić 3, 51000 Rijeka, Croatia,
info@intechopen.com or visit our website at intechopen.com.

Meet the editor

Alexander Vakhrushev received his PhD degree in Technical Sciences from Izhevsk Institute of Mechanical Engineering, Russia, in 1982, and his degree of Doctor of Sciences in Physics & Mathematics from the Institute of Continuous Media Mechanics of Ural Branch of Russian Academy of Sciences, Russia, in 2003. Now, he is the head of the department "Nanotechnology and Microsystems Technic" of Kalashnikov Izhevsk State Technical University and head of the department "Mechanics of Nanostructures" of the Institute of Mechanics, Udmurt Federal Research Center, Ural Branch of the Russian Academy of Sciences. He has over 400 publications to his name. His research interests include multiscale mathematical modeling of physical-chemical processes into the nanoheterosystems.

Contents

Preface

Photonic crystals (PCs) are structures characterized by periodic change in dielectric constant in the space. PCs allow to obtain the permitted and forbidden areas for photon energy, thanks to periodic change of the refractive index. Therefore, PC is an optical filter. PCs have a great perspective for use: the creation of light-conducting channels, spectral separation of channels, devices that protect against interference of light channels, etc. That is why, issues related to the methods of forming a PC, developing methods for calculating their parameters, and developing various devices based on PCs are of importance and need to be dealt with urgency.

This book is devoted to the description of research on these problems. Topics included in the book cover a wide range of research in the field of theoretical analysis and experimental investigation: the electromagnetic field in the photonic crystal, propagation of TM and TE waves in the one-dimensional gyrotropic magnetophotonic crystals, low one-photon absorption, one- and two-dimensional ultratransparent photonic crystals, colloidal assembly, photonic crystal application for development of all-optical computational system, design of novel photonic crystal devices as artificially engineered materials, design strategies for PC devices, self-organization of liquid crystalline nanostructures, and optical diodes and the basic formulas for their calculations.

A collection of 10 scientific chapters on theoretical and practical advances of PC research is presented. Each chapter represents a comprehensive study. According to the predominance of theoretical or practical research, the article is placed in two sections of the book. The first part of the book is entitled "Theory and Calculations," and the second part is called "Design and Application."

Part 1 "Theory and Calculations" includes:

Chapter 1 presents the complete theoretical investigations of modification of the electromagnetic field in the photonic crystal medium. The authors discuss the origins of the effect of the change in the electron mass caused by the modification of the electromagnetic interaction in a PC and its possible applications.

Chapter 2 presents the theoretical analysis of propagation of TM and TE waves in the one-dimensional gyrotropic magnetophotonic crystals with ferrite and plasma-like layers. Research methods are based on the transfer matrix method of F. Abeles for special cases of two-dimensional model of wave propagation in periodic layered media.

In Chapter 3, the crucial conditions for low one-photon absorption (LOPA)-based direct laser writing (DLW) were theoretically investigated and then experimentally demonstrated using a simple optical confocal microscope. Various 1D, 2D, and 3D submicrostructures were fabricated in different materials. Research methods are based on diffraction theory of the electromagnetic field of a tightly focused beam.

Chapter 4 presents the recent progress of one- and two-dimensional ultratransparent photonic crystals. Research methods are based on the ray optics and a nonlocal effective medium theory for the homogenization of PC. Many experimental results are presented, and a comparison with simulation results is given.

Chapter 5 presents the summarization about the influence of the substrate wettability on colloidal assembly. Basic assembly principle for colloidal crystals and several examples for the understanding of the influence of wettability on colloidal assembly are described.

Part 2 "Design and Application" includes:

Chapter 6 presents a short overview about photonic crystal application for development of all-optical computational system. A new paradigm is developed for optical computation using photonic crystals. The information presented in this chapter can be considered as an introduction to the application of photonic crystals for an all-optical computational system with logic gates, memory units, and networks, which can be constructed using these structures.

Chapter 7 presents the complete investigations of the design of novel photonic crystal devices as artificially engineered materials.

The author demonstrates that the synthetic periodic dielectric materials can exhibit various anomalous transmission properties, such as negative refraction, self-focusing, zero phase delay, or effective-zero-index properties, which are determined by the characteristics of their band structures and equal frequency contours.

Chapter 8 presents a short description of the fundamentals of 1D photonic crystals and the overview of the applications of these crystals in silicon photonics. Special attention of the author is given to providing insight into the design strategies for these devices.

Chapter 9 presents the review of photoswitching of photonic band gap (PBG) based on phase transition and orientation behavior of liquid crystalline materials (LCs) and self-organization of LC nanostructures. PBG tuning of 3D photonic crystals and self-organized LC nanostructures is considered.

Chapter 10 presents the review of the different kinds of schemes for reciprocal and nonreciprocal optical diodes based on two-dimensional PC. The authors describe main parameters of the different schemes of optical diodes and the basic formulas for their calculations.

This book will be useful for engineers, technologists, researchers, and postgraduate students interested in the research, design, fabrication processes, and applications of photonic crystals.

I would like to express my appreciation to all the contributors of this book. My special thanks go to the Publishing Process Manager, **Ms. Renata Sliva,** and other staff of InTechOpen publishing for their kind support and great efforts in bringing the book to completion.

Prof. Alexander Vakhrushev
Head of Department "Mechanics of Nanostructures"
Institute of Mechanics, Udmurt Federal Research Center
Ural Branch of the Russian Academy of Sciences, Russia

Head of Department "Nanotechnology and Microsystems"
Kalashnikov Izhevsk State Technical University, Russia

Section 1

Theory and Calculation

Modification of the Electromagnetic Field in the Photonic Crystal Medium and New Ways of Applying the Photonic Band Gap Materials

Renat Gainutdinov, Marat Khamadeev,
Albert Akhmadeev and Myakzyum Salakhov

Additional information is available at the end of the chapter

http://dx.doi.org/10.5772/intechopen.71367

Abstract

Photonic crystals (PCs) are periodic systems that consist of dielectrics with different refractive indices. Photonic crystals have many potential technological applications. These applications are mainly based on the photonic bang gap effect. However the band gap is not only effect that follows from the periodic changing of the refractive index in the photonic crystal. The periodic change of the photon-matter interaction in photonic crystal medium gives rise to the fact that the mass of an electron in the photonic crystal must differ from its mass in vacuum. Anisotropy of a photonic crystal results in the dependence of the electromagnetic mass correction on the orientation of the electron momentum in a photonic crystal. This orientation dependence in turn gives rise to the significant correction to the transition frequencies in an atom placed in air voids of a photonic crystal. These corrections are shown to be comparable to the atomic optical frequencies. This effect allows one to control the structure of the atomic energy levels and hence to control resonance processes. It can serve as the basis for new line spectrum sources. The effect provides new ways of realization of quantum interference between decay channels that can be important for quantum information science.

Keywords: photonic crystals, electron mass, anisotropic vacuum, electromagnetic field, Lamb shift

1. Introduction

Photonic crystals (PCs) are a major field of research having many potential applications [1–15]. These applications are mainly based on the photonic bang gap effect in the photonic crystal. In Ref. [16], it has been shown that a strong modification of the electromagnetic interaction in photonic crystals results in the fact that the electron mass changes its value. Actually in this

INTECH
open science | open minds

© 2018 The Author(s). Licensee InTech. This chapter is distributed under the terms of the Creative Commons Attribution License (http://creativecommons.org/licenses/by/3.0), which permits unrestricted use, distribution, and reproduction in any medium, provided the original work is properly cited.

case, we deal with a quantum electrodynamical (QED) effect that does not manifest itself in the free space. In fact, the interaction of an electron with its own radiation field gives rise to a contribution to its physical mass m_{ph} known as the electromagnetic mass of the electron m_{em}. Nonrenormalizable ultraviolet divergences do not allow one to calculate the electron electromagnetic mass. However, fortunately, only physical mass m_{ph} is observable, and hence m_{em} can be included into it. On the other hand, the modification of the electromagnetic interaction in PC medium gives rise to a correction to the electromagnetic mass m_{em}. This correction δm_{pc} cannot be hidden in the physical mass of the electron and hence is an observable. Thus in PC medium, the novel observable δm_{pc} comes into play. A remarkable feature of δm_{pc} is its dependence on the orientation of the electron momentum in a PC, and this dependence gives rise to significant corrections to the transition frequencies in an atom placed in air voids of a photonic crystal, being comparable to the ordinary atomic frequencies. Such an effect is a consequence of the fact that in the case of atoms in the PC medium, the most contribution comes from the self-energy of electrons associated with mass correction m_{em}^{pc} rather than from the self-energy of atoms associated with the Lamb shift being the QED corrections to the nucleus-electrons coupling. In this chapter, we discuss the origins of the effect of the change in the electron mass caused by the modification of the electromagnetic interaction in a PC and its possible applications.

2. Lamb shift in hydrogen atom in the free space

The processes of the interaction of charged particles with their own radiation field play the important role in the modern physics. These processes give rise to the fact that actually we deal with the particles dressed by a cloud consisting of virtual particles (photons, electron-positron pairs, and so on). In the case of electrons or muons bound to an atomic nucleus, the self-interaction results in the Lamb shift of the atomic energy levels. The results of the recent measurements of the Lamb shift in muonic hydrogen [17, 18] have allowed to determine the value of the root-mean-square charge radius of the proton r_p which is 4% smaller than the radius determined by electron-proton experiments [19, 20] and precision spectroscopy of the ordinary atomic hydrogen [21–27]. This discrepancy known as the "proton radius puzzle" has not been explained yet. Solving the puzzle may require new insights into the problem of the description of the self-energy of the electron and the Lamb shift.

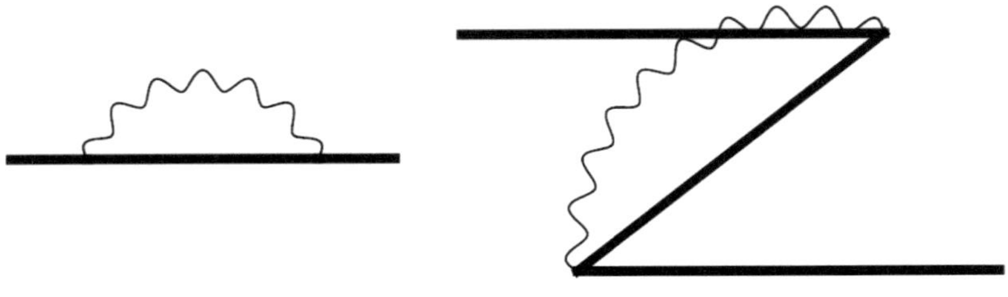

Figure 1. The time-ordered diagrams describing the dominant contribution to the Lamb shift. The thick line denotes the electron (positron) propagating in the Coulomb field; the wavy line denotes emission and reabsorption of a virtual photon.

The Lamb shift consists of the self-energy and vacuum polarization contributions. The modification of the vacuum polarization contribution in the PC medium is negligible, and for this reason, we will focus only on the self-energy one. At leading order self-energy of the electron, which is bound in a hydrogen-like atom, is defined by the process in which a photon is emitted and then is reabsorbed by the electron or positron. This process is described by the time-ordered diagrams in **Figure 1**.

In quantum electrodynamics the corresponding contribution to the Lamb shift in hydrogen-like atoms is given by the term that appears in the second-order perturbation theory and in the Furry picture can be written as

$$\Delta E_{L,n} = \langle n|H_I \frac{1}{E_n^{(0)} - H_0^F} H_I|n\rangle, \tag{1}$$

where H_0^F is the unperturbed Dirac-Coulomb Hamiltonian in the Furry picture ($H_0^F|n\rangle = E_n^{(0)}|n\rangle$), $|n\rangle$ is an atomic state, and

$$H = H_0 + \int d^3x H_I(t = 0, \mathbf{x}), \tag{2}$$

with $H_I(t, \mathbf{x})$ being the interaction Hamiltonian density:

$$H_I(t, \mathbf{x}) = \frac{e}{2} A_\mu(t, \mathbf{x}) \left[\overline{\Psi}(t, \mathbf{x}), \gamma^\mu \Psi(t, \mathbf{x}) \right]. \tag{3}$$

Here $\Psi(x)$ is the Dirac field in the Furry picture. Usually the contributions to the Lamb shift (1) are separated into the low and high energy parts. For the reasons explained bellow, we will focus on the low-energy part of the shift [28]:

$$\Delta E_{L,n}^{<} = \frac{2\pi\alpha}{3m_e^2} \int_0^\Lambda \frac{d^3k}{2|\mathbf{k}|(2\pi)^3} \sum_m \frac{|\langle n|\mathbf{p}|m\rangle|^2}{E_n - |\mathbf{k}| - E_m}, \tag{4}$$

where $\mathbf{p}$ is the operator of the electron momentum and the cutoff Λ limits the energies of virtual photons in the processes of their emission and reabsorption. The cutoff must be much less than typical electron momenta but much larger than the atomic binding energies:

$$(Z\alpha)^2 m_e << \Lambda << (Z\alpha) m_e. \tag{5}$$

Here and below the natural unit system is used, where $\hbar = c = \varepsilon_0 = 1$. This is the reason why one can use the nonrelativistic Hamiltonian:

$$H = \frac{1}{2m_e} [\mathbf{p} - e\mathbf{A}]^2 \tag{6}$$

instead of the Hamiltonian defined in Eqs. (2) and (3). Eq. (4) can be rewritten in the form:

$$\Delta E_{L,n} = -\frac{\Delta m_e^<}{2m_e^2}\langle n|\mathbf{p}^2|n\rangle + \Delta E_L^<, \tag{7}$$

where

$$\Delta m_e^< = \frac{\alpha}{p^2\pi^2}\sum_{\lambda=1}^{2}\int_0^\Lambda \frac{d^3k}{2|\mathbf{k}|^2}|\mathbf{p}\cdot\varepsilon_\lambda(\mathbf{k})|^2 \tag{8}$$

is the low-energy electron mass correction caused by its self-interaction [29]. It should be noted that $\Delta E_{L,n}$ does not contain a term describing the electromagnetic correction to the electron mass. This is the result of making use of the nonrelativistic Hamiltonian (6), and for this reason, the mass correction is extracted from the first term on the right-hand part of Eq. (7) describing the correction to the kinetic energy. Thus, in this case the electromagnetic mass correction is regarded to be included into the physical mass of the electron. The first term on the right-hand part of Eq. (7) must be also included into the physical mass. In this way we arrive at the ordinary expression for the low-energy Lamb shift in hydrogen-like atoms:

$$\Delta E_{L,m}^< = \frac{\alpha}{6\pi^2 m_e^2}\sum_m \int_0^\Lambda \frac{d^3k}{2|\mathbf{k}|^2}\frac{|\langle n|\mathbf{p}|m\rangle|^2}{E_n - |\mathbf{k}| - E_m}(E_n - E_m). \tag{9}$$

Adding to $\Delta E_L^<$ the high energy contribution [28]:

$$\Delta E_L^> = \frac{4\alpha}{3}\frac{Z\alpha}{m^2}|\Psi_{nlmj}(0)|^2\left(\ln\frac{m_e}{2\Lambda} + \frac{11}{24} - \frac{1}{5}\right), \tag{10}$$

where n, l, m, j, and $\Psi_{nlmj}(x)$ being, respectively, the main quantum number, orbital quantum number, magnetic quantum number, inner quantum number, and the wave function, we get the expression to the total Lamb shift of the energies of the states of the hydrogen-like atoms. In the S-state it reads

$$\Delta E_{L,n} = \frac{4\alpha(Z\alpha)^4}{3\pi n^2}\left(\ln\frac{m_e}{2\overline{E}} + \frac{11}{24} - \frac{1}{5}\right)m_e + o(Z\alpha)^4, \tag{11}$$

where $\overline{E} = \alpha^2 m_e$.

3. The Lamb shift in atoms placed in a PC

Investigation of the Lamb shift in hydrogen atom placed in a PC attracts much attention for a long time since the Lamb shift is (historically and in practice) the most important phenomenon of quantum electrodynamics. Interestingly, the calculation results obtained in different works differed strongly in order of magnitude, and the significance of interaction with vacuum, depending on which model of the dispersion of a photon in a photonic crystal, was used.

The first attempt was made by John and Wang [4] by using the solution of the scalar wave equation in one dimension. Thus, the photon dispersion relation was chosen to be isotropic and satisfy the transcendental equation:

$$4n\cos(kL) = (1+n)^2\cos[(2na+b)\omega_k] - (1-n)^2\cos[(2na-b)\omega_k]. \tag{12}$$

Using this dispersion relation, the authors predicted anomalous Lamb shift affecting the odd-parity $2P_{1/2}$ state and not the even-parity $2S_{1/2}$. Magnitude of the effect makes it detectable using microwave. The fact that the anomalous Lamb shift of the $2P_{1/2}$ state is larger than the ordinary Lamb shift of the $2S_{1/2}$ state originates from the dimension of the phase space occupied by band edge photons of vanishing group velocity. John and Wang overestimated this phase space by assuming that $d\omega_k/dk$ vanishes over the entire sphere $|\mathbf{k}| = \pi/L$. At the same time for the case of real photonic crystals, the shift was expected to be comparable to the ordinary Lamb shift of the $2S_{1/2}$ level.

The authors of work [30] noted that a real photonic crystal in general has an anisotropic structure in momentum space and a three-dimensional dispersion relation is required because the density of states (DOS) in isotropic or one-dimensional case has a singularity near band edge. In this study the atomic transition frequency ω is assumed to be near the band edge ω_c, and the dispersion relation was approximated by the expression.

$$\omega_k = \omega_c + A\left|\mathbf{k} - \mathbf{k}_0^i\right|^2, \tag{13}$$

where A is a model-dependent constant and $\mathbf{k}_0^i$ is a finite set of symmetrically placed points leading to a three-dimensional band structure. Using this model the Schrödinger equation was solved, and analytical expression for the Lamb shift was obtained. The value of the Lamb shift turned out to be smaller than that for a hydrogen atom in an ordinary vacuum. Authors explained this result by the fact that the DOS in the photonic crystals with three-dimensional dispersion relations is much lower than that in the ordinary vacuum. This result is also very different from that from the one-dimensional case where DOS has a singularity or from the two-dimensional case where DOS has a sudden jump.

In paper [31] all previous approaches to calculate Lamb shift in photonic crystal were criticized, because they are basically scalar. Authors of this work demonstrated the rigorous solution of the problem of calculation of the Lamb shift in atomic hydrogen in a 3D photonic crystal and showed that the presence of a photonic band gap (PBG) at optical wavelengths can hardly change the Lamb shift. The correction to the energy of electronic state $|m\rangle$ was calculated in the second order of perturbation theory. The quantization of EM fields in a 3D photonic crystal was made by expanding the EM fields in a set of eigenmodes (Bloch states). These states can be solved numerically by means of a plane-wave expansion method. Finally, it was given an expression for the energy shift containing the local density of states (LDOS):

$$\Delta E = \frac{e^2\hbar}{u_0^2 m_e^2}\sum_n E_{nm}\left|p_{mn}\right|^2\int_0^\infty d\omega \frac{\rho(\omega,\mathbf{r})}{\omega^3(E_{nm}+\hbar\omega)}, \tag{14}$$

with $\rho(\omega, \mathbf{r})$ being LDOS:

$$\rho(\omega, \mathbf{r}) = \frac{u_0^2 c^2}{2\hbar\varepsilon_0 (2\pi)^3 \varepsilon^2(\mathbf{r})} \sum_n \int_{BZ} d^3k \frac{|\nabla \times \mathbf{H}_{n\mathbf{k}}(\mathbf{r})|^2}{3\omega_{n\mathbf{k}}} \delta(\omega - \omega_{n\mathbf{k}}), \tag{15}$$

where u_0 is dipole moment, $\varepsilon(\mathbf{r})$ is dielectric constant function, and $\mathbf{H}_{n\mathbf{k}}(\mathbf{r})$ is magnetic field distribution of the Bloch states with energy $\hbar\omega_{n\mathbf{k}}$. The authors estimated the magnitude of the Lamb shift and concluded that PBG at optical wavelengths will not cause an appreciable variation to the energy-level shift induced by self-interaction for different atom positions and different variations of the LDOS.

Vats with colleagues used the anisotropic band edge model and pseudogap model to calculate the Lamb shift in an atom placed in photonic crystal [32]. In the first case near the band edge, dispersion relation (13) was used and corresponding DOS derived. Calculated Lamb shift was an order of magnitude larger than the free space Lamb shift. Then authors treated the case of a pseudogap, for which the stop band does not extend over all propagation directions, thus resulting in a suppression of the DOS rather than the formation of a full PBG:

$$N(\omega) = \omega^2 \left[1 - h\exp\left(-\frac{(\omega - \omega_0)^2}{\Gamma^2} \right) \right]. \tag{16}$$

Here, h and Γ are parameters describing the depth and width of the pseudogap, respectively, and ω_0 is the central frequency of the pseudogap. Vats with coworkers concluded that for a sufficiently strong pseudogap, the maximal value of Lamb shift may be on the order of 15% of the free space value.

The authors of work [33] using method of Green functions developed a general formalism for calculating the Lamb shift in multilevel atoms. The radiative correction to the bound level l is determined by the expression

$$\omega - \omega_l = \sum_j \frac{\alpha_{lj}}{2\pi} (\omega - \omega_j) \beta(\mathbf{r}, \omega - \omega_j), \tag{17}$$

where

$$\alpha_{lj} = \frac{e^2 \left| \mathbf{p}_{lj} \right|^2}{3\pi m_e^2 \varepsilon_0 \hbar c^3} \tag{18}$$

is the relative linewidth of the atomic radiation from the l state to the j state in vacuum

$$\beta(\mathbf{r}, \omega - \omega_j) = P \int_0^{m_e c^2/\hbar} d\omega' \frac{g(\mathbf{r}, \omega')}{(\omega - \omega_j - \omega')\omega'}. \tag{19}$$

The function $g(\mathbf{r}, \omega)$ is the local spectral response function (LSRF) proportional to the photon LDOS:

$$g(\mathbf{r},\omega)=\frac{c^3V_{pc}}{2\pi\omega}\sum_n\int_{BZ}d^3k|\mathbf{E}_{n\mathbf{k}}(\mathbf{r})|^2\delta(\omega-\omega_{n\mathbf{k}})\tag{20}$$

with V_{pc} being the PC volume and $\mathbf{E}_{n\mathbf{k}}(\mathbf{r})$ being the electromagnetic eigenmodes. Authors revealed that in a 3D PC, real photons make a dominant contribution to the value of the Lamb shift, while the contribution from interaction with virtual photons is small. This differs significantly from the free space case. It was shown that the PC structure can lead to a giant Lamb shift, that is, up to two orders of magnitude larger than that for an ordinary vacuum [34]. The Lamb shift is sensitive to both the position of an atom in PCs and the transition frequency of the related excited level.

4. Photonic crystal medium corrections to the electron rest mass

For a long time in investigations of QED effects in the PC medium, researches focused on study of the Lamb shift in hydrogen atom placed in a PC. In all the listed studies, the subtraction of the modified by PC medium self-energy of the free electron from the modified self-energy of the bound electron was used. This procedure was correct, if this self-energy could be included into the electron physical mass. However this is not the case, because the electromagnetic mass of the electron in a PC differs from that in the free space and cannot be hidden in the physical mass. In fact

$$m_{em}^{pc}=m_{em}+\delta m_{pc}\tag{21}$$

and hence the total electron mass m_e^{pc} in a PC is

$$m_e^{pc}=m_e+\delta m_{pc}.\tag{22}$$

Thus, the modification of the interaction of the electron with its own radiation field in the PC medium results in the change in its mass. Let us now determine the mass correction δm_{pc}. For this we have to generalize our analysis of the electron self-energy to the case where it is in the PC medium. It is natural to start from determining of a quantized vector potential of electromagnetic field inside PC. It could be made by taking into account that photon states in periodic dielectric media have Bloch structure. Photonic Bloch states $|\mathbf{k}n\rangle$ can be obtained by means of the plane-wave expansion method [35]. By introducing the operators $\hat{a}^+_{\mathbf{k}n}$ and $\hat{a}_{\mathbf{k}n}$ that describe the creation and annihilation of the photon in the state $|\mathbf{k}n\rangle$, respectively ($\hat{a}^{\mathbf{k}n+}|0\rangle=|\mathbf{k}n\rangle$ and $\hat{a}_{\mathbf{k}n}|\mathbf{k}n\rangle=|0\rangle$), we can construct a modified vector potential:

$$\mathbf{A}_{pc}(\mathbf{r},t)=\sum_{\mathbf{k}n}\left[\mathbf{A}_{\mathbf{k}n}(\mathbf{r})\hat{a}_{\mathbf{k}n}e^{-i\omega_{\mathbf{k}n}t}+\mathbf{A}^*_{\mathbf{k}n}(\mathbf{r})\hat{a}^+_{\mathbf{k}n}e^{i\omega_{\mathbf{k}n}t}\right],\tag{23}$$

where $\mathbf{A}_{\mathbf{k}n}(\mathbf{r})=\sqrt{1/V\omega_{\mathbf{k}n}}\mathbf{E}_{\mathbf{k}n}(\mathbf{r})$ with $\mathbf{E}_{\mathbf{k}n}(\mathbf{r})$ being the Bloch eigenfunctions satisfying the following orthonormality condition:

$$\int_V d^3r\varepsilon(\mathbf{r})\mathbf{E}_{\mathbf{k}n}(\mathbf{r})\mathbf{E}^*_{\mathbf{k}'n'}(\mathbf{r}) = V\delta_{\mathbf{k}\mathbf{k}'}\delta_{nn'}. \tag{24}$$

Using vector potential (23) we can define nonrelativistic interaction Hamiltonian in the form

$$H_I^{pc} = -\frac{e}{m_e}\mathbf{p}\cdot\mathbf{A}_{pc}. \tag{25}$$

The matrix element $\langle\mathbf{p}';\mathbf{k},n|H_I^{pc}|\mathbf{p}\rangle$ of this Hamiltonian can be represented in the form

$$\langle\mathbf{p}';\mathbf{k},n|H_I^{pc}|\mathbf{p}\rangle = -\frac{e}{m_e}\int d^3r\Psi^*_{\mathbf{p}'}(\mathbf{r})(-i\nabla_\mathbf{r}\mathbf{A}_{\mathbf{k}n}(\mathbf{r}))\Psi_\mathbf{p}(\mathbf{r}) = \frac{e}{m_eV^{3/2}\sqrt{\omega_{\mathbf{k}n}}}\int d^3re^{-i\mathbf{p}'\mathbf{r}}(i\nabla_\mathbf{r}\mathbf{E}_{\mathbf{k}n}(\mathbf{r}))e^{i\mathbf{p}\mathbf{r}} \tag{26}$$

with $\Psi_\mathbf{p}(\mathbf{r})$ being the normalized wave function of the electron state $\Psi_\mathbf{p}(\mathbf{r})=\langle\mathbf{r}|\,\mathbf{p}\rangle$. Here we have taken into account that $\Psi_\mathbf{p} = e^{i\mathbf{p}\mathbf{r}}/\sqrt{V}$ for $\mathbf{r}\in V$ and $\Psi_\mathbf{p}=0$ for $\mathbf{r}\notin V$. Taking also into account that $\mathbf{E}_{\mathbf{k}n}(\mathbf{r})$ can be expanded as

$$\mathbf{E}_{\mathbf{k}n}(\mathbf{r}) = \sum_\mathbf{G}\mathbf{E}_{\mathbf{k}n}(\mathbf{G})e^{i(\mathbf{k}+\mathbf{G})\cdot\mathbf{r}} \tag{27}$$

with $\mathbf{G}$ being the reciprocal lattice vector of the photonic crystal ($\mathbf{G}=N_1\mathbf{b}_1+N_2\mathbf{b}_2+N_3\mathbf{b}_3$ where $\mathbf{b}_i$ is the basis vector of a reciprocal lattice), for $\langle\mathbf{p}';\mathbf{k},n|H_I^{pc}|\mathbf{p}\rangle$ we get

$$\langle\mathbf{p}';\mathbf{k},n|H_I^{pc}|\mathbf{p}\rangle = -\frac{e}{m_e}\frac{1}{\sqrt{V\omega_{\mathbf{k}n}}}\sum_\mathbf{G}\mathbf{p}\cdot\mathbf{E}_{\mathbf{k}n}(\mathbf{G})\delta_{\mathbf{p},\mathbf{q}} \tag{28}$$

with $\mathbf{q}=\mathbf{p}'+\mathbf{k}+\mathbf{G}$. For $\langle\mathbf{p}|H_I^{pc}|\mathbf{p}';\mathbf{k},n\rangle$ we find

$$\langle\mathbf{p}|H_I^{pc}|\mathbf{p}';\mathbf{k},n\rangle = -\frac{e}{m_e}\frac{1}{\sqrt{V\omega_{\mathbf{k}n}}}\sum_\mathbf{G}\mathbf{p}\cdot\mathbf{E}^*_{\mathbf{k}n}(\mathbf{G})\delta_{\mathbf{p},\mathbf{q}}. \tag{29}$$

Using these matrix elements, we can determine the mass correction δm_{pc} as a difference of the electromagnetic masses in PC and free space:

$$\delta m_{pc} = -\frac{2e^2}{\mathbf{p}^2V}\left(\sum_\mathbf{G}\sum_{\mathbf{k}n}\frac{1}{\omega_{\mathbf{k}n}}\frac{|\mathbf{p}\cdot\mathbf{E}_{\mathbf{k}n}(\mathbf{G})|^2}{\frac{\mathbf{p}^2}{2m_e}-\frac{(\mathbf{p}-\mathbf{k}-\mathbf{G})^2}{2m_e}-\omega_{\mathbf{k}n}} - \sum_\mathbf{k}\sum_{\lambda=1}^{2}\frac{1}{2|\mathbf{k}|}\frac{|\mathbf{p}\cdot\varepsilon_\lambda(\mathbf{k})|^2}{\frac{\mathbf{p}^2}{2m_e}-\frac{(\mathbf{p}-\mathbf{k})^2}{2m_e}-|\mathbf{k}|}\right). \tag{30}$$

It should be noted that this expression has a natural cutoff because dielectric constant vanishes at higher optical energies. Taking into account that electron momentum is much higher that photon momentum, Eq. (30) can be rewritten in the form

$$\delta m_{pc} = \frac{2e^2}{\mathbf{p}^2V}\left(\sum_\mathbf{G}\sum_{\mathbf{k}n}\frac{|\mathbf{p}\cdot\mathbf{E}_{\mathbf{k}n}(\mathbf{G})|^2}{\omega^2_{\mathbf{k}n}} - \sum_\mathbf{k}\sum_{\lambda=1}^{2}\frac{|\mathbf{p}\cdot\varepsilon_\lambda(\mathbf{k})|^2}{2\mathbf{k}^2}\right). \tag{31}$$

Now in the expression of δm_{pc}, we can replace the discreet sums by integrals:

$$\int d^3k \sum_{\mathbf{k}n} \to \frac{V}{(2\pi)^3} \sum_n \int d^3k, \quad \sum_{\mathbf{k}} \to \frac{V}{(2\pi)^3} \int d^3k. \tag{32}$$

In this way we get

$$\delta m_{pc} = \frac{\alpha}{\pi^2} \left[\sum_n \int_{FBZ} \frac{d^3k}{\omega_{\mathbf{k}n}^2} \sum_{\mathbf{G}} \left| \frac{\mathbf{p}}{|\mathbf{p}|} \cdot \mathbf{E}_{\mathbf{k}n}(\mathbf{G}) \right|^2 - \int \frac{d^3k}{2\mathbf{k}^2} \sum_{\lambda=1}^{2} \left| \frac{\mathbf{p}}{|\mathbf{p}|} \cdot \varepsilon_\lambda(\mathbf{k}) \right|^2 \right]. \tag{33}$$

Accounting for the effect under study for the energy of an electron in the PC medium, we get

$$E_p = m_e + \delta m_e \left(\widehat{\mathbf{p}/|\mathbf{p}|} \right) + \frac{\mathbf{p}^2}{2m_e} \delta m_{pc} \left(\widehat{\mathbf{p}/|\mathbf{p}|} \right) + o\left(\frac{|\mathbf{p}|^4}{m_e^4} \right) m_e. \tag{34}$$

In dealing with an atomic electron, we have also to take into account that its momentum should be described by the momentum operator $\widehat{\mathbf{p}}$ and hence δm_{pc} should be described by the corresponding operator $\delta m_{pc}\left(\widehat{\mathbf{p}/|\mathbf{p}|}\right)$. In this way we arrive at the following expression for the mass correction ΔE_i^{mc} to energies of the states of a hydrogen-like atom:

$$\Delta E_i^{mc} = \left\langle i \left| \delta m_e \left(\widehat{\mathbf{p}/|\mathbf{p}|} \right) \right| i \right\rangle + \left\langle i \left| \frac{|\mathbf{p}|^2}{2m_e} \delta m_e \left(\widehat{\mathbf{p}/|\mathbf{p}|} \right) \right| i \right\rangle + o(\cdots) m_e. \tag{35}$$

In the ground S-state $|S\rangle$, the mean value of the operator $\delta m_{pc}\left(\widehat{\mathbf{p}/|\mathbf{p}|}\right)$ is

$$\langle \delta m_{pc} \rangle_S = \frac{4\alpha}{3\pi} \int d\omega \frac{N(\omega) - \omega^2}{\omega^2}, \tag{36}$$

where $N(\omega) = N_{DOS}(\omega) D(\omega)$ and $N_{DOS}(\omega)$ is the photon density of states

$$N_{DOS}(\omega) = \frac{1}{4\pi} \sum_n \int_{FBZ} d^3k \delta(\omega - \omega_{\mathbf{k}n}) \tag{37}$$

and

$$D(\omega) = \sum_{\mathbf{G}} |\mathbf{E}_{\mathbf{k}n}(\mathbf{G})|^2_{|\omega_{\mathbf{k}n} = \omega}. \tag{38}$$

The function $N(\omega)$ is closely associated with DOS of the PC. The exact calculation of this function is challenging for 3D PC; therefore we will use a model having the form

$$N(\omega) = \omega^2 n_{eff}^3 \left[1 - h\exp\left(-\frac{(\omega - \omega_0)^2}{\sigma^2} \right) \right] F(\omega), \tag{39}$$

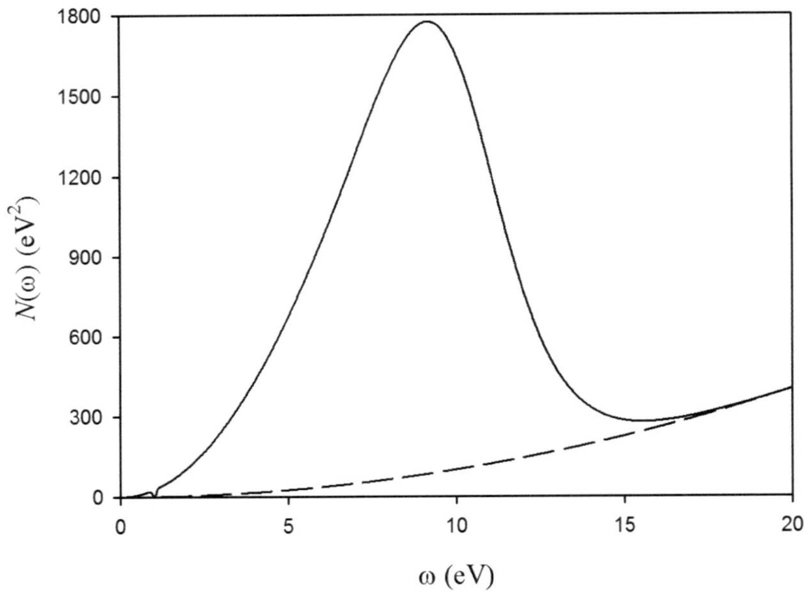

Figure 2. The model $N(\omega)$ determined by the Eq. (39) with $n_{eff}=3$, $h=0.96$, $\sigma=0.07$ eV, $\mu=15$ eV, $\tau=0.01$ eV, and $\omega_0=1$ eV. Dashed line denotes the free space DOS.

where the factor $F(\omega) = n_{eff}^{-3} + \left(1 - n_{eff}^{-3}\right)/\left(\exp\{(\omega-\mu)/\tau\}+1\right)$ with $n_{eff} \equiv \sqrt{\bar{\varepsilon}}$. $\bar{\varepsilon} = \varepsilon \cdot f +$ $(1-f)$ is an average dielectric constant with ε being the dielectric constant of the host material and f being the dielectric fraction in the PC. This model can recapture the existence of photonic band gap, optical density of dielectric host of PC sample, and the fact that at high enough photon energies, $N(\omega)$ must approach the free space DOS (**Figure 2**). For the parameters which were used in **Figure 2**, our calculations have given $\langle \delta m_{pc} \rangle_S = 2.4 \cdot 10^{-6} m_e$.

Let us now consider the effect of the change in the electron mass on the energies of the atomic states and the transition frequencies. Here we will restrict ourselves to the hydrogen-like atoms. In the free space, the energy of the atoms in the state $|a\rangle = |n, j, l, m\rangle$ is the sum of the energy derived from the solution of the Dirac equation $E^D = m_e R_{nj}$ and the Lamb shift of the energy in this state:

$$E_{njl} = m_e R_{nj} + \Delta E_{L,a}, \tag{40}$$

where

$$R_{nj} = \left[1 + \left(\frac{Z\alpha}{n - (j+1/2) + \sqrt{(j+1/2)^2 - \alpha^2}}\right)^2\right]^{-1/2} \tag{41}$$

and $\Delta E_{L,a}$ is the Lamb shift of the energy of the state $|a\rangle$. The transition frequency between this state and the state $|b\rangle = |n', j', l', m\rangle$ is given by

http://dx.doi.org/10.5772/intechopen.71367

$$\omega_{ab} = m_e\left(R_{nj} - R_{n'j'}\right) + \left(\Delta E_{L,a} - \Delta E_{L,b}\right). \tag{42}$$

When the atom is placed in the void of a PC, the transition frequencies ω_{ab}^{PC} are modified as follows:

$$\omega_{ab}^{PC} = \left(m_e + \left\langle a\left|\delta m_e^{PC}\left(\widehat{\mathbf{p}/|\mathbf{p}|}\right)\right|a\right\rangle\right)R_{nj} + \Delta E_{L,a}^{PC} - \left(m_e + \left\langle b\left|\delta m_e^{PC}\left(\widehat{\mathbf{p}/|\mathbf{p}|}\right)\right|b\right\rangle\right)R_{n'j'} - \Delta E_{L,b}^{PC}. \tag{43}$$

In the case when the atom is light, Eq. (43) is reduced to the following expression:

$$\begin{aligned}\omega_{ab}^{PC} &= \left\langle a\left|\delta m_e^{PC}\left(\widehat{\mathbf{p}/|\mathbf{p}|}\right)\right|a\right\rangle\left(1 - \frac{Z^2\alpha^2}{2n^2}\right) - \left\langle b\left|\delta m_e^{PC}\left(\widehat{\mathbf{p}/|\mathbf{p}|}\right)\right|b\right\rangle^2\left(1 - \frac{Z^2\alpha^2}{2n^2}\right)\\ &+ \frac{m_e Z\alpha^2}{2}\left(\frac{1}{n^2} - \frac{1}{(n')^2}\right) + \Delta E_{L,a}^{PC} - \Delta E_{L,b}^{PC} = \frac{m_e Z\alpha^2}{2}\left(\frac{1}{n^2} - \frac{1}{(n')^2}\right) + \Delta\omega_{ab}^{PC} + o\left(Z^2\alpha^4\right)m_e,\end{aligned} \tag{44}$$

where $\Delta\omega_{ab}^{PC}$ is the correction to the transition frequency in the PC medium given by

$$\Delta\omega_{ab}^{PC} = \left\langle a\left|\delta m_e^{PC}\left(\widehat{\mathbf{p}/|\mathbf{p}|}\right)\right|a\right\rangle - \left\langle b\left|\delta m_e^{PC}\left(\widehat{\mathbf{p}/|\mathbf{p}|}\right)\right|b\right\rangle. \tag{45}$$

As we have shown, the values of the mass corrections $\left\langle i\left|\delta m_{PC}\left(\widehat{\mathbf{p}/|\mathbf{p}|}\right)\right|i\right\rangle$ may be of order $10^{-6}m_e$, and hence the corrections to the transition frequencies are comparable to the atomic optical frequencies.

5. Experimental observation

Since spectra remain discrete when the PC medium affects interaction between atoms and their own emission fields, it would be logical to conduct an experiment in which we could observe this effect. This could be accomplished by observing the classical spectra of the atoms in the gas phase, pumped into PC cavities. From a theoretical point of view, it would be best to conduct the experiment with hydrogen atoms, since they are the simplest physical system. However, the handling of atomic hydrogen creates a number of technical difficulties; from a practical point of view, the best candidates for the role of such atoms are those of the noble gases, for example, helium. With respect to the requirements for a PC sample, it is first of all obvious that it should have cavities that are sufficiently interconnected to ensure the possibility of pumping gas. Second, the material of the PC sample should have the largest possible refractive index in the widest possible range of energies, since the effect depends strongly on the optical contrast [36]. Finally, the larger the amount of material filling the PC volume, the greater the effect. At the same time, the cavities must remain large enough to meet the condition that the atoms are free to move. It should be noted that an increase in the relative shift of the lines $\delta\omega/\omega$, along with an increase in the main quantum number n, is unequivocal confirmation of the effect, since the predicted shift of the lines does not depend on it.

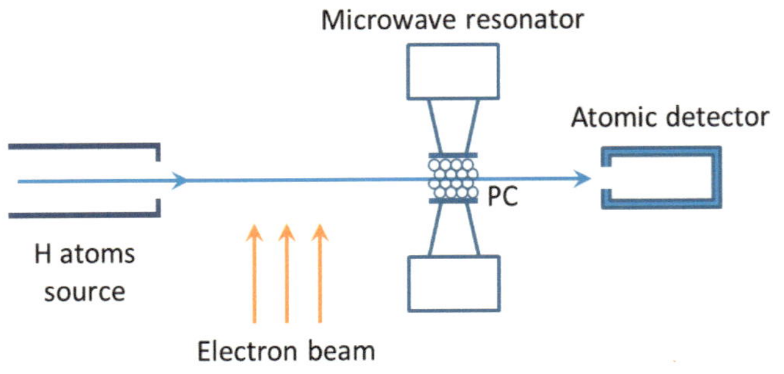

Figure 3. Scheme of modified Lamb shift experiment.

As a simple and natural way to confirm the considered effect, we propose to use a modified experiment to measure Lamb shift in hydrogen atom placed in the voids of photonic crystal (**Figure 3**). In the experiment the hydrogen atoms are exposed to electromagnetic radiation of a certain frequency, and if this frequency corresponds to the difference between the $2S_{1/2}$ and $2P_{1/2}$ energy levels (~1058 MHz without PC medium), no excited atoms will reach the detector. However taking into account the influence of the photonic crystal on the energy levels of atoms the Lamb shift will differ from 1058 MHz, the excited atoms will appear on the detector which will confirm the effect. Then we can measure new Lamb shift by adjusting the frequency of electromagnetic radiation.

There are a number of technical issues which need to be resolved. First, all exposed atoms must be within the photonic crystal, that is, electromagnetic radiation should be concentrated in a relatively small volume of a photonic crystal using antennas or waveguides. Second, as already noted, there are many requirements to the sample of photonic crystal, including the quality of the structure and possibility of free passage of hydrogen atoms through the PC medium. To solve the last one, we propose to use photonic crystals with inverted opal structure [37], the volume fraction of air voids which is approximately 74%. Such structures are fabricated from synthetic opals by filling voids between spherical particles with any desired material. After that initial particles are removed leaving a framework with spherical air voids. However, the resulting structures have a large number of defects and have significant limitations in linear dimensions.

6. Prospects of applications of the effect

The most surprising feature of the effect under study is that the electromagnetic mass of the electron comes into play when an atom is placed in the voids of a PC. There are no analogs of such QED effect in the free space. The correction to the electromagnetic mass caused by the modification of the electromagnetic interaction strongly changes the character of processes of the spontaneous emission and the absorption of atoms placed in the PC medium, and this can open up new possibilities for applying PCs. For the first time, one can change the transitions on the value comparable to the ordinary atomic transition frequencies. This effect becomes possible due to the dependence of the electromagnetic mass correction on the orientation of the electron momentum in the PC medium. This provides a way to control the structure of the atomic energy levels. In this way, in particular, light sources with the line spectrum of a new type could be developed.

The line spectrum sources such as He-Ne laser play an important role in physics and technologies. However, the corresponding transition frequencies in the optical range are limited. The mass-change effect under study opens possibilities to tune the energy levels of He and Ne and, as a consequence, to increase the slope efficiency. It allows one to create the new He-Ne-like lasers.

One of the most perspective applications of the effect is a realization of quantum interference. Quantum interference among different decay channels caused by the anisotropic vacuum is the major field of research. Several ways have been proposed to create the anisotropy and to provide interference between atomic levels in such materials as negative-index materials [38–43], metasurfaces [44], hyperbolic metamaterials [45], metallic nanostructures [46, 47], topological insulators [48], and external fields [49–51]. The possibility for making use of anisotropy in the PC medium for these purposes has been investigated in Refs. [52–55]. The authors of the listed papers based themselves on the idea voiced by Agarwal [56] who pointed that the anisotropy of the vacuum can cause the quantum interference between nearest energy levels (e.g., Zeeman sublevels) having orthogonal dipole moments. The effect of the change in the electron mass in a PC provides new possibilities to create conditions at which quantum interference becomes possible via nonradiative transitions between atomic levels with breaking the strict selection rules.

7. Conclusion

The QED effects on which we focused play an important role in the physics of PCs. The Lamb shift in atoms that is one of the most important phenomena of the QED becomes larger in the case when the atom is placed in the air voids of PCs. But what is especially important is that in the case where an atom is placed in the artificial PC medium, we face a phenomenon that does not manifest itself in vacuum. This phenomenon consists in the fact that the part of the electromagnetic mass m_{em} of the electron that together with the bare mass m_0 constitutes the physical mass $m_{ph}=m_0+m_{em}$ becomes observable. In vacuum only m_{ph} is observable. This fact is used in the renormalization theory that is of the central importance in QED. The renormalization procedure implies that the terms describing the self-energy of the free electron should be removed from any expressions describing the processes in which the electron takes place. This is an explanation of the fact that for long time, this subtraction procedure was used in describing the Lamb shift in atoms placed in PCs despite that the electromagnetic interaction in the PC medium is significantly modified. The correction $\delta m_{pc} = m_{em}^{pc} - m_{em}$ to the electromagnetic mass of the electron caused by this modification cannot be hidden in the physical mass of the electron and for this reason is observable. Thus, in the case of the artificial PC medium, the electromagnetic mass (more precisely its part δm_{pc}) comes into play. In contrast to the Lamb shift that is relatively small correction to the atomic energy levels, the electromagnetic mass correction δm_{pc} can have a significant effect not only on the energy levels of atoms placed in the PC medium but also on the physical processes in these atoms. The key point is that δm_{pc} depends on the orientation of the electron momentum in a PC and actually is an operator $\delta m_{pc}\left(\widehat{\mathbf{p}/|\mathbf{p}|}\right)$ whose diagonal matrix elements determine the corrections to the transition frequencies that are comparable to the atomic frequencies in the free space. The nondiagonal matrix elements determine nonradiative transitions between the states with breaking the strict selection rules. These transitions give rise to the quantum interference between the different decay channels. The possibility of controlling these quantum-interference processes can be important for quantum information science.

Author details

Renat Gainutdinov[1]*, Marat Khamadeev[1,2], Albert Akhmadeev[1,2] and Myakzyum Salakhov[1,2]

*Address all correspondence to: renat.gainutdinov@kpfu.ru

1 Institute of Physics, Kazan Federal University, Kazan, Russian Federation

2 Tatarstan Academy of Sciences, Kazan, Russian Federation

References

[1] Yablonovitch E. Inhibited spontaneous emission in solid-state physics and electronics. Physical Review Letters. 1987;**58**(20):2059-2062. DOI: 10.1103/PhysRevLett.58.2059

[2] John S. Strong localization of photons in certain disordered dielectric superlattices. Physical Review Letters. 1987;**58**(23):2486-2489. DOI: 10.1103/PhysRevLett.58.2486

[3] Quang T, Woldeyohannes M, John S, Agarwal GS. Coherent control of spontaneous emission near a photonic band edge: A single-atom optical memory device. Physical Review Letters. 1997;**79**(26):5238-5241. DOI: 10.1103/PhysRevLett.79.5238

[4] John S, Wang J. Quantum electrodynamics near a photonic band gap: Photon bound states and dressed atoms. Physical Review Letters. 1990;**64**(20):2418-2421. DOI: 10.1103/PhysRevLett.64.2418

[5] John S, Wang J. Quantum optics of localized light in a photonic band gap. Physical Review B. 1991;**43**(16):12772-12789. DOI: 10.1103/PhysRevB.43.12772

[6] Bay S, Lambropoulos P, Mølmer K. Fluorescence into flat and structured radiation continua: An atomic density matrix without a master equation. Physical Review Letters. 1997;**79**(14):2654-2657. DOI: 10.1103/PhysRevLett.79.2654

[7] Bay S, Lambropoulos P, Mølmer K. Atom-atom interaction in strongly modified reservoirs. Physical Review A. 1997;**55**(2):1485-1496. DOI: 10.1103/PhysRevA.55.1485

[8] Busch K, Vats N, John S, Sanders BC. Radiating dipoles in photonic crystals. Physical Review E. 2000;**62**:4251-4260. DOI: 10.1103/PhysRevE.62.4251

[9] Zhu S-Y, Chen H, Huang H. Quantum interference effects in spontaneous emission from an atom embedded in a photonic band gap structure. Physical Review Letters. 1997;**79**(2):205-208. DOI: 10.1103/PhysRevLett.79.205

[10] John S, Quang T. Localization of superradiance near a photonic band gap. Physical Review Letters. 1995;**74**(17):3419-3422. DOI: 10.1103/PhysRevLett.74.3419

[11] Lopez C. Materials aspects of photonic crystals. Advanced Materials. 2003;**15**:1679-1704. DOI: 10.1002/adma.200300386

[12] Mateos L, Molina P, Galisteo J, López C, Bausá LE, Ramírez MO. Simultaneous generation of second to fifth harmonic conical beams in a two dimensional nonlinear photonic crystal. Optics Express. 2012;**20**:29940-29948. DOI: 10.1364/OE.20.029940

[13] Pinto AMR, Lopez-Amo M. Photonic crystal fibers for sensing applications. Journal of Sensors. 2012;**2012**:598178. DOI: 10.1155/2012/598178

[14] Tuyen LD, Liu AC, Huang C-C, Tsai PC, Lin JH, C-W W, Chau L-K, Yang TS, Minh LQ, Kan H-C, Hsu CC. Doubly resonant surface-enhanced Raman scattering on gold nanorod decorated inverse opal photonic crystals. Optics Express. 2012;**20**:29266-29275. DOI: 10.1364/OE.20.029266

[15] Callahan DM, Munday JN, Atwater HA. Solar cell light trapping beyond the ray optic limit. Nano Letters. 2012;**12**:214-218. DOI: 10.1021/nl203351k

[16] Gainutdinov RK, Khamadeev MA, Salakhov MK. Electron rest mass and energy levels of atoms in the photonic crystal medium. Physical Review A. 2012;**85**(5):053836(1-7). DOI: 10.1103/PhysRevA.85.053836

[17] Pohl R, Gilman R, Miller GA, Pachucki K. Muonic hydrogen and the proton radius puzzle. Annual Review of Nuclear and Particle Science. 2013;**63**:175-204. DOI: 10.1146/annurev-nucl-102212-170627

[18] Carlson CE. The proton radius puzzle. Progress in Particle and Nuclear Physics. 2045;**82**:59-77. DOI: 10.1016/j.ppnp.2015.01.002

[19] Sick I. On the rms-radius of the proton. Physics Letters B. 2003;**576**:62-67. DOI: 10.1016/j.physletb.2003.09.092

[20] Bernauer JC et al. (A1 collaboration). High-precision determination of the electric and magnetic form factors of the proton. Physical Review Letters. 2010;**105**:242001. DOI: 10.1103/PhysRevLett.105.242001

[21] Niering M, Holzwarth R, Reichert J, Pokasov P, Udem T, Weitz M, Hänsch TW, Lemonde P, Santarelli G, Abgrall M, Laurent P, Salomon C, Clairon A. Measurement of the hydrogen 1S-2S transition frequency by phase coherent comparison with a microwave Cesium fountain clock. Physical Review Letters. 2000;**84**:5496. DOI: 10.1103/PhysRevLett.84.5496

[22] Fischer M, Kolachevsky N, Zimmermann M, Holzwarth R, Udem T, Hänsch TW, Abgrall M, Grünert J, Maksimovic I, Bize S, Marion H, Pereira dos Santos F, Lemonde P, Santarelli G, Laurent P, Clairon A, Salomon C, Haas M, Jentschura UD, Keitel CH. New limits on the drift of fundamental constants from laboratory measurements. Physical Review Letters. 2004;**92**:230802. DOI: 10.1103/PhysRevLett.92.230802

[23] Parthey CG, Matveev A, Alnis J, Bernhard B, Beyer A, Holzwarth R, Maistrou A, Pohl R, Predehl K, Udem T, Wilken T, Kolachevsky N, Abgrall M, Rovera D, Salomon C, Laurent P, Hänsch TW. Improved measurement of the hydrogen 1S-2S transition frequency. Physical Review Letters. 2011;**107**:203001. DOI: 10.1103/PhysRevLett.107.203001

[24] Schwob C, Jozefowski L, de Beauvoir B, Hilico L, Nez F, Julien L, Biraben F, Acef O, Zondy JJ, Clairon A. Optical frequency measurement of the 2S-12D transitions in hydrogen and deuterium: Rydberg constant and lamb shift determinations. Physical Review Letters. 1999;**82**:4960. DOI: 10.1103/PhysRevLett.82.4960

[25] de Beauvoir B, Nez F, Julien L, Cagnac B, Biraben F, Touahri D, Hilico L, Acef O, Clairon A, Zondy JJ. Absolute frequency measurement of the 2S-8S/D transitions in hydrogen and deuterium: New determination of the Rydberg constant. Physical Review Letters. 1997;**78**:440. DOI: 10.1103/PhysRevLett.78.440

[26] de Beauvoir B, Schwob C, Acef O, Jozefowski L, Hilico L, Nez F, Julien L, Clairon A, Biraben F. Metrology of the hydrogen and deuterium atoms: Determination of the Rydberg constant and lamb shifts. European Physical Journal D: Atomic, Molecular, Optical and Plasma Physics. 2000;**12**:61-93. DOI: 10.1007/s100530070043

[27] Arnoult O, Nez F, Julien L, Biraben F. Optical frequency measurement of the 1S-3S two-photon transition in hydrogen. European Physical Journal D: Atomic, Molecular, Optical and Plasma Physics. 2010;**60**:243-256. DOI: 10.1140/epjd/e2010-00249-6

[28] Bjorken JD, Drell SD. Relativistic Quantum Mechanics. Vol. 1. New York: McGraw-Hill; 1964. 311 p

[29] Schweber SS. An Introduction to Relativistic Quantum Field Theory. New York: Dover; 2005. 928 p

[30] Zhu S-Y, Yang Y, Chen H, Zheng H, Zubairy MS. Spontaneous radiation and lamb shift in three-dimensional photonic crystals. Physical Review Letters. 2000;**84**(10):2136-2139. DOI: 10.1103/PhysRevLett.84.2136

[31] Li Z-Y, Xia Y. Optical photonic band gaps and the lamb shift. Physical Review. B, Condensed Matter and Materials physics. 2001;**63**(12):121305(1-4). DOI: 10.1103/PhysRevB.63.121305

[32] Vats N, John S, Busch K. Theory of fluorescence in photonic crystals. Physical Review A. 2002;**65**(4):43808(1-13). DOI: 10.1103/PhysRevA.65.043808

[33] Wang X-H, Kivshar YS, Gu B-Y. Giant lamb shift in photonic crystals. Physical Review Letters. 2004;**93**(7):073901(1-4). DOI: 10.1103/PhysRevLett93.073901

[34] Wang X-H, B-Y G, Kivshar YS. Spontaneous emission and lame shift in photonic crystals. Science and Technology of Advanced Materials. 2005;**6**(7):814-822. DOI: 10.1016/j.stam.2005.06.025

[35] Sakoda K. Optical Properties of Photonic Crystals. Berlin: Springer; 2001. 227 p. DOI: 10.1007/978-3-662-14324-7

[36] Gainutdinov RK, Salakhov MK, Khamadeev MA. Optical contrast of a photonic crystal and the self energy shift of the energy levels of atoms. Bulletin of the Russian Academy of Sciences: Physics. 2012;**76**(12):1301-1305. DOI: 10.3103/S106287381212012X

[37] Khokhlov PE, Sinitskii AS, Tretyakov YD. Inverse photonic crystals based on silica. Doklady Chemistry. 2006;**408**(1):61-64. DOI: 10.1134/S0012500806050028

[38] Li G-X, Evers J, Keitel CH. Spontaneous emission interference in negative-refractive-index waveguides. Physical Review. B, Condensed Matter and Materials Physics. 2009;**80**(4):045102 (1-7). DOI: 10.1103/PhysRevB.80.045102

[39] Zeng X, Xu J, Yang Y. Spontaneous emission interference enhancement with a μ-negative metamaterial slab. Physical Review A. 2011;**84**(3):033834(1-5). DOI: 10.1103/PhysRevA.84.033834

[40] Yang Y, Xu J, Chen H, Zhu S. Quantum interference enhancement with left-handed materials. Physical Review Letters. 2008;**100**(4):043601(1-4). DOI: 10.1103/PhysRevLett.100.043601

[41] Xu J, Chang S, Yang Y, Al-Amri M. Casimir-polder force on a v-type three-level atom near a structure containing left-handed materials. Physical Review A. 2016;**93**(1):012514 (1-8). DOI: 10.1103/PhysRevA.93.012514

[42] Zeng X, Yu M, Wang D, Xu J, Yang Y. Spontaneous emission spectrum of a V-type three-level atom in a fabry-perot cavity containing left-handed materials. Journal of the Optical Society of America B: Optical Physics. 2011;**28**(9):2253-2259. DOI: 10.1364/JOSAB.28.002253

[43] Xu J-P, Yang Y-P. Quantum interference of V-type three-level atom in structures made of left-handed materials and mirrors. Physical Review A. 2010;**81**(1):013816(1-8). DOI: 10.1103/PhysRevA.81.013816

[44] Jha PK, Ni X, Wu C, Wang Y, Zhang X. Metasurface-enabled remote quantum interference. Physical Review Letters. 2015;**115**(2):025501(1-5). DOI: 10.1103/PhysRevLett.115.025501

[45] Sun L, Jiang C. Quantum interference in a single anisotropic quantum dot near hyperbolic metamaterials. Optics Express. 2016;**24**(7):7719-7727. DOI: 10.1364/OE.24.007719

[46] Yannopapas V, Paspalakis E, Vitanov NV. Plasmon-induced enhancement of quantum interference near metallic nanostructures. Physical Review Letters. 2009;**103**(6):063602(1-4). DOI: 10.1103/PhysRevLett.103.063602

[47] Evangelou S, Yannopapas V, Paspalakis E. Simulating quantum interference in spontaneous decay near plasmonic nanostructures: Population dynamics. Physical Review A. 2011;**83**(5):055805(1-4). DOI: 10.1103/PhysRevA.83.055805

[48] Fang W, Yang Z-X, Li G-X. Quantum properties of an atom in a cavity constructed by topological insulators. Journal of Physics B: Atomic, Molecular and Optical Physics. 2015;**48**(24):245504(1-10). DOI: 10.1088/0953-4075/48/24/245504

[49] Mortezapour A, Saleh A, Mahmoudi M. Birefringence enhancement via quantum interference in the presence of a static magnetic field. Laser Physics. 2013;**23**(6):065201(1-7). DOI: 10.1088/1054-660X/23/6/065201

[50] Ficek Z, Swain S. Simulating quantum interference in a three-level system with perpendicular transition dipole moments. Physical Review A. 2004;**69**(2):023401(1-10). DOI: 10.1103/PhysRevA.69.023401

[51] Tan H-T, Xia H-X, Li G-X. Quantum interference and phase-dependent fluorescence spectrum of a four-level atom with antiparallel dipole moments. Journal of Physics B: Atomic, Molecular and Optical Physics. 2009;**42**(12):125502(1-6). DOI: 10.1088/0953-4075/42/12/125502

[52] J-P X, Wang L-G, Yang Y-P, Lin Q, Zhu S-Y. Quantum interference between two orthogonal transitions of an atom in one-dimensional photonic crystals. Optics Letters. 2008;**33**(17):2005-2007. DOI: 10.1364/OL.33.002005

[53] Li G-X, Li F-L, Zhu S-Y. Quantum interference between decay channels of a three-level atom in a multilayer dielectric medium. Physical Review A. 2001;**64**(1):013819(1-10). DOI: 10.1103/PhysRevA.64.013819

[54] Zhang HZ, Tang SH, Dong P, He J. Spontaneous emission spectrum from a V-type three-level atom in a double-band photonic crystal. Journal of Optics B: Quantum and Semiclassical Optics. 2002;**4**(5):300-307. DOI: 10.1088/1464-4266/4/5/312

[55] Zhang HZ, Tang SH, Dong P, He J. Quantum interference in spontaneous emission of an atom embedded in a double-band photonic crystal. Physical Review A. 2002;**65**(6A):063802(1-8). DOI: 10.1103/PhysRevA.65.063802

[56] Agarwal GS. Anisotropic vacuum-induced interference in decay channels. Physical Review Letters. 2000;**84**(24):5500-5503. DOI: 10.1103/PhysRevLett.84.5500

Chapter 2

Anomalous Transmission Properties Modulated by Photonic Crystal Bands

Guoyan Dong

Additional information is available at the end of the chapter

http://dx.doi.org/10.5772/intechopen.71403

Abstract

Photonic crystals (PhCs) can be utilized to control the propagation behaviors of light within a frequency band (i.e., conduction band or stop band) for their periodic arrangements of dielectric. Utilizing the effect of band gap one may introduce defects in a photonic crystal to limit or guide the electromagnetic wave with the frequency in stop gaps. Furthermore, based on synthetic periodic dielectric materials photonic crystals can exhibit various anomalous transmission properties, such as negative refraction, self-focusing, zero phase delay or effective-zero-index properties that are determined by the characteristics of their band structures and equal frequency contours (EFC). These extraordinary results contributing to the design of novel PhC devices and the development of PhC application are demonstrated in this chapter.

Keywords: photonic crystal, photonic band gap, negative refraction, zero phase delay, Dirac-like point

1. Introduction

Photonic crystals (PhCs) are structures in which the dielectric constant is periodically modulated on a length scale comparable to the desired wavelength of operation [1, 2], and the resultant photonic dispersion may exhibit photonic band gaps (PBGs) and anomalous propagation behaviors which are useful in controlling light behavior according to different theoretical principles. Based on the PBG effect we may introduce a line defect in a photonic crystal to guide the electromagnetic waves with the frequency in stop gap [3]. The line defect is called a photonic crystal waveguide (PCW), which is very compact (with the typical width is around half-wavelength) and allows sharp bends without losses for its all-dielectric structure [4]. Many attempts have been made to fabricate materials with complete photonic

INTECH
open science | open minds

© 2018 The Author(s). Licensee InTech. This chapter is distributed under the terms of the Creative Commons Attribution License (http://creativecommons.org/licenses/by/3.0), which permits unrestricted use, distribution, and reproduction in any medium, provided the original work is properly cited. (cc) BY

band gaps (PBGs) at near-infrared [5] and visible frequencies [6], such as semiconductor microfabrication [7], self-assembly of colloidal particles [8], electron-beam lithography [9], multiphoton polymerization [10], holographic lithography (HL) [11], and so on. Among them, HL is a very promising technique for the inexpensive fabrication of high-quality two- and three-dimensional (2D and 3D) PhC templates with the unique advantages of one-step recording, the ability to obtain an inverse lattice by using a template, inexpensive volume recording and rapid prototyping. The PhCs formed by HL which usually have irregular "atoms" or columns. Since the PBG property of resultant structure varies with the shape of "atoms" or columns, thus the PBGs and propagation properties for PhCs made by HL will be different from those of regular structures.

Veselago predicted a kind of materials with negative refractive index in 1968 [12] which has attracted lots of attention in recent years. Such materials are generally referred to as left-handed materials (LHMs), double negative materials [13], or backward-wave media et al. [14], whose best known characteristic is to refract light in opposite direction. Shelby et al. have accomplished one of the first experiments to demonstrate the negative-index behavior [15]. However, the absorption loss of the metal limits its potential optical applications. Different from LHMs, PhCs made of periodic all-dielectric materials can exhibit an fascinating dispersion such as negative refraction and self-focusing properties which are determined by the properties of their band structures and equal frequency contours (EFC) [16, 17].

In this chapter, the maximal complete relative photonic band gaps of 22.1% and 25.1% are introduced how to be achieved in 2D- and 3D photonic crystals formed by HL method. The symmetry mismatch between the incident wave and the Bloch modes of PhC can be used to guide light efficiently. The unique features of negative refraction, dual-negative refraction, triple refraction phenomena, and special collimation effects of symmetrical positive-negative refraction have been verified by numerical simulations or experiments. Effective measurement method has been proposed to identify the Dirac-like point of finite PhC arrays accurately. A mechanism for generation of efficient zero phase delay of electromagnetic wave propagation based on wavefront modulation is investigated with parallel wavefronts (or phasefronts) extending along the direction of energy flow. The band structures of PBG are calculated by the plane-wave expansion method [18], and a finite-difference time-domain (FDTD) method [19] is used to calculate the transmission property of the guided mode. The method of wave-vector diagram is generally used to predict the properties of beam propagation in PhCs.

2. Characteristics of photonic band gap in 2D- and 3D-PhCs

2.1. Photonic band gap in 2D-PhCs

Since the shape and size of lattice columns in 2D case or atoms in 3D case are usually determined by the isointensity surfaces of the interference field, the columns or atoms formed by this way often are of irregular shapes [20, 21]. Consequently, the PBG property of resultant structure is closely related to the fabrication process. Therefore the complete PBG can be obtained and improved by proper designs of the shape and size of lattice cells [22]. Compared with the 3D case, the 2D PhCs are easier to fabricate and study theoretically for the fact that, the wave propagation of 2D PhCs can be investigated separately for two orthogonal polarizations known as TM and TE. They have important practical significance because they offer the possibility of guiding and manipulating light in planar defect circuits [23], photonic crystal fibers [24] and controlling polarization through their anisotropic band structures.

A kind of 2D 3-fold PhCs conforming to the hybrid triangular configuration formed by HL was proposed [20]. This kind of hybrid triangular lattice is formed by two sets of triangular lattices, as depicted in **Figure 1**, with the big dots indicating the triangular lattice sites with lattice constant a, while the small dots with the same lattice structure shifting $\sqrt{3}a/3$ in y direction. Different from the 6-fold rotational symmetry of regular triangular lattice structure, this hybrid structure has 3-fold rotational symmetry.

The basis vectors of the hybrid triangular lattice may be chosen as $\boldsymbol{a}_1 = a(1/2, \sqrt{3}/2)$ and $\boldsymbol{a}_2 = a(-1, 0)$. To produce this 2D hybrid triangular lattice holographically, we may use the following intensity of interference field,

$$I = 3 + \cos\left[\frac{2\pi}{a}\left(x - \frac{y}{\sqrt{3}}\right)\right] + \cos\left[\frac{2\pi}{a}\left(x + \frac{y}{\sqrt{3}}\right)\right] + \cos\left(\frac{4\pi}{\sqrt{3}a}y\right)$$
$$+c\left\{\cos\left[\frac{2\pi}{a}\left(x - \left(\frac{y}{\sqrt{3}} - \frac{1}{3}\right)\right)\right] + \cos\left[\frac{2\pi}{a}\left(x + \left(\frac{y}{\sqrt{3}} - \frac{1}{3}\right)\right)\right] + \cos\left(\frac{4\pi}{\sqrt{3}a}\left(y - \frac{\sqrt{3}}{3}\right)\right)\right\}, \tag{1}$$

where the first constant has no essential effect, which can be changed by adjusting the light intensity threshold I_t. The former and latter three cosine terms in this formula define the above mentioned two sets of triangular lattices, respectively. The constant c is a modulation coefficient which has a considerable effect on the final structure and thus the corresponding PBGs.

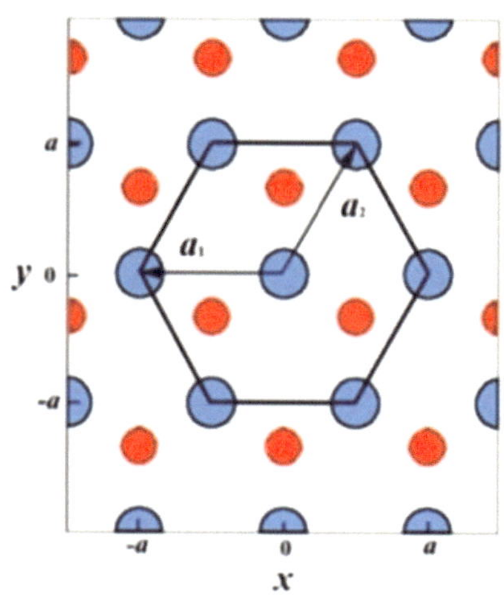

Figure 1. Hybrid triangular lattice with the big dots define a triangular lattice while the small dots compose a same one with a shift of $\sqrt{3}a/3$ in y direction.

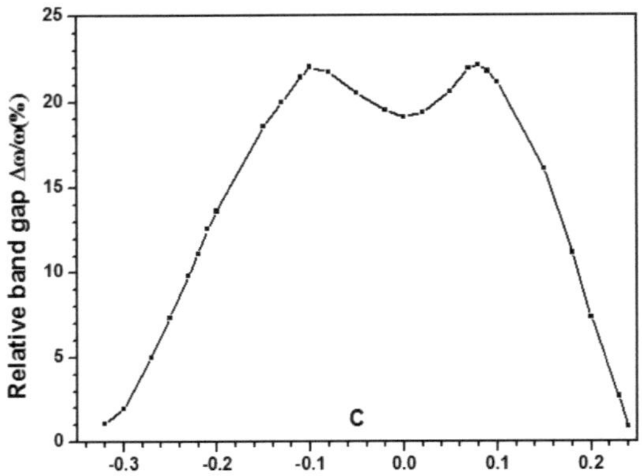

Figure 2. Relation between the value of c and the corresponding maximum relative band width when $\varepsilon = 13.6$. The two peaks of relative band gap occur at $c = -0.1$ and 0.08.

It is clear that the factors of c and I_t can be modulated to control the PBGs. For a given c, a certain lattice structure can be formed by washing away the region of $I < I_t$. If we fill this structure with a material of high refractive index, such as GaAs, and then remove the template, an inverse structure can be achieved as has been successfully demonstrated in Ref. [25]. **Figure 2** represents the relative gap size of optimized structures as a function of c for the dielectric constant contrast of 13.6 to 1 corresponding to GaAs in air. Obviously, the band gaps of PhCs made by HL may be widened from 18.9 to 22.1% by introducing irregularity of the columns and lowering the symmetry of the structure.

In addition, another kind of 2D 6-fold hybrid triangular configuration formed holographically is proposed [21], in which the complete PBGs can be found even with much lower dielectric constant ($\varepsilon = 3.8$). This 2D periodic structure is a hybrid triangular lattice combining two sets of triangular sublattices as depicted in **Figure 3**, where the red dots denote a triangular sublattice with lattice constant a, while the blue dots denote another set with the lattice constant of $a/\sqrt{3}$ which is rotated by an angle of 30° with respect to the former.

For the triangular structure formed holographically the complete PBGs always appear in inverse structures (air columns in dielectric material) with high dielectric constant instead of normal structures [26]. However, the complete PBG may be obtained for normal structure (dielectric column in air background) with lower dielectric contrast ($\varepsilon = 3.8$). **Figure 4** indicates the relations of relative band gap $\Delta\omega/\omega$ with maximum peak value to filling ratio f for different dielectric contrasts. Computations show that the peak value of optimum relative PBG not always augments with the increase of dielectric constant ε for normal structure, instead, the peak value reaches the maximum at about $\varepsilon = 8.9$, such as 9.9% with $f = 15.9\%$ for $\varepsilon = 8.9$ and $c = 1.2$, which is larger than the result (8.8%) of best designed pincushion columns with the same dielectric constant [27]. Specially, in this kind of normal structures, the required minimum permittivity to open a complete photonic band gap with $\Delta\omega/\omega > 1\%$ is near 3.8, which is much lower than the value of 6.4 in the case of pincushion columns [27] and lower than all the results of 2D photonic crystals ever reported before. In addition, the complete PBGs of this

http://dx.doi.org/10.5772/intechopen.71403

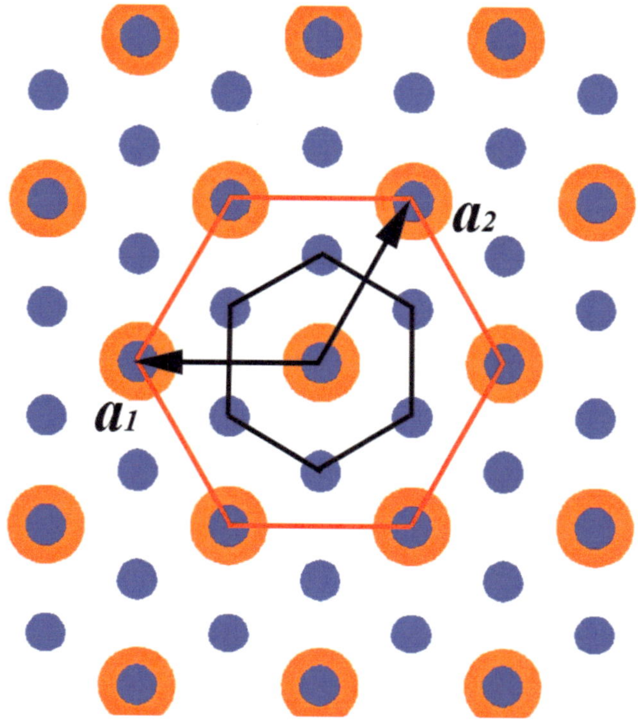

Figure 3. The hybrid triangular lattice, where red dots define triangular sublattice, blue dots define another group with lattice constant $a/\sqrt{3}$ rotated 30°.

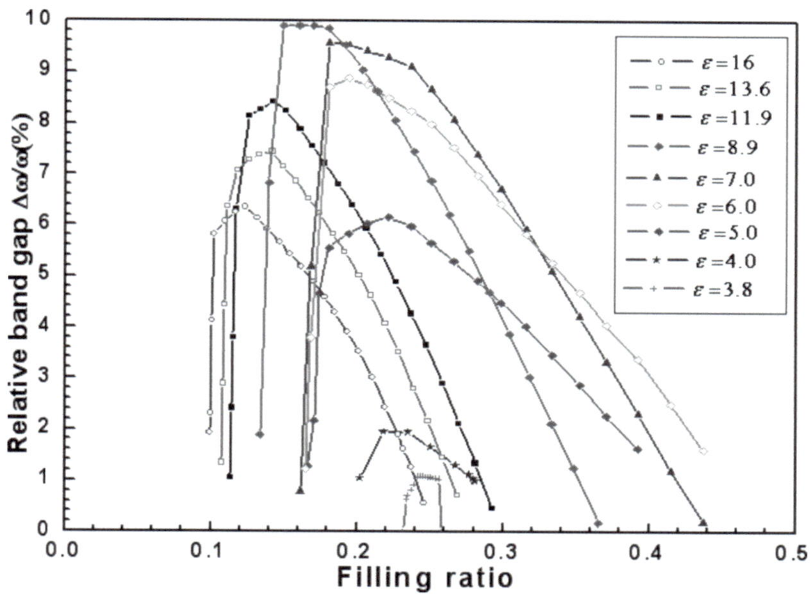

Figure 4. The relations of relative band gaps to different filling ratios *f*.

normal structure exist over wide ranges of coefficient c, filling ratio f and dielectric constant, which can relax the experimental conditions greatly.

2.2. Photonic band gap in 3D-PhCs

An important way to make 3D PhCs by HL is the interference of four umbrellalike beams (IFUB) where three ambient beams (A-beams) form the same apex angle θ with a central beam (C-beam), as shown in **Figure 5** [28, 29]. The possible lattices that IFUB may produce and the polarization optimization in the formation of different lattices have been discussed in the previous works [30, 31]. As a special case of IFUB, the symmetric umbrella configuration, where any two of the three A-beams also form the same angle, is widely used in HL since it can be conveniently realized with the use of a diffraction beam splitter (DBS). The continuous increasing of apex angle θ leads to continuous variation of primitive vectors, reciprocal vectors and the irreducible Brillouin zone of the resultant structure.

The interference of four noncoplanar plane waves of the same frequency will result in an intensity distribution

$$I = I_0 + 2\Delta I(\boldsymbol{r}), \tag{2}$$

where

$$I_0 = \sum_{j=1}^{4} E_j^2, \quad \Delta I(\boldsymbol{r}) = \sum_{i<j} E_i E_j e_{ij} \cos\left[(\boldsymbol{K}_i - \boldsymbol{K}_j)\cdot \boldsymbol{r} + \phi_{i0} - \phi_{j0} + \sigma_{ij}\right], \tag{3}$$

I_0 is the background intensity, ΔI is the spatial variation of intensity, φ_{j0} is initial phase of the jth wave, $e_{ij} = |\boldsymbol{e}_i \cdot \boldsymbol{e}_j^*|$, and $\sigma_{ij} = \arg(\boldsymbol{e}_i \cdot \boldsymbol{e}_j^*)$. The four wave vectors are expressed as functions of the apex angle θ,

$$\begin{aligned} \boldsymbol{K}_1 &= (2\pi/\lambda)\big((\sqrt{3}/2)\sin\theta, -(1/2)\sin\theta, \cos\theta\big), \\ \boldsymbol{K}_2 &= (2\pi/\lambda)\big(-(\sqrt{3}/2)\sin\theta, -(1/2)\sin\theta, \cos\theta\big), \\ \boldsymbol{K}_3 &= (2\pi/\lambda)(0, \sin\theta, \cos\theta), \\ \boldsymbol{K}_4 &= (2\pi/\lambda)(0, 0, 1). \end{aligned} \tag{4}$$

With the angle θ increasing from zero to 180°, the shape of Brillouin zone changes from a small hexangular plane spreading out on xy plane to a simple cubic at $\theta = 70.53°$ and finally to a long hexangular pillar along z axis. **Figure 6** gives the lattice structure and the Brillouin zone calculated at the special angles θ. The special case of $\theta = 38.94°$ corresponds to the fcc structure which belongs to space group No. 166 ($R\bar{3}m$), the structure of θ = 109.47° is approximate to Schwarz's triply periodic minimal D surface [32].

The plane-wave expansion method [18] was used to study the PBG properties of structures of this kind and search for the corresponding optimum volume filling ratio f yielding maximum relative PBGs. **Figure 7** shows the maximum size of the relative PBG for the optimum filling ratio at different apex angle θ. The complete PBGs exist over a very

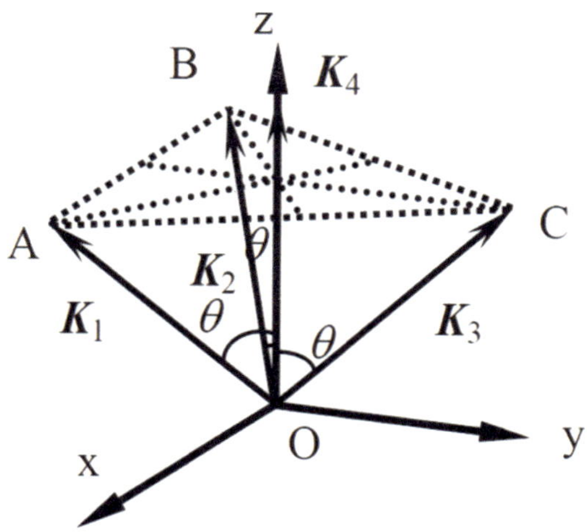

Figure 5. Symmetric umbrella recording geometry and the coordinate system used for calculations.

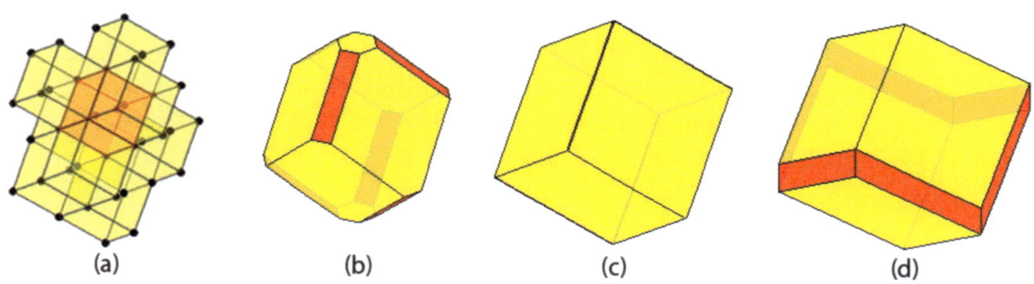

Figure 6. (a) The lattice structure at $\theta = 70.53°$ and (b) the irreducible Brillouin zones of rhombohedral structures for $\theta = 60°$, (c) $\theta = 70.53°$ and (d) $\theta = 80°$.

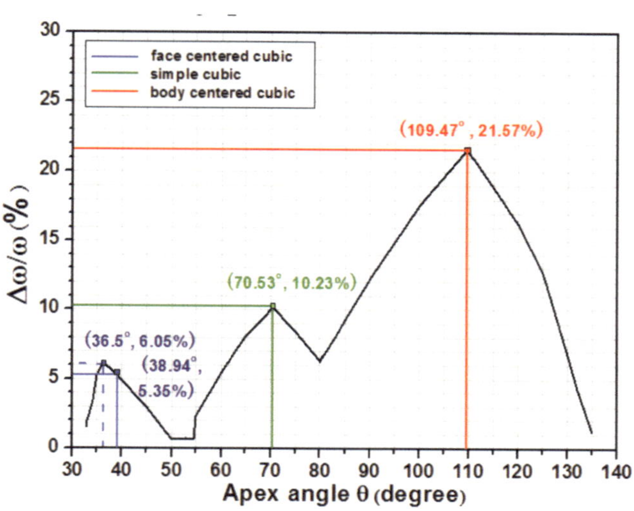

Figure 7. Relative band gap of optimized structures as function of apex angle for $33° < \theta < 135°$ when $\varepsilon = 11.9$.

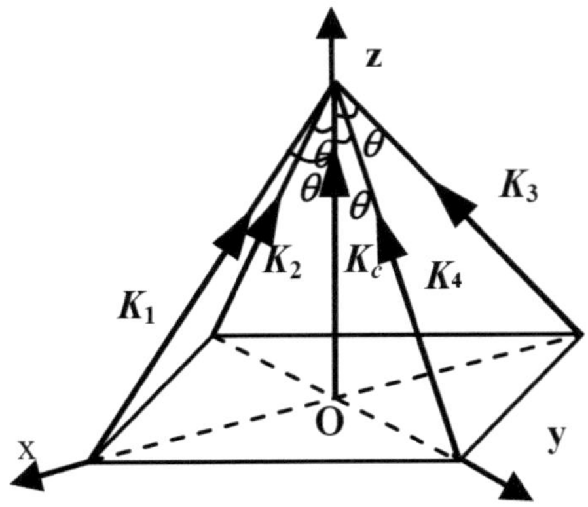

Figure 8. Symmetric umbrella recording geometry and the coordinate system used for calculations.

wide range of apex angle from 33 to 135° except the narrow region from 49 to 55°, and the resultant PBGs of simple cubic-sc, face centered cubic-fcc and body centered cubic-bcc accord with the previous works nicely [32]. The relative PBG width is a continuous function of apex angle with three peaks, and the PBG gradually decreases to zero at $\theta = 33°$ and 135°. The first peak of 6.05% corresponds to the rhombohedral structure appears at $\theta = 36.5°$. The second of 10.23% corresponding to the sc structure occurs at $\theta = 70.53°$; and the third one of 21.57% corresponds to the bcc structure at $\theta = 109.47°$.

As mentioned above, the fcc lattice can be obtained by the interference of one central beam and three ambient beams symmetrically scattered around the former, but the structure made in this geometry has only quite a narrow PBG of 5.35%. An alternative beam design was proposed to fabricate the fcc lattice with a large complete PBG, but it requires four beams incident from two opposite surfaces of a sample [33], making it difficult to realize in practice. So a five-beam symmetric umbrella configuration is proposed to make 3D PhCs with large complete PBGs. The proposed recording geometry of five-beam symmetric umbrella configuration is shown in **Figure 8**, where the central beam (C-beam) is set along the z direction, while the four ambient beams (A-beams) are in the *yoz* and *xoz* planes, respectively, to form the same apex angle θ. The above mentioned five wave vectors can be expressed as,

$$\begin{aligned} &\boldsymbol{K}_1 = (2\pi/\lambda)(-\sin\theta,\ 0,\ \cos\theta), \quad \boldsymbol{K}_2 = (2\pi/\lambda)(0,\ \sin\theta,\ \cos\theta), \\ &\boldsymbol{K}_3 = (2\pi/\lambda)(\sin\theta,\ 0,\ \cos\theta), \quad \boldsymbol{K}_4 = (2\pi/\lambda)(0,\ -\sin\theta,\ \cos\theta), \\ &\boldsymbol{K}_c = (2\pi/\lambda)(0,\ 0,\ 1). \end{aligned} \tag{5}$$

This geometry can be realized by using a DBS to obtain a zero-order diffracted beam and four symmetric first-order diffracted beams. The polarization of the central beam is circularly polarized and all the four A-beams are linearly polarized. The unit polarization vectors of five beams are

$$\boldsymbol{e}_1 = \boldsymbol{e}_3 = (0,\ 1,\ 0)\ ,\quad \boldsymbol{e}_2 = \boldsymbol{e}_4 = (1,\ 0,\ 0)\ ,\quad \boldsymbol{e}_c = \frac{1}{\sqrt{2}}(1,\ -i,\ 0), \text{respectively} \tag{6}$$

In general, the lattice structures are tetragonal symmetric structures. The continuous increase of apex angle θ leads to continuous variation of primitive vectors, reciprocal vectors and the irreducible Brillouin zone of the resultant structure. At $\theta = 70.53°$, the structure is fcc symmetry with respect to diamond structure. Around 70.53°, the lattice is face-centered-tetragonal (fct) symmetry. **Figure 9(a)** and **(b)** show the real fcc structure and its primitive cell fabricated by five-beam symmetric umbrella configuration at $\theta = 70.53°$, which obviously differ from the rhombohedral structure of fcc symmetry and its primitive cell formed by four-beam symmetric umbrella configuration at $\theta = 38.94°$, as shown in **Figure 9(c)** and **(d)**. At the apex angle of $\theta = 90°$, the body-centered-cubic (bcc) lattice can be obtained. For value of θ is near 90°, the structure has body-centered-tetragonal (bct) symmetry.

Full photonic band gaps exist over a very wide range of apex angle with a relatively low refractive index contrast. **Figure 10** represents the relative gap sizes of optimized structures for the apex angle range of $50° < \theta < 115°$ with a dielectric constant contrast of 11.9 to 1 corresponding to silicon in air. It is clear that there are complete PBGs above 10% in the range of $52° < \theta < 112°$, and even larger PBGs above 20% for $59° < \theta < 92°$. The maximum relative gap size of 25.1% appears at $\theta = 70.53°$ corresponding to fcc structure, and the relative gap size of

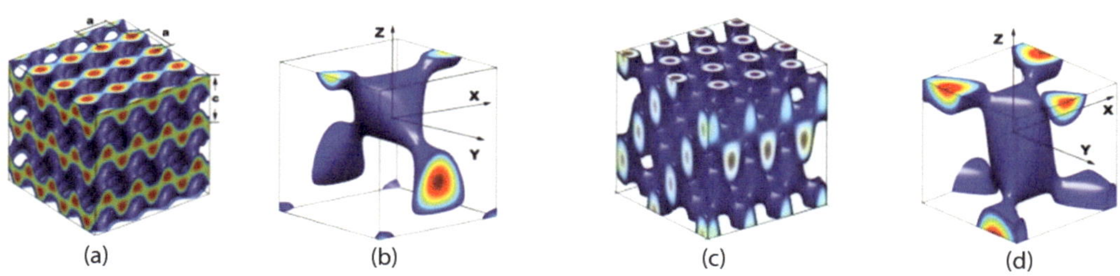

Figure 9. (a) The real fcc structure formed by five-beam symmetric umbrella configuration at $\theta = 70.53°$ and $I_t = 1.39$; (b) the primitive cell of fcc structure; (c) the rhombohedral structure of fcc symmetry constructed by four-beam symmetric umbrella configuration for $\theta = 38.94°$; and (d) the primitive cell of the lattice structure of (c).

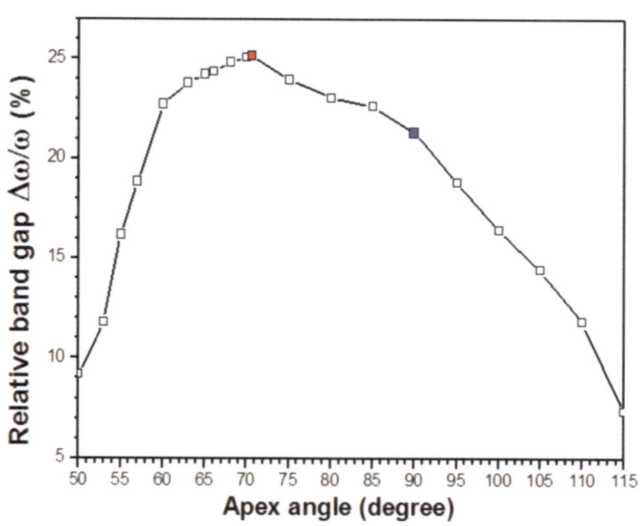

Figure 10. Relative band gap for $50° < \theta < 115°$ when $\varepsilon = 11.9$. The solid symbols are the date for fcc and bcc structures at $\theta = 70.53°$ and 90°, respectively.

bcc structure is 21.3% at $\theta = 90°$. **Figure 11** gives the band structure of the fcc lattice with a large PBG from 0.330 to 0.425$\omega a/2\pi c$ between the second and third bands. Comparing with the four-beam symmetric umbrella configuration, one can find that two results have the similar PBGs of 21.3% for the bcc lattice structures, and the biggest PBG of 25.1% for fcc lattice structure formed by five umbrellalike beams is much larger than the value of 5.35% formed by four umbrellalike beams for the shape reason of PhC lattice cell.

2.3. Photonic crystal waveguide

A waveguide can be created in the PhC slab by introducing a linear defect in the in-plane 2D periodic structure [34, 35]. Since the ability to guide light waves around sharp corners with high efficiency is crucial for photonic integrated circuits, many studies have been carried out concerning waveguide bends through sharp bends in 2D PhC slabs [36, 37]. However, all these works limit the studies to the structures formed by air rods with regular circular cross sections. PBGs for PhCs made by HL may be different from those of regular structures, so will the propagation properties.

When the 2D periodic structure is a triangular Bravais lattice formed by the interference technique of three noncoplanar beams [26], the structure was filled with a material of high refractive index and then removing the template, an inverse structure can be obtained. When the intensity threshold I_t changes from 3.0 to 2.1, the shape of air holes of this inverse structure changes gradually from a circle to a hexagon approximately. a waveguide with two 60° bends are shown in **Figure 12(a)** with the distance between two bends of 9a.

Some peaks of transmission may result from the resonance between two bends in the waveguide of ordinary PhCs with circular air holes [38]. The band diagram of the PhC configuration has been calculated with the intensity threshold of $I_t = 2.5$ and filling ratio of $f = 48.7\%$ for the TE-like mode. **Figure 12(b)** indicates the transmission and reflection spectra of the waveguide.

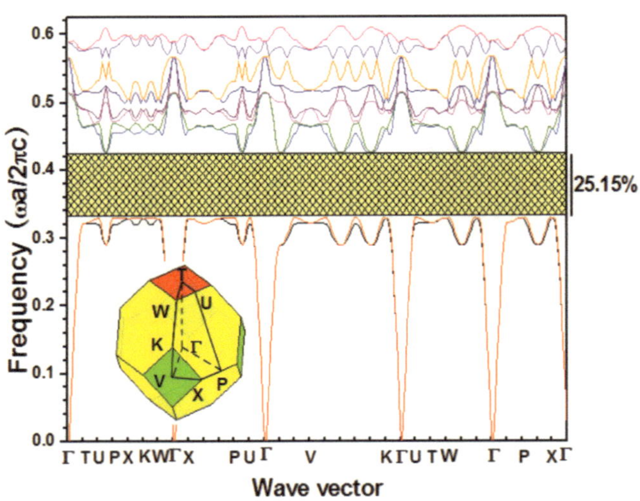

Figure 11. Photonic band structure for the fcc structure with $\theta = 70.53°$. The position of the high symmetry points together with the irreducible Brillouin zone are shown in the inset.

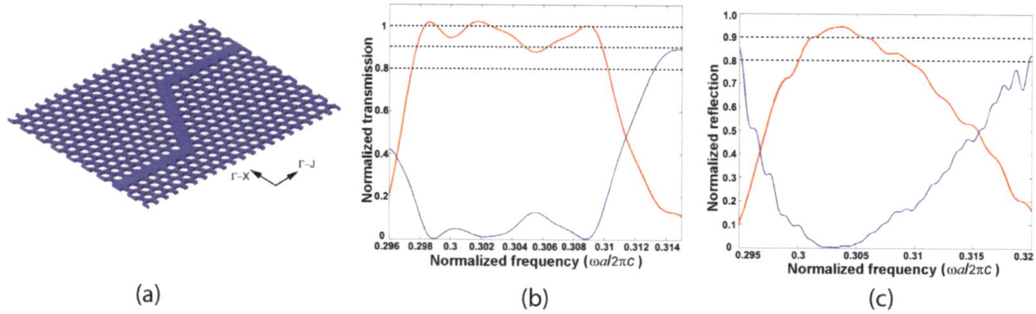

Figure 12. (a) Schematic of a 2D PhC waveguide with two 60° bends and transmission (red curve) and reflection spectra (blue curve) of the PhC waveguide (b) with two bends and (c) one bend.

The red curve here corresponds to the transmission spectra and the blue one to the reflection spectra. High transmission of more than 90% can be obtained in a wide frequency range from 0.298 to 0.310 ($\omega a/2\pi c$). The transmission spectrum of similar PCW with sole 60° bend is shown in **Figure 12(b)**. It is clear that the frequency range of high transmission (>90%) of **Figure 12(b)** is much wider than **Figure 12(c)**. This difference convincingly demonstrates that the existing resonance between two bends induce strong effect on the transmission property of the PCW, which can be used to optimize the PCW design effectively.

Different from the total internal reflection and photonic crystal fibers (PCF) with full 2D PBGs by introducing line defect, a 2D photonic crystal waveguide (PCW) formed by an air core and two identical semi-infinite layers of left-handed holographic PhC is proposed to confine light in air waveguide. As shown in **Figure 13**, The EFCs plot and wave-vector diagram of TM2 band in the HL photonic crystal indicate the PhC is left-handed. Considering the symmetry of Bloch modes of this PhC, the incident interface can be placed in ΓM or ΓK directions. Simulations have demonstrated that the incident beam can readily travel through the PhC slab with the input surface interface normal to ΓM direction, but restrained in ΓK direction, which may originate from the symmetry mismatch between the external incident wave and the Bloch mode of this PhC structure [39].

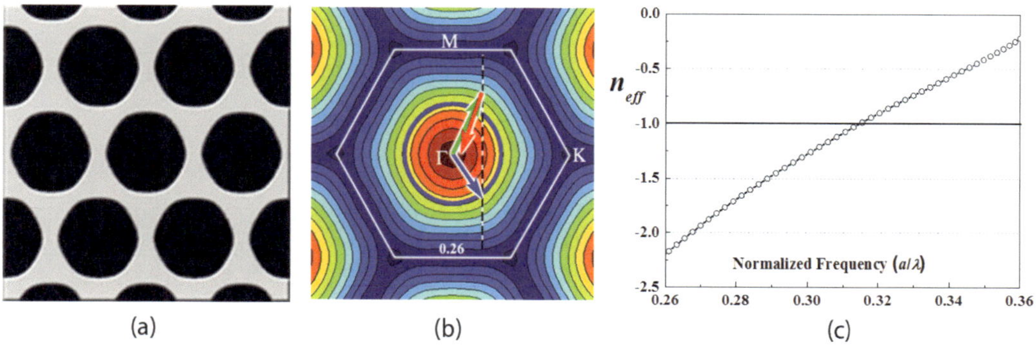

Figure 13. (a) The dielectric pattern of cross section with $I_t = 2.33$; (b) EFC plot and wave-vector diagram of TM2 band; (c) effective index of TM2 band vs. the normalized frequency.

An air waveguide is introduced in the PhC along ΓK direction, as shown in **Figure 14(a)**, with the length 50*a* of PCW. The power flow can become negative in the PhC cladding when $\varepsilon_x < 0$, but remain positive in the air core. By reducing the air width to a critical thickness for TM mode at certain frequency, the group velocity decreases to zero due to the energy flow in the air core was offset by the energy flow of the PhC cladding [39]. According to the result of Ref. [40] the wave vector of guided waves can be given by $k_z = m\pi/d$ with $m = 1,2,\ldots$, and the guided waves can pass through the super waveguide composed of one air layer and LHM for the width of air waveguide $d > \lambda_0/2$ (λ_0 is the wavelength in air). The larger d is, the more modes are guided in the super waveguide. To satisfy the guided condition the width of air layer d is chosen to be equal or greater than that of 2 layers. A Gaussian pulse with the frequency spanning from 0.26 to 0.38 is excited at the input (left) side of the waveguide to investigate the transmission properties of this holographic PCW. The transmission spectrums of different widths of air waveguides are shown in **Figure 14(b)**. Obviously, a high transmission (>90%) happened in the frequency regime from 0.315 to 0.365 through the air waveguide with the width of 4 layers, corresponding to $0 < |n_{eff}| < 1$, which verifies the oscillating modes has a real propagation constant [40]. With the width of air waveguide reducing from 4 to 2 layers, the group velocities decrease gradually and backscattering loss becomes the dominant factor [41], the critical excitation frequency changes from 0.315 to 0.334 and the transmittance decreases gradually for the backscattering loss weakening the guided Bloch mode.

Figure 15 shows TM field in the waveguide with 3 layers of air width for the cases when frequencies are (a) 0.28, (b) 0.315 and (c) 0.33 respectively, light attenuation can be seen clearly in **Figure 15(a)** as a result of vertical scattering loss. Since air thickness decreases from 4 layers to 3 layers, the group velocities slow down and the backscattering loss becomes a dominant loss factor. The light with the frequency of 0.315 display a low transmission in the waveguide as shown in **Figure 15(b)**. **Figure 15(c)** verifies that the incident wave with frequency beyond the minimum critical excitation frequency can be well confined to the air waveguide. Based on the symmetry mismatch between the incident wave and the Bloch modes of the holographic PhC, a holographic PCW without PBGs can also efficiently guide light in a wide frequency region with high transmission efficiency.

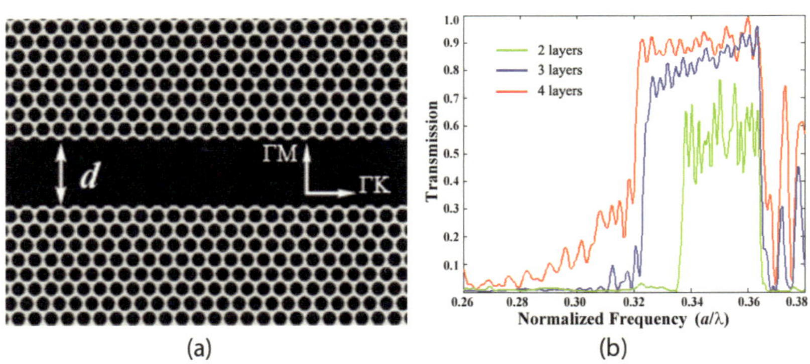

Figure 14. (a) Schematic of the holographic PhC waveguide with a 4 layers width; (b) variation of transmission spectrum for the holographic PCW with different air width.

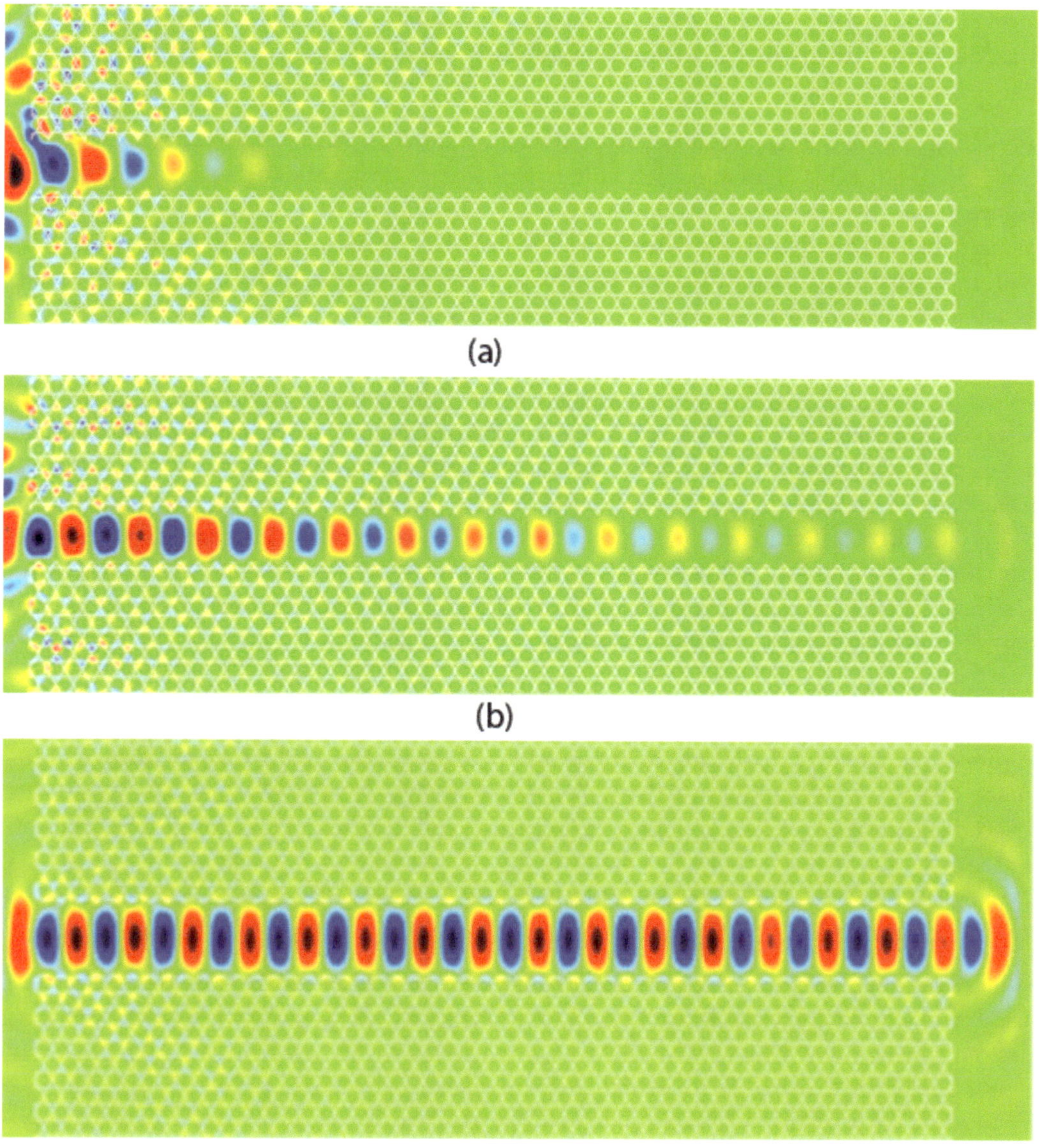

(a) (b) (c)

Figure 15. Snapshots of the propagations of TM polarization monochromatic wave at the frequencies of (a) 0.28, (b) 0.315 and (c) 0.33 in the holographic PCW of $I_t = 2.33$ with 3 layers width.

3. Anomalous refraction behaviors modulated by PhC bands

Photonic crystals can exhibit an extraordinarily high, nonlinear dispersion such as negative refraction and self-focusing properties that are solely determined by the characteristics of their band structures and equal frequency contours (EFC) [42–44]. The 2D honeycomb structures formed by single-exposure interference fabrication methods is used to investigate a series of HL PhC structures in order to obtain comprehensive understanding of the anomalous refractive properties in PhCs.

The filling ratio of the HL PhC is determined by the ratio of the total exposure dose. Silicon with $\varepsilon = 11.56$ (i.e., n = 3.4) is used to analyze the dispersion characteristics of the holographic PhCs. **Figure 16(a)** gives the cross section of the HL PhC sample. The EFCs plot and wave-vector diagram of TM2 band in **Figure 16(b)** illustrated the EFCs around point Γ are convex and shrink with increasing frequency, indicating the PhC is left-handed. Due to the symmetry mismatch between the external plane wave at normal incidence and the Bloch modes of this PhC as mentioned above, the interface between PhC and free space is arranged along the ΓK direction.

As shown in **Figure 16(a)**, the surface of dielectric PhC slab with the trigonal flange (cut 0.4a) was disposed to reduce the reflection and scattering losses effectively [45], because of the effective index gradually varying to match with free space. A continuous monochromatic TM polarization plane wave at the desired frequency $\omega = 0.348$ incidents on the PhC slab with the incident angles of $\theta = 30°$ and 60°. The wave patterns are shown in **Figure 16(c)** and **(d)**, respectively, with the refracted beams and incident beams symmetrically located on the same side of normals, which illustrate the effective refractive index of this PhC is $n_{\text{eff}} = -1$, and the phenomenon of negative refraction in this PhC is an absolute left-handed behavior with $\boldsymbol{K}_{\text{r}} \cdot \boldsymbol{V}_{\text{gr}} < 0$.

For a continuous point source of $\omega = 0.348$ located on the upper side of the PhC slab with the distance of $d_{\text{o1}} = 8.0a$ away from the upper interface (i.e., the object distance), as shown in **Figure 17(a)**, the image point approximately locates at the edge of the lower surface with the image distance $d_{\text{i1}} = 0$. In **Figure 17(b)**, the relevant image distance becomes $d_{\text{i2}} \approx 4.6a$ for the object distance of $d_{\text{o2}} = 3.5a$. Obviously, the sum of d_{o} and d_{i} in this PhC slab is nearly a constant and satisfies the Snell's law of a flat lens with $n_{\text{eff}} = 1$. In addition, note that there is an internal focus inside the PhC slab of **Figure 17(b)**, which is a clear evidence of LHMs following the geometric optics rules.

Multi-refraction effects in the 2D triangular PhC have also been found [43]. The special EFC distributions of different bands can be used to predicate the propagating properties of incident electromagnetic wave (EMW). The EFC plot of the second band is shown in **Figure 18(a)** with almost straight EFC in the ΓK direction at the frequency of 0.26 a/λ. The group velocity v_{gr} of refracted waves ought to be perpendicular to the incident surface among the incident angle region from 0 to 35°, which have been demonstrated by the simulation results in **Figure 18(b)–(d)**. This unusual collimation effect has a series of exciting potential applications, such as spatial light modulator and optical collimator.

The k-conservation relation is observed in wave-vector diagrams [44]. Traditional EMWs propagate in media with their wavefronts perpendicular to the energy flow direction. Here, the EMW of 0.36 a/λ incident upon the ΓK surface at $\theta_{inc} = 25°$, the wavefronts of refracted wave are modulated by the periodic PhC to parallel to the energy flow direction with $\boldsymbol{k} \cdot \boldsymbol{v}_{gr} = 0$. **Figure 19(a)** gives the analysis of EFC plot and wave-vector diagram. FDTD simulations of electric field distribution in **Figure 19(b)** prove with the certainty of theory analysis results with the parallel wavefronts extending along the transmission line.

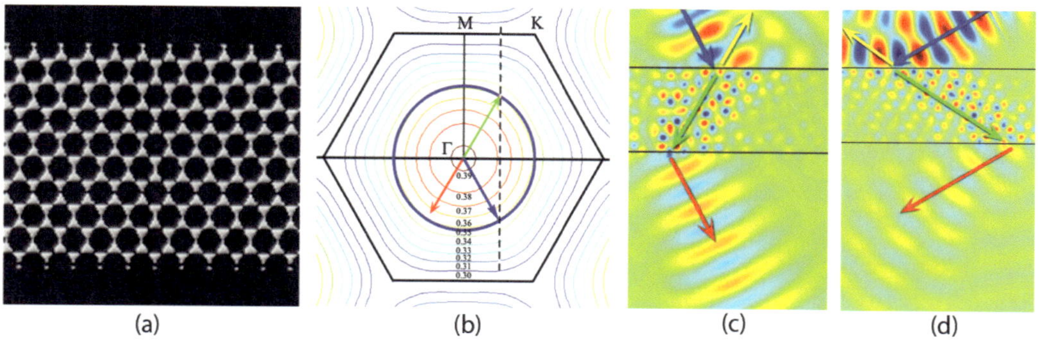

Figure 16. (a) Schematic view of the HL PhC slab with the trigonal dielectric flange; and (b) the wave patterns of negative refractions for different incident angles of (c) $\theta = 30°$ and (d) $\theta = 60°$.

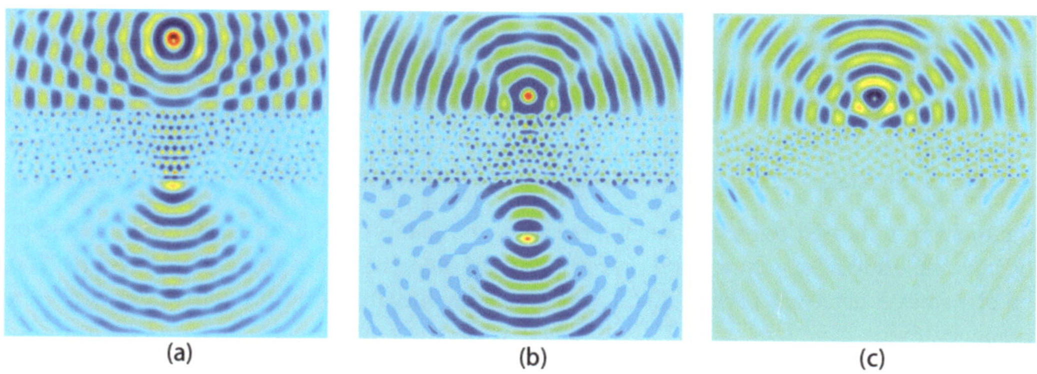

Figure 17. Field patterns for the flat superlens. The object distance is 8.5*a* (a) and 3.5*a* (b) for the interface normal to ΓM direction, and (c) 3.5*a* for the interface normal to ΓK direction.

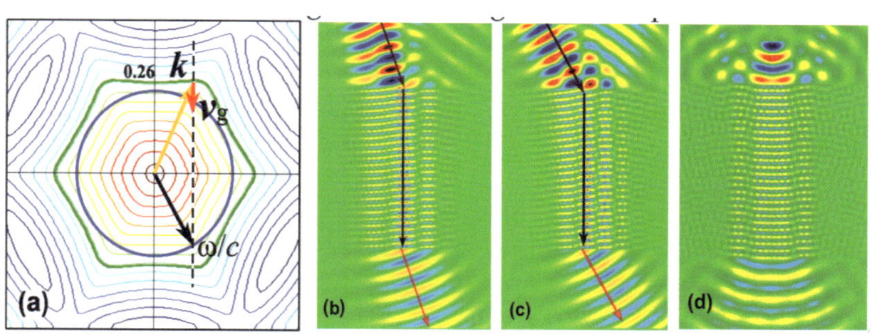

Figure 18. (a) EFC plot of the second band with the wave-vector diagrams at 0.26a/λ and the FDTD simulations of electric field distributions with different incident angles of (b) 20°, (c) 30° and (d) point source incident.

For the normal HL structure with triangular lattice symmetry of $I_t = 1.90$ (or $f = 82.8\%$), the EFC plots of TM2 and TM3 bands are calculated [46]. Different from the EFCs of inverse HL structure, TM2 and TM3 bands intersect in the frequencies regime from 0.225 to 0.343.

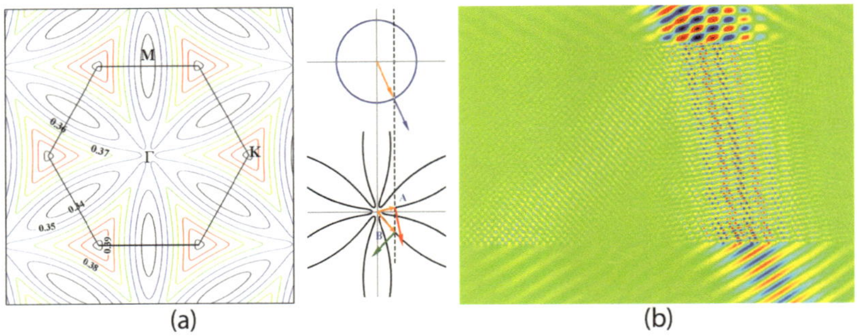

Figure 19. (a) EFC plot of the fourth band and wave-vector diagram at $0.36a/\lambda$ with $\theta_{inc} = 25°$; (b) the FDTD simulations of electric field distributions.

When a Gaussian beam of $\omega = 0.31$ incidents on the ΓK surface at $\theta = 25°$, the wave-vector diagram is shown in **Figure 20(a)** with the blue circle representing the EFC in air, the brown ring denoting the EFC of TM2 band, and the green hexagram corresponding to the EFC of TM3 band. Obviously, the dashed conservation line intersects with EFCs of TM2 and TM3 simultaneously to excite two beams of left-handed negative refraction with different refractive angles, because of the unique EFC features, as shown in **Figure 20(b)** to verify this result.

The more complicated refraction behaviors can be excited in the higher band regions based on the intricate undulation of one band or the overlap of different bands. As shown in **Figure 21(a)**, the sixth band has dual parallel EFCs with opposite curvatures by the red rings within the frequency range from 0.44 to 0.47 a/λ within a wide scope of incident angle. As an example, when the working frequency is chosen to be 0.46a/λ, the incident wave at $\theta_{inc} = 30°$ can excite positive and negative refracted waves have the symmetrical refractive angles of ±30°. The seventh band has more frequency undulations circled by the blue rings which lead to the more intricate triple refraction within the frequency scope from 0.488 to 0.50 a/λ (**Figure 21(b)**).

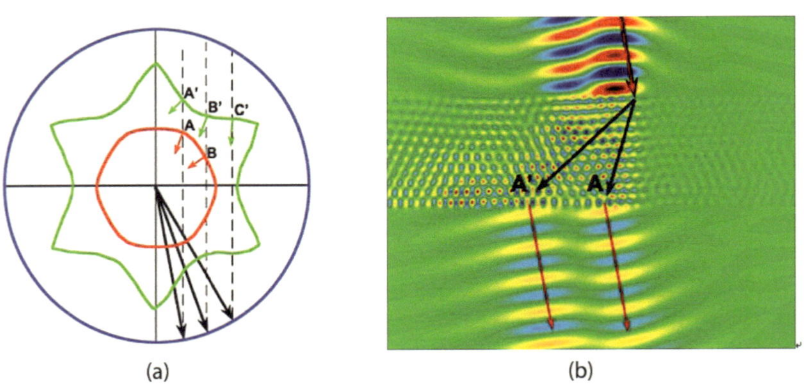

Figure 20. (a) Wave-vector diagram of $\omega = 0.31$ for TM2 (brown ring) and TM3 (green hexagram); (b) field pattern with the incident beam of $\omega = 0.31$ at the incident angle of 25°.

http://dx.doi.org/10.5772/intechopen.71403

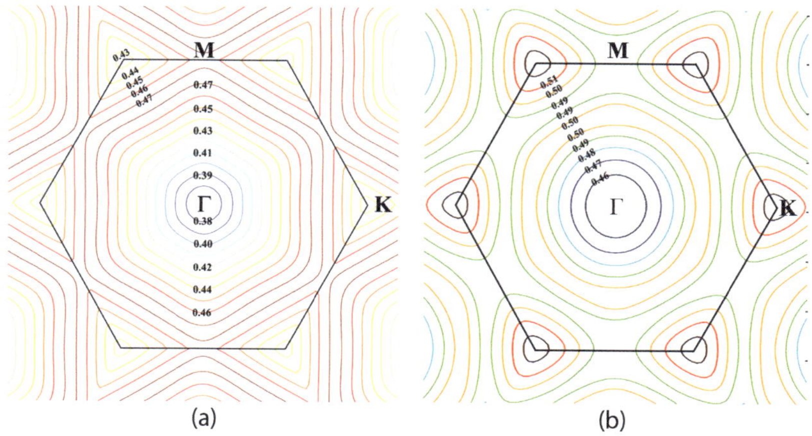

Figure 21. EFC plots of (a) the sixth band and (b) the seventh band.

4. Phenomena of zero phase delay in PhCs and application

Recently, materials with zero or near-zero-n has attracted great focus for the characteristics of uniform phase and infinite wavelength [47–50]. A series of exciting potential applications have been found in zero-n materials, such as wavefront reshaping [51], beam self-collimation [52], extremely convergent lenses [53], etc. One of their best known applications is optical links in lumped nanophotonic circuits over hundreds of wavelengths without introducing phase shifts so as to reduce the unwanted frequency dispersion. Different strategies have been provided to realize zero permittivity ε. One is to use metallic metamaterial structures with effective zero permittivity and/or permeability [54, 55]. However, these metamaterials suffer from the strong absorption loss of the metals and hence with greatly deteriorated transmittance. Some alternative approaches have been provided, include the combination of negative- and positive-index materials [52], the microwave waveguides below cutoff [56] or the periodic superlattice formed by positive index homogeneous dielectric media and negative index photonic crystals (PhCs) [57].

A plane wave can be described as $\tilde{E}(\boldsymbol{r}) = A\exp\left[i(\boldsymbol{k}\cdot\boldsymbol{r}+\varphi_0)\right]$, where the symbols of A, $\boldsymbol{k}$, $\boldsymbol{r}$, ϕ_0 denote wave amplitude, wave vector, position vector and initial phase. The spatial phase shift is determined by the spatial phase factor $\boldsymbol{k}\cdot\boldsymbol{r}$, with the traditional wave vector $\boldsymbol{k}$ pointing to the direction of energy flow (i.e. the direction of group velocity $\boldsymbol{v}_{gr}$). Assuming the condition of $\boldsymbol{k}\cdot\boldsymbol{r} = 0$ is satisfied, the wave vector $\boldsymbol{k}$ is perpendicular to the energy flow with the wavefronts extending along the propagation direction with the stationary spatial phase along the direction of energy flow $\boldsymbol{S}$. In general, it is difficult to modulate the direction of wavefronts in homogeneous materials. Since this formula of $\boldsymbol{k}\cdot\boldsymbol{r} = 0$ can also be expressed as $\boldsymbol{k}\cdot v_{gr} = 0$, by the definition of group velocity $\boldsymbol{v}_{gr} = \nabla_{\boldsymbol{k}}\omega$ in PhCs, the group velocity vector is perpendicular to EFCs pointing to the frequency-increasing direction. Hence, by adjusting the EFC distribution, the condition of zero phase delay with $\boldsymbol{k}\cdot\boldsymbol{r} = 0$ can be satisfied.

The PhC sample of triangular lattice is composed of dielectric rods with $\varepsilon_r = 10$, diameter $d = 10$ mm, height $h = 10$ mm and lattice constant a = 10 mm. When the beam of the frequency $\omega = 0.376$ is incident upon the ΓK surface with the incident angle of $\theta_{inc} = 30°$. The wave-vector diagram of the fourth band is shown in **Figure 22**. According to the condition of boundary conservation, two refracted waves can be excited in the PhC. The condition of $\boldsymbol{k} \cdot \boldsymbol{r} = 0$ is satisfied for the positive refracted wave. Simulations and experiments in a near-field scanning system have demonstrated that the wavefronts exactly are parallel to the energy flow with zero phase delay, as shown in **Figure 23(a)** and **(b)**. The measured phase contrast image is shown in **Figure 23(c)** at 11.213 GHz. Obviously, the incident wave and the exit wave can be connected directly as if the PhC slab did not exist. Since the PhC structure can be engineered readily with the large design flexibility, the frequency and incident angle of zero phase delay can be adjustable in a relaxed PhC configuration design.

Another approach to realize zero phase delay in PhCs is based on the accidental degeneracy of two dipolar modes and a single monopole mode generates at the Dirac-like point (DLP) with the linear dispersions of Dirac cone at the Brillouin zone center of PhCs [58]. At the DLP shown in **Figure 24(a)**, the PhC can mimic the zero-index medium (ZIM) with the characteristics of uniform field distribution. Linear dispersions near the center point induced by the triple degeneracy display many unique scattering properties, such as conical diffraction [59], wave shaping and cloaking [60]. The dispersion properties obeying the 1/L scaling law near the DLP in the normal propagation direction have been verified theoretically and experimentally [61] through the dielectric PhC ribbons with the finite thicknesses, however, it is difficult to distinguish the precise crossing point from the wide extremum range just relying on the transmittance spectrum.

A Gaussian pulse source of TM mode was placed in front of a PhC square-lattice array with the incident angle of 90°, thus three transmission spectrums can be measured outside the other three output interface of the PhC array with one in parallel direction and two in perpendicular upward and downward directions. As long as the size is large enough, the PhC array can be

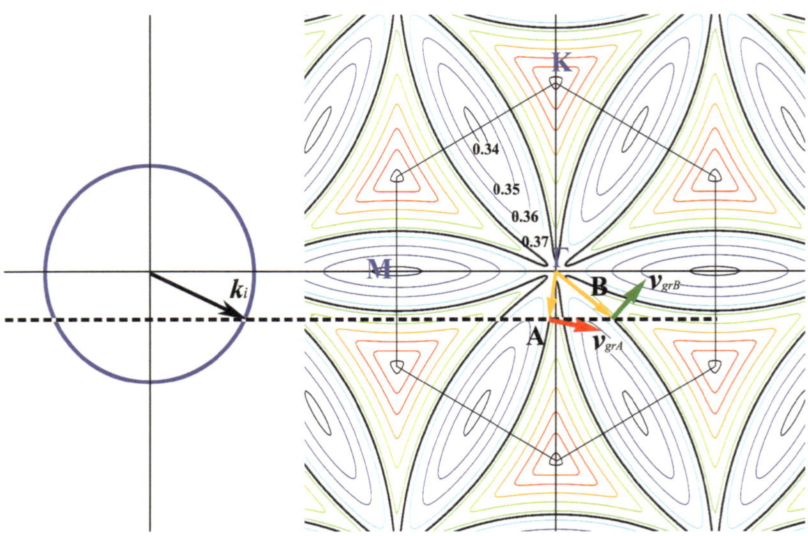

Figure 22. EFCs plot of the fourth band with the wave-vector diagram at $\omega = 0.376a/\lambda$ with $\theta_{inc} = 30°$.

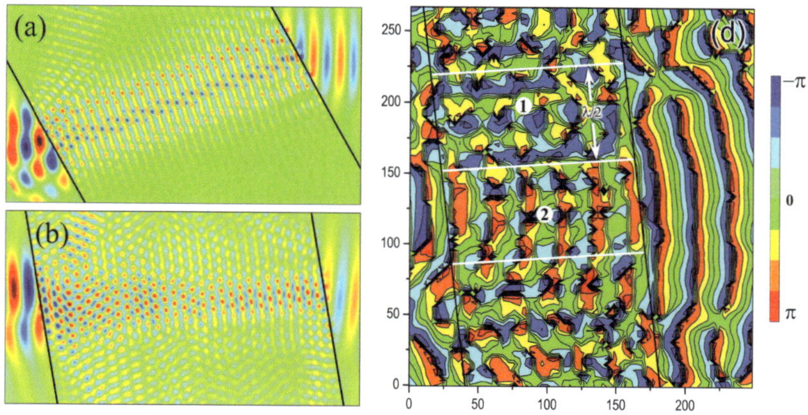

Figure 23. (a, b) Wavefront distribution in the PhC slab with different incident angles; (c) measured phase contrast image.

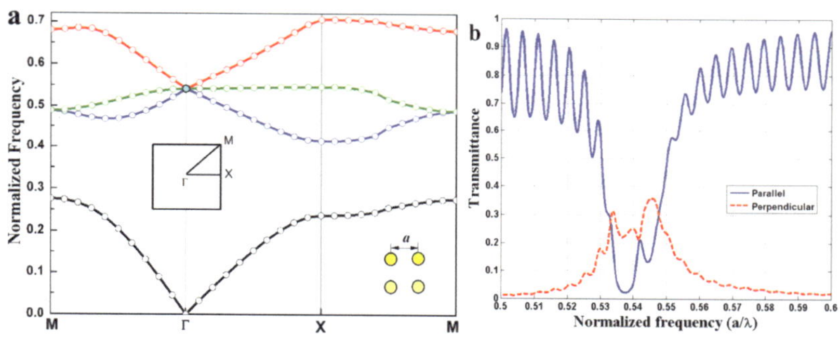

Figure 24. (a) Band structure of TM waves for the 2D square-lattice PhC at the normalized frequency $\omega_D = 0.541$. (b) The simulated transmission spectra in the directions parallel and perpendicular to the incident propagation direction.

regarded as an infinite PhC with the similar transmittance at the different exit boundaries with the same interface structure. As shown in **Figure 24(b)**, the upward and downward sharp cusps embedded in the parallel and perpendicular extremum transmittance spectra intersect at the DLP frequency, therefore, the DLP can be identified accurately due to the property of uniform field distribution of PhC with effective zero index at DLP.

5. Conclusion

In summary, some novel kinds of 3- and 2-D lattices formed by holographic lithography have been investigated to find that the complete PBG over wide ranges of system parameters can be achieved by proper design. The beam design for making this structure is also derived. By using the interference of symmetric four umbrellalike beams one can obtain complete PBGs over a very wide range of apex angle, three special lattice structures of fcc, sc and bcc were achieved with different complete PBGs of 5.35%, 10.23% and 21.57%. The similar PBG of 21.3% for the bcc lattice structures, and the biggest PBG of 25.1% for fcc lattice structure were achieved by five umbrellalike beams, which is much larger than the value of 5.35% formed by four umbrellalike beams for the shape reason of PhC lattice cell. The holographic PCW can efficiently guide light in a wide range of frequency around sharp corners and the resonance between the two bends has a close relation with the configuration of PhC which can be regarded as an important factor to improve the transmission property of 2D PCW. Since the absolute refractive index 1 of the air guiding core is larger than the effective index of PhC cladding, the total internal reflections can be achieved approximately with a high transmittance in the PhC waveguide with the effective refraction index near zero. For PhC waveguide when the width of air layer is greater than the critical thickness, the problems of scattering loss can be avoided. The proposed left-handed holographic PCW is more suitable for many applications in the area of photonic integrated circuits, and the idea and method of analysis here open a new freedom for PCW engineering.

Different from the anomalous phenomena of PhCs in the high bands, the PhCs of honeycomb-lattice in the lower band region are more suitable to realize the effect of all-angle left-handed negative refraction. For the straight EFCs distribution in special directions, the regular PhC can be applied to the design of optical collimator. Moreover, multi-refraction has been found in the higher bands. There are two ways to realize multi-refractions, one arises from a single band with intricate undulations and another originates from the overlap of multi-bands. The unique phenomena of dual-negative refraction and positive-negative refraction have been found in the higher bands of PhC. Based on the design flexibility of PhC, the transmission properties of PhC can be engineered with a large freedom. These results are important and useful for the design of PhC device.

An efficient way to achieve EMW propagation with zero phase delay is to modulate the wavefronts paralleling to the direction of energy flow. This method can be extended from 2D to 3D cases or other artificially engineered materials, which opens a new door to obtain perfect zero-phase-delay propagation of EMW and has significant potential in many applications. Furthermore, the research found that the Dirac-like point of PhC can be identified accurately by measuring the transmission spectra through a finite photonic crystal square array.

All these results can be extended to the fabrication of other artificially engineered materials and provide guidelines to the design of new type optical devices.

Acknowledgements

This work was supported by National Natural Science Foundation of China (11574311, 51532004, 61275014)

Author details

Guoyan Dong

Address all correspondence to: gydong@ucas.ac.cn

College of Materials Science and Opto-Electronic Technology, University of Chinese Academy of Sciences, Beijing, China

References

[1] Yablonovitch E. Inhibited spontaneous emission in solid-state physics and electronics. Physical Review Letters. 1987;**58**:2059. DOI: 10.1103/PhysRevLett.58.2059

[2] Joannopoulos JD. et al. Photonic Crystals: Molding the Flow of Light. NJ, USA: Princeton University Press; 2008. DOI: 10.1016/S0038-1098(96)00716-8

[3] Johnson S, Villeneuve P, Fan S, Joannopoulos J. Linear waveguides in photonic-crystal slabs. Physical Review B. 2000;**62**:8212-8222. DOI: 10.1103/PhysRevB.62.8212

[4] Mekis A, Chen J, Kurland I, Fan S, Villeneuve P, Joannopoulos J. High transmission through sharp bends in photonic crystal waveguides. Physical Review Letters. 1996;**77**:3787. DOI: 10.1103/PhysRevLett.77.3787

[5] Krauss T, Rue R, Brand S. Two-dimensional photonic-bandgap structures operating at near-infrared wavelengths. Nature. 1996;**383**:699-702. DOI: 10.1038/383699a0

[6] Zhang X, Jackson T, Lafond E, Deymier P, Vasseur J. Evidence of surface acoustic wave band gaps in the phononic crystals created on thin plates. Applied Physics Letters. 2006;**88**:041911. DOI: 10.1063/1.2167794

[7] Aoki K, Miyazaki H, Hirayama H, Inoshita K, Baba T, Sakoda K, Shinya N, Aoyagi Y. Microassembly of semiconductor three-dimensional photonic crystals. Nature Materials. 2003;**2**:117-121. DOI: 10.1038/nmat802

[8] Moroz A. Three-dimensional complete photonic-band-gap structures in the visible. Physical Review Letters. 1999;**83**:5274-5277. DOI: 10.1103/PhysRevLett.83.5274

[9] Liu K, Avouris P, Bucchignano J, Martel R, Sun S, Michl J. Simple fabrication scheme for sub-10 nm electrode gaps using electron-beam lithography. Applied Physics Letters. 2002;**80**:865-867. DOI: 10.1063/1.1436275

[10] Farsari M, Vamvakaki M, Chichkov B. Multiphoton polymerization of hybrid materials. Journal of Optics. 2010;**12**:124001. DOI: 10.1088/2040-8978/12/12/124001

[11] Miklyaev Y, Meisel D, Blanco A, Freymann G. Three-dimensional face-centered-cubic photonic crystal templates by laser holography: Fabrication, optical characterization, and band-structure calculations. Applied Physics Letters. 2003;**82**:1284-1286. DOI: 10.1063/1.1557328

[12] Veselago V, Lebedev P. The electrodynamics of substances with simultaneously negative values of ε and μ. Soviet Physics Uspekhi. 1968;**10**:509-514. DOI: 10.1070/PU1968v010n04ABEH003699

[13] Ziolkowski R, Heyman E. Wave propagation in media having negative permittivity and permeability. Physical Review E. 2001;**64**:056625. DOI: 10.1103/PhysRevE.64.056625

[14] Lindell I, Tretyakov S, Nikoskinen K, Ilvonen S. BW media—media with negative parameters, capable of supporting backward waves. Microwave & Optical Technology Letters. 2001;**31**:129-133. DOI: 10.1002/mop.1378

[15] Shelby R, Smith D, Schultz S. Experimental verification of a negative index of refraction. Science. 2001;**292**:77-79. DOI: 10.1126/science.1058847

[16] Rakich P, Dahlem M, Tandon S, et al. Achieving centimetre-scale supercollimation in a large-area two-dimensional photonic crystal. Nature Materials. 2006;**5**:93. DOI: 10.1038/005093a0

[17] Vasić B, Gajić R. Self-focusing media using graded photonic crystals: Focusing, Fourier transforming and imaging, directive emission, and directional cloaking. Journal of Applied Physics. 2011;**110**:1771-1180. DOI: 10.1063/1.3630116

[18] Cao Y, Hou Z, Liu Y. Convergence problem of plane-wave expansion method for phononic crystals. Physics Letters A. 2004;**327**:247-253. DOI: 10.1016/j.physleta.2004.05.030

[19] Bondeson A, Rylander T, Ingelström P. The finite-difference time-domain method. In: Computational Electromagnetics. Vol. 51. New York, NY: Springer; 2001. pp. 57-86. DOI: 10.1007/0-387-26160-5_5

[20] Dong G, Yang X, Cai L, Shen X, Meng X, Xu X, Wang Y. Band gap analysis and holographic design of 3-fold hybrid triangular photonic crystals of irregular columns with large full band gaps. Journal of Optics A: Pure and Applied Optics. 2007;**9**:531. DOI: 10.1088/1464-4258/9/5/017

[21] Dong G, Yang X, Cai L, Shen X, Meng X, Xu X, Zhang H. Six-fold hybrid photonic crystal formed holographically with full band gap for low refractive index. EPL. 2007;**80**:14006. DOI: 10.1209/0295-5075/80/14006

[22] Cai L, Dong G, Feng C, Yang X, Shen X, Meng X. Holographic design of a two-dimensional photonic crystal of square lattice with a large two-dimensional complete bandgap. Journal of the Optical Society of America B. 2006;**23**:1708-1711. DOI: 10.1364/JOSAB.23.001708

[23] Joannopoulos J, Villeneuve P, Fan S. Erratum: Photonic crystals: Putting a new twist on light. Nature. 1997;**386**:143-149. DOI: 10.1038/386143a0

[24] Russell P. Photonic crystal fibers. Science. 2003;**299**:358. DOI: 10.1126/science.1079280

[25] Sharp D, Campbell M, Dedman E, Harrison M, Denning R, Turberfield A. Photonic crystals for the visible spectrum by holographic lithography. Nature. 2000;**404**:53. DOI: 10.1038/35003523

[26] Yang X, Cai L, Liu Q. Theoretical bandgap modeling of two-dimensional triangular photonic crystals formed by interference technique of three-noncoplanar beams. Optics Express. 2003;**11**:1050-1055. DOI: 10.1364/OE.11.001050

[27] Cai L, Feng C, He M, Yang X, Meng X, Dong G, Yu X. Holographic design of a two-dimensional photonic crystal of square lattice with pincushion columns and large complete band gaps. Optics Express. 2005;**13**:4325. DOI: 10.1364/OPEX.13.004325

[28] Dong G, Cai L, Yang X, Shen X, Meng X, Xu X, Wang Y. Analysis of structure and band gap evolution of photonic crystals formed holographically by symmetric umbrella configuration

with varying apex angles. Journal of Physics D: Applied Physics. 2006;**39**:3566. DOI: 10.1088/0022-3727/39/16/007

[29] Dong G, Cai L, Yang X, Shen X, Meng X, Xu X, Wang Y. Holographic design and band gap evolution of photonic crystals formed with five-beam symmetric umbrella configuration. Optics Express. 2006;**14**:8096-8102. DOI: 10.1364/OE.14.008096

[30] Cai L, Liu Q, Yang X. Polarization optimization in the interference of four umbrellalike symmetric beams for making three-dimensional periodic microstructures. Applied Optics. 2002;**41**:6894. DOI: 10.1364/AO.41.006894

[31] Cai L, Yang X, Liu Q, Wang Y. What kind of Bravais lattices can be made by the interference of four umbrellalike beams? Optics Communications. 2003;**224**:243-246. DOI: 10.1016/j.optcom.2003.07.004

[32] Meisel D, Wegener M, Busch K. Three-dimensional photonic crystals by holographic lithography using the umbrella configuration: Symmetries and complete photonic band gaps. Physical Review B. 2004;**70**:165104. DOI: 10.1103/PhysRevB.70.165104

[33] Chan T, Toader O, John S. Photonic band gap templating using optical interference lithography. Physical Review E. 2005;**71**:046605. DOI: 10.1103/PhysRevE.71.046605

[34] Dong G, Yang X, Cai L, Shen X, Wang Y. Improvement of transmission properties through two-bend resonance by holographic design for a two-dimensional photonic crystal waveguide. Optics Express. 2008;**16**:15375. DOI: 10.1364/OE.16.015375

[35] Dong G, Yang X, Cai L. Transmission properties of an air waveguide with left-handed holographic photonic crystal cladding. 中国物理快报:英文版. 2011;**28**:014210. DOI: 10.1088/0256-307X/28/1/014210

[36] Cui X, Hafner C, Vahldieck R, Robin F. Sharp trench waveguide bends in dual mode operation with ultra-small photonic crystals for suppressing radiation. Optics Express. 2006;**14**:4351. DOI: 10.1364/OE.14.004351

[37] Smajic J, Hafner C, Erni D. Design and optimization of an achromatic photonic crystal bend. Optics Express. 2003;**11**:1378. DOI: 10.1364/OE.11.001378

[38] Chutinan A, Noda S. Waveguides and waveguide bends in two-dimensional photonic crystal slabs. Physical Review B. 2000;**62**:4488-4492. DOI: 10.1103/PhysRevB.62.4488

[39] Sakoda K. Optical Properties of Photonic Crystals. Berlin, Heidelberg: Springer; 2005. DOI: 10.1007/978-3-662-14324-7

[40] Cheng Q, Cui T. High-power generation and transmission through a left-handed material. Physical Review B. 2005;**72**:113112. DOI: 10.1103/PhysRevB.72.113112

[41] Jiang T, Zhao J, Feng Y. Stopping light by an air waveguide with anisotropic metamaterial cladding. Optics Express. 2009;**17**:170. DOI: 10.1364/OE.17.000170

[42] Dong G, Cai L, Yang X. Anomalous refractive effects in honeycomb lattice photonic crystals formed by holographic lithography. Optics Express. 2010;**18**:16302. DOI: 10.1364/OE.18.016302

[43] Dong G, Zhou J, Yang X, Meng X. Multi-refraction with same polarization state in two dimensional triangular photonic crystals. Progress in Electromagnetics Research. 2012;**128**: 91-103. DOI: 10.2528/PIER12040306

[44] Notomi M. Theory of light propagation in strongly modulated photonic crystals: Refractionlike behavior in the vicinity of the photonic band gap. Physical Review B. 2000; **62**:10696-10705. DOI: 10.1103/PhysRevB.62.10696

[45] Matsumoto T, Eom K, Baba T. Focusing of light by negative refraction in a photonic crystal slab superlens on silicon-on-insulator substrate. Optics Letters. 2006;**31**:2786-2788. DOI: 10.1364/OL.31.002786

[46] Dong G, Zhou J, Yang X, Cai L. Dual-negative refraction in photonic crystals with hexagonal lattices. Optics Express. 2011;**19**:12119-12124. DOI: 10.1364/OE.19.012119

[47] Dong G, Zhou J, Yang X, Meng X. Precise identification of Dirac-like point through a finite photonic crystal square matrix. Scientific Reports. 2016;**6**:36712. DOI: 10.1038/srep36712

[48] Dong G, Zhou J, Cai L. Zero phase delay induced by wavefront modulation in photonic crystals. Physical Review B. 2012;**87**:125107. DOI: 10.1103/PhysRevB.87.125107

[49] Yun S, Jiang Z, Xu Q, Liu Z, Werner D, Mayer T. Low-loss impedance-matched optical metamaterials with zero-phase delay. ACS Nano. 2012;**6**:4475. DOI: 10.1021/nn3012338

[50] Zhang F, Houzet G, Lheurette E, Lippens D, Chaubet M, Zhao X. Negative-zero-positive metamaterial with omega-type metal inclusions. Journal of Applied Physics. 2008;**103**:084 312. DOI: 10.1063/1.2910831

[51] Alù A, Silveirinha MG, Salandrino A, Engheta N. Epsilon-near-zero metamaterials and electromagnetic sources: Tailoring the radiation phase pattern. Physical Review B. 2007;**75**: 155410. DOI: 10.1103/PhysRevB.75.155410

[52] Mocella V, Cabrini S, Chang A, et al. Self-collimation of light over millimeter-scale distance in a quasi-zero-average-index metamaterial. Physical Review Letters. 2009;**102**:133902. DOI: 10.1103/PhysRevLett.102.133902

[53] Silveirinha M, Engheta N. Tunneling of electromagnetic energy through subwavelength channels and bends using epsilon-near-zero materials. Physical Review Letters. 2006;**97**:157403. DOI: 10.1103/PhysRevLett.97.157403

[54] Belov P, Zhao Y, Tse S, et al. Transmission of images with subwavelength resolution to distances of several wavelengths in the microwave range. Physical Review B. 2008;**77**:193 108. DOI: 10.1103/PhysRevB.77.193108

[55] Podolskiy V, Pollard R, Murphy A, et al. Optical nonlocalities and additional waves in epsilon-near-zero metamaterials. Physical Review Letters. 2009;**102**:127405. DOI: 10.1103/PhysRevLett.102.127405

[56] Edwards B, Alù A, Young M, Silveirinha M, Engheta N. Experimental verification of epsilon-near-zero metamaterial coupling and energy squeezing using a microwave waveguide. Physical Review Letters. 2008;**100**:033903. DOI: 10.1103/PhysRevLett.100.033903

[57] Kocaman S, Aras M, Hsieh P, et al. Zero phase delay in negative-refractive-index photonic crystal superlattices. Nature Photonics. 2011;**5**:499-505. DOI: 10.1038/nphoton.2011.129

[58] Huang X, Lai Y, Hang Z, Zheng H, Chan C. Dirac cones induced by accidental degeneracy in photonic crystals and zero-refractive-index materials. Nature Materials. 2011;**10**:582. DOI: 10.1038/nmat3030

[59] D'Aguanno G, Mattiucci N, Conti C, Bloemer M. Field localization and enhancement near the Dirac point of a finite defectless photonic crystal. Physical Review B. 2013;**87**:085135. DOI: 10.1103/PhysRevB.87.085135

[60] Chan C, Hang Z, Huang X. Dirac dispersion in two-dimensional photonic crystals. Advances in OptoElectronics. 2012;**2012**:313984. DOI: 10.1155/2012/313984

[61] Zandbergen S, de Dood M. Experimental observation of strong edge effects on the pseudodiffusive transport of light in photonic graphene. Physical Review Letters. 2010;**104**: 043903. DOI: 10.1103/PhysRevLett.104.043903

Chapter 3

Dispersion Properties of TM and TE Modes of Gyrotropic Magnetophotonic Crystals

Alexander A. Shmat'ko, Viktoria N. Mizernik,
Eugene N. Odarenko and Viktor T. Lysytsya

Additional information is available at the end of the chapter

http://dx.doi.org/10.5772/intechopen.71273

Abstract

This chapter discusses the propagation of TM and TE waves in the one-dimensional gyrotropic magnetophotonic crystals with ferrite and plasma-like layers. Elements of the transfer matrix are calculated in closed analytical form on the base of electrodynamic problem rigorous solution for arbitrary location of the gyrotropic elements on the structure period. Dispersion equation of the layered periodic structure with gyrotropic elements is obtained. Dispersion properties of the structure for TE and TM modes are analyzed for different configurations of magnetophotonic crystals (ferrite and plasma-like layers). Existence areas of transmission bands for surface and bulk waves are obtained. The effect of problem parameters on the dispersion properties of magnetophotonic crystals for TM and TE modes is investigated. Regimes of complete transmission of wave through limited magnetophotonic crystal are analyzed for bulk and surface waves.

Keywords: magnetophotonic crystal, gyrotropic media, dispersion diagrams, TE and TM modes, bulk and surface waves

1. Introduction

Photonic crystals (PCs) are artificial periodic structures with spatially modulated refractive index in one or more coordinates [1, 2]. Their outstanding optical properties are due to the existence of frequency band gaps where the propagation of electromagnetic waves is impossible. Application of these structures became very attractive for modern optoelectronics which uses the various waveguides, resonators, sensors, and other devices on the basis of PC [3, 4]. Moreover, the control of the PC structure characteristics is the important problem that is usually solved using external electric or magnetic fields. These methods of providing controllability are based on the variation of refractive index of special materials such as liquid crystals

INTECH
open science | open minds

© 2018 The Author(s). Licensee InTech. This chapter is distributed under the terms of the Creative Commons Attribution License (http://creativecommons.org/licenses/by/3.0), which permits unrestricted use, distribution, and reproduction in any medium, provided the original work is properly cited. (cc) BY

and magnetic materials [5, 6]. Since these sensitive materials are anisotropic, then theoretical analysis of their properties is more complicated.

When at least one of the PCs' unit cell components is a magnetically sensitive (gyrotropic) material, they exhibit unique magneto-optical properties and identified as magnetophotonic crystals (MPCs). Investigations of the MPCs are begun for simplest one-dimensional structures [7, 8]. However, one-dimensional MPCs are the basis elements for various active field-controlling applications so far [9, 10]. Changing of the permeability by external magnetic field is one of the main phenomena that allow developing electronically tuned devices in different frequency bands: filters, circulators and so on [11–13].

Along with the properties inherent in conventional PCs, these structures have additional optical and magneto-optical properties which considerably expand their functionality. Kerr effect, Faraday rotation and optical nonlinearity can be enhanced in MPC due to light localization within magnetic multilayer. Magneto-optical system with large Faraday or Kerr rotation can be used for effective optical isolators [14, 15], spatial light-phase modulators [16] and magnetic field and current sensor [17] development. Furthermore, one can obtain stronger enhancement of the magneto-optical phenomena due to resonant effects in the MPCs [18], which characterized by specific polarization properties. Using PCs with magneto-optical layers provides possibility of control of optical bistability threshold in structure based on graphene layer [19]. It should be noted that not only magnetic materials are suitable for MPC. Namely, one-dimensional PC with plasma layers can be tunable by external magnetic field [20].

A number of applications of the MPCs are inspired by their nonreciprocal properties. For example, special spatial structure of the MPC layers provides the asymmetry of dispersion characteristics and, as a result, the effect of unidirectional wave propagation [21]. This phenomenon allows enhancing field amplitude in the MPC without any periodicity defects. In this case, the so-called frozen mode regime occurs instead the defect mode one.

One of the unique properties of gyrotropic materials is the possibility of negative values of material parameters under the certain conditions. Usually, these are so-called single-negative media that are divided into epsilon-negative media (plasma) and mu-negative ones (gyrotropic magnetic materials). The term "double-negative media" or "left-handed materials" is used for media with negative values of both permittivity and permeability and often replaced by term "metamaterials." Application of metamaterials in one-dimensional PC systems results in unusual regularities of bulk and surface wave propagation and is the subject of experimental and theoretical research [22, 23].

Theoretical description of the various types of one-dimensional PCs is usually based on the transfer-matrix method of Abelès [24] that was applied by Yeh et al. to periodic layered media [25]. This method cannot be applied in general case for anisotropic multilayer structures because of mode coupling. However, this is possible in special cases, namely, in two-dimensional model of wave propagation in periodic layered media [26]. This case is considered in this chapter. Such an approach makes it possible to simplify significantly the analysis of physical phenomena in complex layered media with various combinations of gyrotropic and isotropic elements. Moreover using well-known permutation duality principle of Maxwell's equations results in a reduction of unique combinations number. In turn, this allows better understanding of regularities of bulk and surface wave propagation in one-dimensional MPC and finding new modes for applications in modern microwave, terahertz and optical devices.

2. Formulation and solution of the problem for modes of gyrotropic periodic structures

2.1. Basic relationships

We study electromagnetic wave propagation in periodic structure in general case with bigyrotropic layers (one-dimensional MPC) (**Figure 1**). Each of two layers on the structure period $L=a+b$ is an anisotropic medium (plasma or ferrite or their combinations). Their permittivity and permeability are characterized by tensor values of standard form [26]:

$$\overset{\leftrightarrow}{\varepsilon}_j = \begin{Vmatrix} \varepsilon_j & -i\varepsilon_{aj} & 0 \\ i\varepsilon_{aj} & \varepsilon_j & 0 \\ 0 & 0 & \varepsilon_{\parallel j} \end{Vmatrix}, \quad \overset{\leftrightarrow}{\mu}_j = \begin{Vmatrix} \mu_j & -i\mu_{aj} & 0 \\ i\mu_{aj} & \mu_j & 0 \\ 0 & 0 & \mu_{\parallel j} \end{Vmatrix}. \tag{1}$$

For plasma media, the value of the permittivity is tensor, whereas the value of the permeability is scalar. Such media are called electrically gyrotropic. It is opposite for the ferrite media; the permeability is tensor, whereas permittivity is scalar. Such media are usually called magnetically gyrotropic. If the permittivity and permeability are simultaneously described by the tensors (Eq. (1)), such media are called gyrotropic or bigyrotropic. The material parameters included in tensors $\overset{\leftrightarrow}{\varepsilon}_j$ and $\overset{\leftrightarrow}{\mu}_j$ are defined by the value of the external bias magnetic field $\vec{H}_0 = \vec{z}_0 H_0$, which is directed along Oz axis.

The study of the general case of gyrotropic media with material parameters of form (Eq. (1)) is reasonable, primarily because it allows using the permutation duality principle when obtaining main equations for fields and characteristic Eqs. [26]. According to this principle generalized for gyrotropic media, namely, when simultaneously the substitution of fields

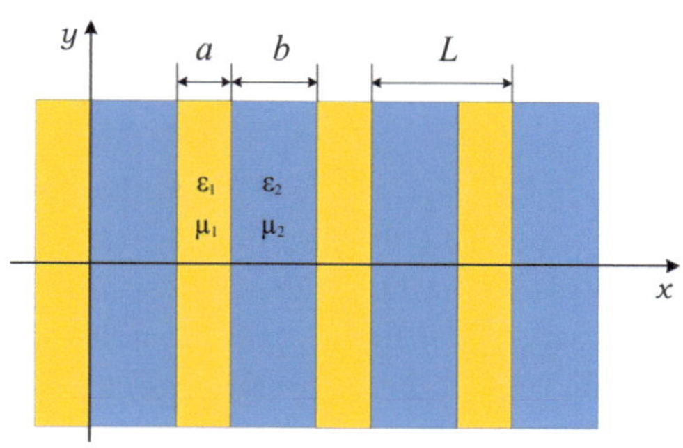

Figure 1: Schematic of the periodic structure.

$\vec{E}\leftrightarrow\vec{H}$ and material parameters $\overleftrightarrow{\varepsilon}\leftrightarrow-\overleftrightarrow{\mu}$ is done, the receiving of general equations, from which the equations for magneto-gyrotropic media (ferrite), electro-gyrotropic media (plasma), and gyrotropic media (bianisotropic media) can be obtained easily, turns to be more simple than in each of the mentioned particular cases separately.

Indeed, it follows from Maxwell equations:

$$rot\ \vec{E}=-\frac{1}{c}\frac{\partial\vec{B}}{\partial t},\quad rot\ \vec{H}=\frac{1}{c}\frac{\partial\vec{D}}{\partial t},\tag{2}$$

where $\vec{D}=\overleftrightarrow{\varepsilon}\vec{E}$ is inductance vector of electric field and $\vec{B}=\overleftrightarrow{\mu}\vec{H}$ is inductance vector of magnetic field. Components of these vectors can be written as:

$$D_x=\left(\overleftrightarrow{\varepsilon}\vec{E}\right)_x=\varepsilon E_x-i\varepsilon_a E_y,\quad D_y=\left(\overleftrightarrow{\varepsilon}\vec{E}\right)_y=i\varepsilon_a E_x+\varepsilon E_y,\quad D_z=\left(\overleftrightarrow{\varepsilon}\vec{E}\right)_z=\varepsilon_{\parallel}E_z,$$

$$B_x=\left(\overleftrightarrow{\mu}\vec{H}\right)_x=\mu H_x-i\mu_a H_y,\quad B_y=\left(\overleftrightarrow{\mu}\vec{H}\right)_y=i\mu_a H_x+\mu H_y,\quad B_z=\left(\overleftrightarrow{\mu}\vec{H}\right)_z=\mu_{\parallel}H_z.$$

In general case, one can obtain two connected differential equations for longitudinal components of electromagnetic fields E_z and H_z (along Oz axis and the direction of bias magnetic field $\vec{H}_0=\vec{z}_0H_0$). In two-dimensional case ($\partial/\partial z=0$), these equations can be broken up into two independent Helmholtz equations with respect to the selected longitudinal field components E_z and H_z (TM and TE waves) [26]. Indeed, we can show it for two-dimensional case and harmonic dependence $\exp(-i\omega t)$ of the fields on time t.

Using the relation between field components E_z and H_z via transverse field components, one can obtain the Helmholtz equation for two polarizations E_z and H_z, respectively:

$$\frac{\partial^2 E_z}{\partial x^2}+\frac{\partial^2 E_z}{\partial y^2}+k^2\mu_{\perp}\varepsilon_{\parallel}E_z=0,\quad \frac{\partial^2 H_z}{\partial x^2}+\frac{\partial^2 H_z}{\partial y^2}+k^2\varepsilon_{\perp}\mu_{\parallel}H_z=0.\tag{3}$$

Here:

$$\mu_{\perp j}=\mu_j\left(1-\frac{\mu_{aj}^2}{\mu_j^2}\right),\quad \varepsilon_{\perp j}=\varepsilon_j\left(1-\frac{\varepsilon_{aj}^2}{\varepsilon_j^2}\right).$$

Eq. (3) describes TM waves (H_x, H_y, E_z), E_z-polarization (s-polarization) and TE waves (E_x, E_y, H_z), H_z-polarization (p-polarization). Therefore, the vector of the electric field for s-polarization is directed perpendicular to the xy plane and the vector of the electric field for the p-polarization is parallel to this plane. It is necessary to use the boundary conditions for the tangential components of the electric and magnetic fields at the media interfaces to solve the boundary electrodynamic problem for the eigenvalues and the eigenfunctions of the Laplace operator. We use the conditions of continuity of the components E_z and H_y for TM waves and H_z and E_y for TE waves.

This analysis shows the important conclusion that in the general case of a gyrotropic medium, the fields of TM and TE waves with respect to the direction of the bias magnetic field $\vec{H}_0$ in the two-dimensional case are divided into two independent solutions of the Maxwell equations. Moreover, Eq. (3) and the expressions for the tangential components of fields E_y and H_y yield the permutation duality principle for the TM and TE waves. These equations are transformed each other after replacing the component of field E_z by H_z, and simultaneously, the effective permeability $\mu_\perp$ should be replaced by the effective permittivity $(-\varepsilon_\perp)$ and $\varepsilon_\parallel$ by $(-\mu_\parallel)$. The boundary conditions for the components E_y and H_y also satisfy the permutation duality principle. Hence, it is possible to simplify this general electrodynamic problem, and we can consider only one type of TM or TE waves for any kind of media in order to obtain a solution for another type of wave. Below, we consider the propagation of E_z-polarized (TM) waves in a gyrotropic PC structure.

To determine the eigenvalues and the corresponding eigenfunctions of the two-layer gyromagnetic MPC, we consider Helmholtz equation for the E_z-polarization with the corresponding boundary conditions for the tangential components of the fields E_z and H_y at the interfaces of the periodic structure layers. For the case of H_z-polarization (TE waves), it is necessary to use the permutation duality principle in the solution for TM waves.

The solution of the Helmholtz equation for TM waves for both tangential components of the fields E_z and H_y can be written in the following form:

$$\begin{aligned} E_z^1(x,y) &= \left(a_{n-1}e^{i\xi_1(x-(n-1)L)} + b_{n-1}e^{-i\xi_1(x-(n-1)L)}\right)e^{i\beta y}, \quad b < x-(n-1)L < L \\ H_y^1(x,y) &= \frac{\xi_1}{-k\mu_{\perp 1}}\left(a_{n-1}g_1^+ e^{i\xi_1(x-(n-1)L)} - b_{n-1}g_1^- e^{-i\xi_1(x-(n-1)L)}\right)e^{i\beta y}, \end{aligned} \tag{4}$$

$$\begin{aligned} E_z^2(x,y) &= \left(c_n e^{i\xi_2(x-nL)} + d_n e^{-i\xi_2(x-nL)}\right)e^{i\beta y}, \quad 0 < x-nL < b \\ H_y^2(x,y) &= \frac{\xi_2}{-k\mu_{\perp 2}}\left(c_n g_2^+ e^{i\xi_2(x-nL)} - d_n g_2^- e^{-i\xi_2(x-nL)}\right)e^{i\beta y}. \end{aligned} \tag{5}$$

Here, $n = 1, 2, \ldots$ is a number of the period; $\xi_j = \sqrt{k^2\mu_{\perp j}\varepsilon_{\parallel j} - \beta^2}$ are the transverse wave numbers in gyrotropic layers in the direction of the Ox axis; β is the longitudinal wave number of the MPC; a_n, b_n, c_n, and d_n are the wave amplitudes in the layers; and $g_j^\pm = 1 \pm i\frac{\mu_{aj}}{\mu_j\xi_j}\beta$.

Let us note one feature in the expressions for the electromagnetic fields (Eqs. (4) and (5)). The presence of gyrotropy in the layers ($\mu_{aj} \neq 0$) leads to the fact that the distributions of the amplitude of the fields in the layers differ by factors $g_j^\pm$ for the forward and backward waves, which propagating in the layers along the direction of periodicity.

To find the dispersion equation that relates the longitudinal wave number β with the structure parameters for a given frequency ω, it is necessary to use the boundary conditions on the interfaces of the layers and the Floquet-Bloch theorem [27, 28] for the periodic structure. Using the conditions of continuity of the E_z and H_y components of the fields at the interfaces $x-(n-1)L = 0$ and $x = nL$, we get matrix equations:

$$\begin{pmatrix} a_{n-1} \\ b_{n-1} \end{pmatrix} = \begin{pmatrix} a_{11} & a_{12} \\ a_{21} & a_{22} \end{pmatrix} \begin{pmatrix} c_n \\ d_n \end{pmatrix}, \quad \begin{pmatrix} c_n \\ d_n \end{pmatrix} = \begin{pmatrix} b_{11} & b_{12} \\ b_{21} & b_{22} \end{pmatrix} \begin{pmatrix} a_n \\ b_n \end{pmatrix}. \tag{6}$$

where

$$a_{11} = \frac{1}{2}\left(g_1^- + \frac{\xi_2}{\xi_1}\frac{\mu_{\perp 1}}{\mu_{\perp 2}} g_2^+\right) e^{-i\xi_2 L}, \quad a_{12} = \frac{1}{2}\left(g_1^- - \frac{\xi_2}{\xi_1}\frac{\mu_{\perp 1}}{\mu_{\perp 2}} g_2^-\right) e^{i\xi_2 L}, \tag{7}$$

$$a_{21} = \frac{1}{2}\left(g_1^+ - \frac{\xi_2}{\xi_1}\frac{\mu_{\perp 1}}{\mu_{\perp 2}} g_2^+\right) e^{-i\xi_2 L}, \quad a_{22} = \frac{1}{2}\left(g_1^+ + \frac{\xi_2}{\xi_1}\frac{\mu_{\perp 1}}{\mu_{\perp 2}} g_2^-\right) e^{i\xi_2 L}. \tag{8}$$

$$b_{11} = \frac{1}{2}\left(g_2^- + \frac{\xi_1}{\xi_2}\frac{\mu_{\perp 2}}{\mu_{\perp 1}} g_1^+\right) e^{i\xi_2 a} e^{-i\xi_1 a}, \quad b_{12} = \frac{1}{2}\left(g_2^- - \frac{\xi_1}{\xi_2}\frac{\mu_{\perp 2}}{\mu_{\perp 1}} g_1^-\right) e^{i\xi_2 a} e^{i\xi_1 a}, \tag{9}$$

$$b_{21} = \frac{1}{2}\left(g_2^+ - \frac{\xi_1}{\xi_2}\frac{\mu_{\perp 2}}{\mu_{\perp 1}} g_1^+\right) e^{-i\xi_2 a} e^{-i\xi_1 a}, \quad b_{22} = \frac{1}{2}\left(g_2^+ + \frac{\xi_1}{\xi_2}\frac{\mu_{\perp 2}}{\mu_{\perp 1}} g_1^-\right) e^{-i\xi_2 a} e^{i\xi_1 a}. \tag{10}$$

Eliminating the coefficients c_n, d_n in the matrix equations (Eq. (6)), we obtain the relation for the coefficients in identical layers for two neighboring periods of the structure:

$$\begin{pmatrix} a_{n-1} \\ b_{n-1} \end{pmatrix} = \begin{pmatrix} a_{11} & a_{12} \\ a_{21} & a_{22} \end{pmatrix} \begin{pmatrix} c_n \\ d_n \end{pmatrix} = \begin{pmatrix} a_{11} & a_{12} \\ a_{21} & a_{22} \end{pmatrix} \begin{pmatrix} b_{11} & b_{12} \\ b_{21} & b_{22} \end{pmatrix} \begin{pmatrix} a_n \\ b_n \end{pmatrix} = \begin{pmatrix} A & B \\ C & D \end{pmatrix} \begin{pmatrix} a_n \\ b_n \end{pmatrix}. \tag{11}$$

The elements of the ABCD matrix are calculated by the rule of multiplying two matrices. Using the Eqs. (7)–(10), we find the elements of the given transfer matrix, namely:

$$A = \left\{ \cos\xi_2 b - i\frac{1}{2}\left[\frac{\xi_1}{\xi_2}\frac{\mu_{\perp 2}}{\mu_{\perp 1}} + \frac{\xi_2}{\xi_1}\frac{\mu_{\perp 1}}{\mu_{\perp 2}} + \frac{\beta^2}{\xi_1\xi_2}\frac{\mu_{\perp 2}}{\mu_{\perp 1}}\left(\frac{\mu_{a1}}{\mu_1} - \frac{\mu_{\perp 1}}{\mu_{\perp 2}}\frac{\mu_{a2}}{\mu_2}\right)^2\right] \sin\xi_2 b \right\} e^{-i\xi_1 a}, \tag{12}$$

$$D = \left\{ \cos\xi_2 b + i\frac{1}{2}\left[\frac{\xi_1}{\xi_2}\frac{\mu_{\perp 2}}{\mu_{\perp 1}} + \frac{\xi_2}{\xi_1}\frac{\mu_{\perp 1}}{\mu_{\perp 2}} + \frac{\beta^2}{\xi_1\xi_2}\frac{\mu_{\perp 2}}{\mu_{\perp 1}}\left(\frac{\mu_{a1}}{\mu_1} - \frac{\mu_{\perp 1}}{\mu_{\perp 2}}\frac{\mu_{a2}}{\mu_2}\right)^2\right] \sin\xi_2 b \right\} e^{i\xi_1 a}, \tag{13}$$

$$B = i\frac{1}{2}\sin\xi_2 b\left\{ -\frac{\xi_2}{\xi_1}\frac{\mu_{\perp 1}}{\mu_{\perp 2}} + \frac{\xi_1}{\xi_2}\frac{\mu_{\perp 2}}{\mu_{\perp 1}}\left[1 - i\frac{\beta}{\xi_1}\left(\frac{\mu_{a1}}{\mu_1} - \frac{\mu_{a2}}{\mu_2}\frac{\mu_{\perp 1}}{\mu_{\perp 2}}\right)\right]^2 \right\} e^{i\xi_1 a}, \tag{14}$$

$$C = -i\frac{1}{2}\sin\xi_2 b\left\{ -\frac{\xi_2}{\xi_1}\frac{\mu_{\perp 1}}{\mu_{\perp 2}} + \frac{\xi_1}{\xi_2}\frac{\mu_{\perp 2}}{\mu_{\perp 1}}\left[1 + i\frac{\beta}{\xi_1}\left(\frac{\mu_{a1}}{\mu_1} - \frac{\mu_{a2}}{\mu_2}\frac{\mu_{\perp 1}}{\mu_{\perp 2}}\right)\right]^2 \right\} e^{-i\xi_1 a}. \tag{15}$$

An important property of the *ABCD* matrix is the unimodularity property, when the ratio between the elements of the matrix is fulfilled: $AD - BC = 1$. Using the expressions for the elements of the matrix *ABCD*, one can show that this condition is satisfied. We note that when ξ_1 is a real number, then the elements of the matrix $A = D^*$ и $B = C^*$ are pairwise conjugate.

The resulting matrix equation (Eq. (11)), which determines the relationship of the unknown coefficients in two identical layers of different periods of the periodic structure and the

Floquet-Bloch theorem, allows us to find the characteristic (dispersion) equation for determining the previously introduced unknown longitudinal wave number β for a wave propagating along gyrotropic layers (along the Oy axis) and the Floquet-Bloch wave number K. According to the Floquet-Bloch theorem in its matrix formulation, one can obtain

$$\begin{pmatrix} A & B \\ C & D \end{pmatrix}\begin{pmatrix} a_n \\ b_n \end{pmatrix} = e^{-iKL}\begin{pmatrix} a_n \\ b_n \end{pmatrix}. \tag{16}$$

The phase factor e^{-iKL} is the eigenvalue of the transfer-matrix ABCD, which is determined from the characteristic equation:

$$e^{-iKL} = \frac{1}{2}(A+D) \pm i\sqrt{1-\left[\frac{1}{2}(A+D)\right]^2}. \tag{17}$$

The unknown real values of the roots of the characteristic equation have the form:

$$K_{TM}(\beta) = \frac{1}{L}\arccos\left\{ \cos\xi_2 b\cos\xi_1 a - \frac{1}{2}\left[\begin{array}{l} \frac{\xi_1}{\xi_2}\frac{\mu_{\perp 2}}{\mu_{\perp 1}} + \frac{\xi_2}{\xi_1}\frac{\mu_{\perp 1}}{\mu_{\perp 2}} + \\ + \frac{\beta^2}{\xi_1\xi_2}\frac{\mu_{\perp 2}}{\mu_{\perp 1}}\left(\frac{\mu_{a1}}{\mu_1} - \frac{\mu_{\perp 1}}{\mu_{\perp 2}}\frac{\mu_{a2}}{\mu_2}\right)^2 \end{array}\right]\sin\xi_2 b\sin\xi_1 a \right\}. \tag{18}$$

It is easy to show that this expression is transformed to the well-known solution of the dispersion equation for the case of two magnetodielectric layers ($\mu_{a1} = \mu_{a2} = 0$) [25].

Note that the longitudinal wave number enters into equation as β squared. This indicates that its absolute value is the same for opposite directions of wave propagation along the Oy axis. That is, the dispersion is the same for forward and backward waves propagating along the layers. However, the field distributions for the case of gyrotropic media in the direction of periodicity for forward and backward waves are different.

Using the permutation duality principle, we found the solutions of the electrodynamic problem for TE wave propagation in the gyrotropic MPC. For this case, we change the material parameters according to the rule $\overleftrightarrow{\mu} \leftrightarrow -\overleftrightarrow{\varepsilon}$ in the transfer-matrix elements (Eqs. (12)–(15)), in the dispersion relation, and in the solution (Eq. (18)). Then, we obtain

$$A_{TE} = \left\{ \cos\xi_2 b - i\frac{1}{2}\left[\frac{\xi_1}{\xi_2}\frac{\varepsilon_{\perp 2}}{\varepsilon_{\perp 1}} + \frac{\xi_2}{\xi_1}\frac{\varepsilon_{\perp 1}}{\varepsilon_{\perp 2}} + \frac{\beta^2}{\xi_1\xi_2}\frac{\varepsilon_{\perp 2}}{\varepsilon_{\perp 1}}\left(\frac{\varepsilon_{a1}}{\varepsilon_1} - \frac{\varepsilon_{\perp 1}}{\varepsilon_{\perp 2}}\frac{\varepsilon_{a2}}{\varepsilon_2}\right)^2\right]\sin\xi_2 b \right\} e^{-i\xi_1 a}, \tag{19}$$

$$D_{TE} = \left\{ \cos\xi_2 b + i\frac{1}{2}\left[\frac{\xi_1}{\xi_2}\frac{\varepsilon_{\perp 2}}{\varepsilon_{\perp 1}} + \frac{\xi_2}{\xi_1}\frac{\varepsilon_{\perp 1}}{\varepsilon_{\perp 2}} + \frac{\beta^2}{\xi_1\xi_2}\frac{\varepsilon_{\perp 2}}{\varepsilon_{\perp 1}}\left(\frac{\varepsilon_{a1}}{\varepsilon_1} - \frac{\varepsilon_{\perp 1}}{\varepsilon_{\perp 2}}\frac{\varepsilon_{a2}}{\varepsilon_2}\right)^2\right]\sin\xi_2 b \right\} e^{i\xi_1 a}, \tag{20}$$

$$B_{TE} = i\frac{1}{2}\sin\xi_2 b\left\{ -\frac{\xi_2}{\xi_1}\frac{\varepsilon_{\perp 1}}{\varepsilon_{\perp 2}} + \frac{\xi_1}{\xi_2}\frac{\varepsilon_{\perp 2}}{\varepsilon_{\perp 1}}\left[1 - i\frac{\beta}{\xi_1}\left(\frac{\varepsilon_{a1}}{\varepsilon_1} - \frac{\varepsilon_{a2}}{\varepsilon_2}\frac{\varepsilon_{\perp 1}}{\varepsilon_{\perp 2}}\right)\right]^2 \right\} e^{i\xi_1 a}, \tag{21}$$

$$C_{TE} = -i\frac{1}{2}\sin\xi_2 b\left\{-\frac{\xi_2}{\xi_1}\frac{\varepsilon_{\perp 1}}{\varepsilon_{\perp 2}} + \frac{\xi_1}{\xi_2}\frac{\varepsilon_{\perp 2}}{\varepsilon_{\perp 1}}\left[1 + i\frac{\beta}{\xi_1}\left(\frac{\varepsilon_{a1}}{\varepsilon_1} - \frac{\varepsilon_{a2}}{\varepsilon_2}\frac{\varepsilon_{\perp 1}}{\varepsilon_{\perp 2}}\right)\right]^2\right\}e^{-i\xi_1 a}. \tag{22}$$

Here $\xi_1 = \sqrt{k^2\varepsilon_{\perp 1}\mu_{\parallel 1} - \beta^2}$ and $\xi_2 = \sqrt{k^2\varepsilon_{\perp 2}\mu_{\parallel 2} - \beta^2}$.

It is apparent that in this case, one can write

$$K_{TE}(\beta) = \frac{1}{L}\arccos\left\{\cos\xi_2 b\cos\xi_1 a - \frac{1}{2}\left[\begin{array}{l}\frac{\xi_1}{\xi_2}\frac{\varepsilon_{\perp 2}}{\varepsilon_{\perp 1}} + \frac{\xi_2}{\xi_1}\frac{\varepsilon_{\perp 1}}{\varepsilon_{\perp 2}} + \\ + \frac{\beta^2}{\xi_1\xi_2}\frac{\varepsilon_{\perp 2}}{\varepsilon_{\perp 1}}\left(\frac{\varepsilon_{a1}}{\varepsilon_1} - \frac{\varepsilon_{\perp 1}}{\varepsilon_{\perp 2}}\frac{\varepsilon_{a2}}{\varepsilon_2}\right)^2\end{array}\right]\sin\xi_2 b\sin\xi_1 a\right\}. \tag{23}$$

In the absence of gyrotropy ($\varepsilon_{aj}=0$), this solution is transformed into the well-known expression for the Bloch wave number for TE waves for the magnetodielectric PC [25].

The given elements of the transfer-matrix *ABCD* and the solutions of the dispersion equations (Eqs. (18) and (23)) for both *s*- and *p*-polarizations (TM and TE waves) are suitable for analysis of wide variety of MPCs with different material parameters: two isotropic layers on the crystal period (dielectric, magnetic, magnetodielectric), one isotropic layer with another anisotropic layer, two gyroelectric layers or gyromagnetic ones or a combination of them, and two gyrotropic layers. Such an abundance of variants makes Eq. (17) universal in terms of analyzing the dispersion characteristics of TE and TM waves and establishing the features of their propagation in various one-dimensional MPCs.

2.2. Eigen regimes of MPCs

We perform the analysis of the features for the propagation of the electromagnetic waves in different MPCs for various eigenmodes. We identified 10 variants of such regimes.

1. *The crystal contains two layers of magnetodielectric*: $\varepsilon_{a1}=\varepsilon_{a2}=\mu_{a1}=\mu_{a2}=0$.

2. *The crystal contains magnetodielectric layer,* $\varepsilon_{a1}=\mu_{a1}=0$, *and the layer of a semiconductor plasma*: $\varepsilon_{a2}\neq 0$ and $\mu_{a2}=0$.

3. *The crystal contains magnetodielectric layer,* $\varepsilon_{a1}=\mu_{a1}=0$, *and the ferrite layer:* $\mu_{a2}\neq 0$ and $\varepsilon_{a2}=0$.

Taking into account the permutation duality principle, we obtain equations analogous to variant 2.

4. *The crystal contains magnetodielectric layer and gyrotropic one*: $\varepsilon_{a1}=\mu_{a1}=0$, $\varepsilon_{a2}\neq 0$, and $\mu_{a2}\neq 0$.

5. *The crystal contains layer of semiconductor plasma:* $\varepsilon_{a1}\neq 0$ *and* $\mu_{a1}=0$ *and the ferrite layer*: $\varepsilon_{a2}=0$ and $\mu_{a2}\neq 0$.

6. *The crystal contains two layers of semiconductor plasma*: $\varepsilon_{a1}\neq 0$, $\varepsilon_{a2}\neq 0$ and $\mu_{a1}=0$, $\mu_{a2}=0$.

7. *The crystal contains two ferrite layers*: $\mu_{a1}\neq 0$, $\mu_{a2}\neq 0$ and $\varepsilon_{a1}=0$, $\varepsilon_{a2}=0$.

Taking into account the permutation duality principle, we obtain equations analogous to variant 6.

8. *The crystal contains the plasma layer and gyrotropic layer*: $\varepsilon_{a1} \neq 0, \varepsilon_{a2} \neq 0$ and $\mu_{a1} = 0, \mu_{a2} \neq 0$.

9. *The crystal contains the ferrite layer and gyrotropic layer*: $\varepsilon_{a1} = 0, \varepsilon_{a2} \neq 0$ and $\mu_{a1} \neq 0, \mu_{a2} \neq 0$.

Taking into account the permutation duality principle, we obtain equations analogous to variant 8.

10. *The crystal contains two gyrotropic layers* $\mu_{a1} \neq 0, \mu_{a2} \neq 0$ and $\varepsilon_{a1} \neq 0, \varepsilon_{a2} \neq 0$

In this case, one can use solutions Eqs. (18) and (23) for TM and TE waves, respectively.

3. Analysis of the propagation of TE and TM waves in MPCs

3.1. General aspects of wave propagation in MPCs

Let us do a physical analysis of the obtained results and determine general rules of electromagnetic wave propagation in MPC. If the wave number $K(\beta)$ is real, then the electromagnetic wave propagates in a MPC without attenuation. These particular Floquet-Bloch wave numbers $K(\beta)$ correspond to the transmission bands and satisfy the condition $|\cos K(\beta)L| \leq 1$. On the other hand, a different behavior is observed when the wave number is complex: $K(\beta) = K^{'}(\beta) + iK^{''}(\beta)$. Such waves cannot propagate in a MPC and thus decay along the direction of periodicity. As a result, the forbidden zones are formed that satisfied the condition $|\cos K(\beta) L| > 1$. This condition allows easy determination of the imaginary part $K^{''}(\beta)$ of the wave number. When $K^{''}(\beta) \neq 0$, then $\sin K^{'}(\beta)L = 0$ and $K^{'}(\beta)L = \pi m$ (where $m = 0, 1, 2\ldots$ are forbidden zone numbers). As a result, we can obtain the equation for $K^{''}(\beta)$: $chK''(\beta)L = (-1)^m \frac{1}{2}(A + D)$.

The value $m = 0$ corresponds to the zeroth forbidden zone where the wave number is purely imaginary. It is important to distinguish two cases: the absence ($\varepsilon_{aj} = 0$, $\mu_{aj} = 0$) and presence ($\varepsilon_{aj} \neq 0$, $\mu_{aj} \neq 0$) of gyrotropy in a MPC. Besides of that, if in j^{th} layer the conditions $k^2\varepsilon_{\perp j}\mu_{\parallel j} - \beta^2 < 0$ or $k^2\mu_{\perp j}\varepsilon_{\parallel j} - \beta^2 < 0$ are satisfied, and in one layer $\varepsilon_{\perp 1} < 0$ (or$\mu_{\perp 1} < 0$) and in another layer $\varepsilon_{\perp 2} > 0(\mu_{\perp 2} > 0)$, then the wave in such layer would decay in amplitude along the x-axis. This wave is a surface wave and delayed in respect to the speed of light.

3.2. Analysis of dispersion characteristics

Let us consider first the case of a MPC in the absence of gyrotropy ($\varepsilon_{aj} = 0$ and $\mu_{aj} = 0$). This structure is a periodic sequence of magnetodielectric layers. Dispersion equations for these structures were previously considered in a simplified version by many authors. However, the first investigation of such characteristic (dispersion) equations was carried out as early as the nineteenth century by Rayleigh [29] in the solution of the one-dimensional Hill equation. If the medium is periodic, then the latter equation becomes the traditional one-dimensional Helmholtz equation. Further investigations were performed by various researchers, for example, Brillouin [30], Yeh et al. [25], and Bass [31].

The solutions of the dispersion equations for the magnetodielectric periodic structures written out in this section include all possible combinations of the signs and magnitudes of the material parameters. As an example, let us consider dispersion characteristics of one-dimensional magnetodielectric PCs for several combinations of the parameters.

Figure 2 shows dispersion curves for the value of the longitudinal wave number $\beta=0.9$. In this case, the periodic structure consists of two dielectrics ($\mu_1=\mu_2=\mu_{\|}=1$) with positive values of permittivity $\varepsilon_1=\varepsilon_{\|1}=2$ and $\varepsilon_2=\varepsilon_{\|2}=9$. The normalized width of the layers is $a/L=0.8$ and $b/L=0.2$. Solid and dotted curves denote the real and imaginary parts of the Floquet-Bloch wave number, respectively. The imaginary parts of the Floquet-Bloch wave number characterize the degree of wave decay in the forbidden bands. The wave decay is different in forbidden zones. Moreover, the maximum attenuation is observed for the zeroth zone.

If we consider a MPC consisting of two different magnetodielectrics, then there are no fundamental differences.

Figure 3 shows the dispersion diagrams for PC with two dielectric layers ($\mu_1=\mu_2=1$) on the structure period and with different signs of the permittivities ($\varepsilon_1>0$ and $\varepsilon_2<0$), namely, $\varepsilon_1=\varepsilon_{\|1}=2$ and $\varepsilon_2=\varepsilon_{\|2}=-6$. First the problem of wave propagation at the boundary of two half-spaces from dielectrics with opposite signs of permittivity values was considered by Sommerfeld [32].

The existence of surface waves at the boundaries of PC layers is illustrated in the dispersion diagram (**Figure 3**) for TE waves (p-polarization). The region of these waves existence is located below the light line $k=\beta(\varepsilon_1)^{-1/2}$ (the solid line on the diagram). Taking into account the identical physical nature of the surface wave existence in PC and Zenneck-Sommerfeld wave for two media [33], this regime can be classified as a modified surface Zenneck-Sommerfeld wave. For this wave, the condition $\beta=k\sqrt{\frac{\varepsilon_1\varepsilon_2}{\varepsilon_1+\varepsilon_2}}$ is satisfied approximately, as follows from the dispersion equation. It should be noted that if we consider PC with magnetic layers, then a surface wave will exist for s-polarization.

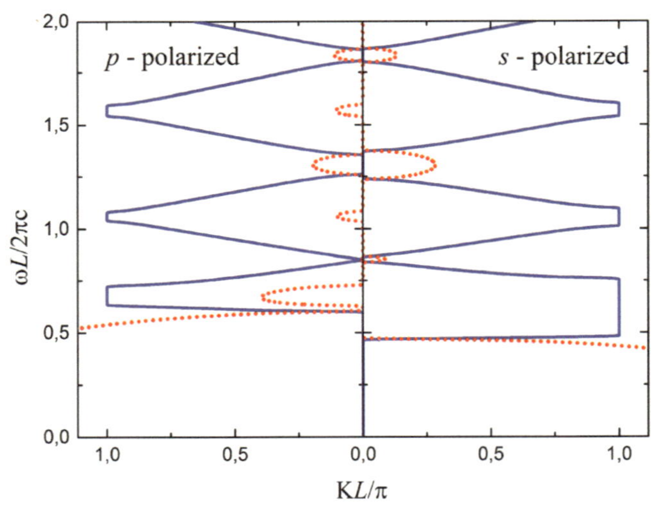

Figure 2: Dispersion characteristics of the one-dimensional photonic crystal.

http://dx.doi.org/10.5772/intechopen.71273

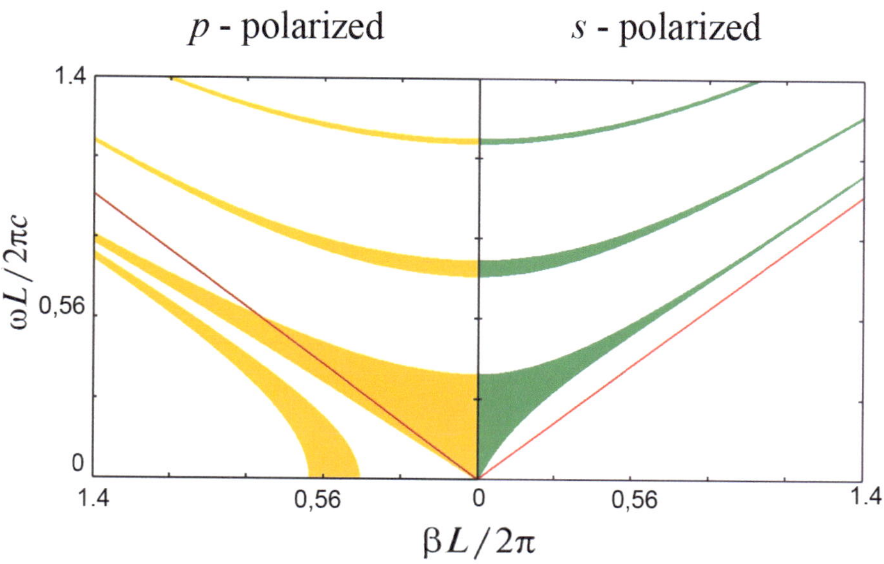

Figure 3. Dispersion diagrams for both polarizations.

With the advent of new artificial media (metamaterials) for which the permittivity and permeability are simultaneously negative, optoelectronics devices with new functionalities are developed. In this connection, it is expedient to consider a MPC, one of whose layers on the structure period is a metamaterial (e.g., $\varepsilon_2 < 0, \mu_2 < 0$).

In **Figure 4**, dispersion diagrams are calculated for the one-dimensional PC with alternative layers of magnetodielectric and metamaterial ($\varepsilon_2 < 0, \mu_2 < 0$). The following parameters were

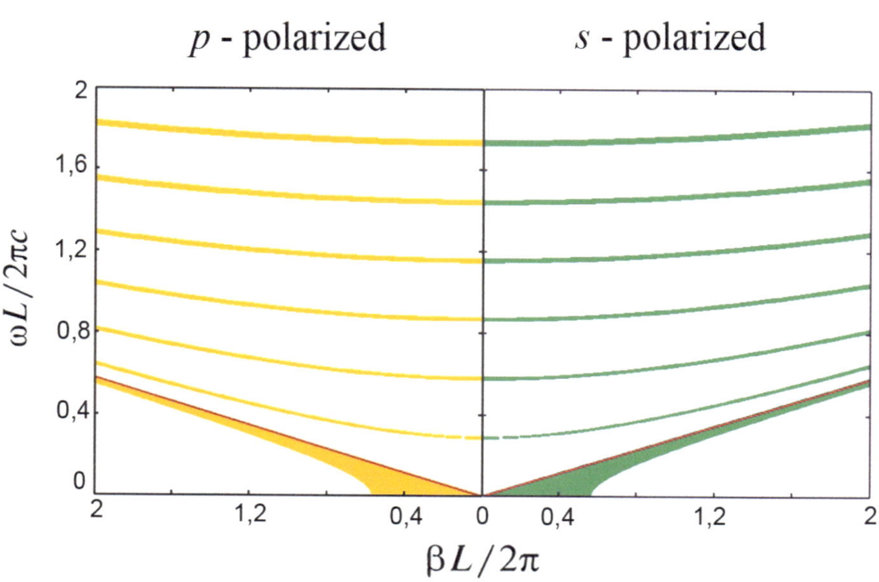

Figure 4. Dispersion diagram of the photonic crystal with metamaterial layer.

selected: $\varepsilon_1 = 4$, $\mu_1 = 3$, $\varepsilon_2 = -2$, $\mu_2 = -6$, and $a/L = 0.5$. **Figure 5a** and **b** shows the dispersion characteristics for such MPC at the values $\beta = 0$ and $\beta = 2\pi/L$, respectively. On the presented graphs, identical dispersion diagrams for both polarizations are seen. Therefore, it is a case of polarization indifference for the forbidden zones and transmission ones. For the presented case, the refractive indices of the layers are the same in absolute value and have phase difference of π, namely, $n_1 = \sqrt{\varepsilon_1 \mu_1} = \sqrt{12}$ and $n_2 = \sqrt{|\varepsilon_2| e^{-i\pi} |\mu_2| e^{-i\pi}} = e^{-i\pi}\sqrt{12}$. For this case, the complete Floquet-Bloch wave transmission in PC for two polarizations is possible.

It is clear that there are transmission bands only for surface slow waves (modified Zenneck-Sommerfeld waves) that are located below the light line $\omega\sqrt{12} = \beta c$. For bulk waves ($k^2 \varepsilon_j \mu_j > \beta^2$), the transmission bands degenerate into curves. These curves are described by equation $(2\pi m)^2 = (k^2 |\varepsilon_j \mu_j| - \beta^2) L^2$ ($m = 1, 2, \ldots$) for both polarizations.

Dispersion curves for real K' and imaginary K'' values (**Figure 5a** and **b**) show that the wave decay for both polarizations is the same. The propagation of waves is observed at discrete points on which the conditions $K'' = 0$ and $|\cos K(\beta) L| = 1$ are satisfied. These points are located outside the transmission bands of surface waves. The phenomenon of polarization indifference makes it possible to develop devices with a finite number of periods of a MPC in which a complete narrow-band propagation of the wave is possible simultaneously for a whole spectrum of discrete frequencies and both polarizations [34].

We now turn to an analysis of the propagation of waves in MPC consisting of a magnetodielectric and a semiconductor plasma layer [35]. It is advisable to consider two cases in the presence of gyrotropy of the medium ($\varepsilon_{a2} \neq 0$), when $\varepsilon_{\perp 2} > 0$ and $\varepsilon_{\perp 2} < 0$ for TE waves. We note that the dispersion characteristics for TM waves are the same as for a conventional magnetodielectric. Let us first consider the features of TE wave propagation in MPC when the plasma layer gyrotropy is such that condition $\varepsilon_{\perp 2} > 0$ ($\varepsilon_{a2} < \varepsilon_2$) is fulfilled. From an analysis of the solutions of the dispersion equation, it follows that in the periodic structure at such a value of the effective permittivity of a plasma medium, there can exist two modes of TE wave propagation. The first

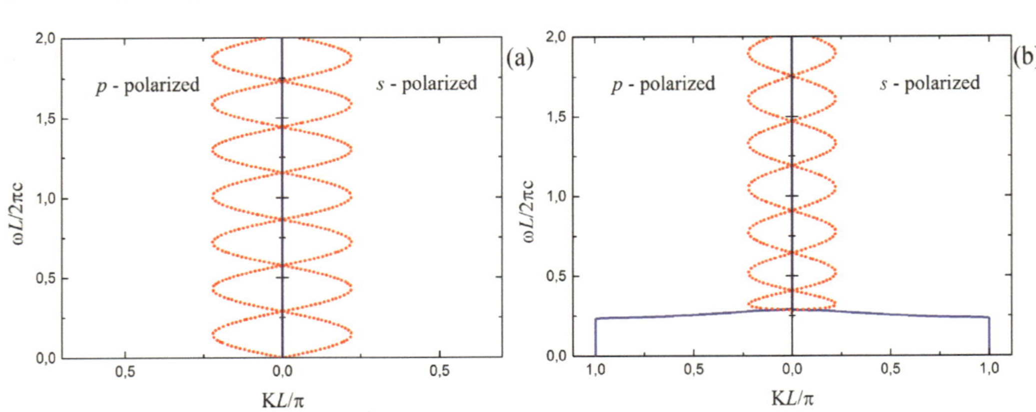

Figure 5. Dispersion characteristics for different values of longitudinal wavenumber: (a) $\beta = 0$; (b) $\beta = 2\pi/L$.

mode is bulk waves whose domain of existence is determined by the simultaneous fulfillment of two conditions: $k^2\varepsilon_{\perp 1}\mu_{\parallel 1}-\beta^2>0$ and $k^2\varepsilon_{\perp 2}\mu_{\parallel 2}-\beta^2>0$. The second mode is the propagation mode of gyrotropic surface waves. This mode is observed when three conditions are fulfilled simultaneously: $k^2\varepsilon_{\perp 1}\mu_{\parallel 1}-\beta^2<0$, $k^2\varepsilon_{\perp 2}\mu_{\parallel 2}-\beta^2<0$, and $\frac{\xi_1}{\xi_2}\frac{\varepsilon_{\perp 2}}{\varepsilon_1}+\frac{\xi_2}{\xi_1}\frac{\varepsilon_1}{\varepsilon_{\perp 2}}+\frac{\beta^2}{\xi_1\xi_2}\frac{\varepsilon_1}{\varepsilon_{\perp 2}}\left(\frac{\varepsilon_{a2}}{\varepsilon_2}\right)^2<0$.

If the above conditions are satisfied, then the existence of real values of the Floquet-Bloch wave number other than zero is possible when the condition $|\cos K(\beta)L|\leq 1$ is fulfilled for pure imaginary values of transverse wave numbers ξ_1 and ξ_2.

The results of calculations of the dispersion diagram are shown in **Figure 6**. The calculation was carried out with the following parameters: $a/L=0.8$, $\mu_1=\mu_2=\mu_{\parallel 1}=\mu_{\parallel 2}=1$, $\varepsilon_1=2$, $\varepsilon_2=4$, $\varepsilon_{a2}=3.2$, and $\varepsilon_{\perp 2}=1.44$. The solid line shows the light line for the first layer on the period of the structure. It can be seen from the figure that when the conditions$k\sqrt{\varepsilon_{\perp 2}\mu_{\parallel 2}}<\beta$, $k\sqrt{\varepsilon_1\mu_{\parallel 1}}<\beta$, and $\frac{\varepsilon_{\perp 2}}{|\varepsilon_{\perp 2}|}\left[\frac{\beta^2}{|\xi_1||\xi_2|}\frac{\varepsilon_1}{|\varepsilon_{\perp 2}|}\left(\frac{\varepsilon_{a2}}{\varepsilon_2}\right)^2-\left(\left|\frac{\xi_1}{\xi_2}\right|\frac{|\varepsilon_{\perp 2}|}{\varepsilon_1}+\left|\frac{\xi_2}{\xi_1}\right|\frac{\varepsilon_1}{|\varepsilon_{\perp 2}|}\right)\right]>0$ are satisfied, there exists a region of real values of the Floquet-Bloch wave number for which the gyrotropic surface wave regime is observed (more shaded area of the transmission zone).

In the case $\varepsilon_{\perp 2}<0$, as for PC with dielectric layers, the regime of a modified Zenneck-Sommerfeld surface wave can exist in MPC. However, in addition to that wave, there is also a gyrotropic surface wave for other parameters of the problem. Indeed, such a wave exists when both transverse wave numbers in two layers are pure imaginary $\left(\xi_1^2<0,\xi_2^2<0\right)$. Also under these conditions and $\varepsilon_{\perp 2}<0$, the condition $\frac{\beta^2}{|\xi_1||\xi_2|}\frac{\varepsilon_1}{|\varepsilon_{\perp 2}|}\left(\frac{\varepsilon_{a2}}{\varepsilon_2}\right)^2<\left|\frac{\xi_1}{\xi_2}\right|\frac{|\varepsilon_{\perp 2}|}{\varepsilon_1}+\left|\frac{\xi_2}{\xi_1}\right|\frac{\varepsilon_1}{|\varepsilon_{\perp 2}|}$ must be fulfilled, that is, opposite to the condition for positive values of the second-layer effective permittivity.

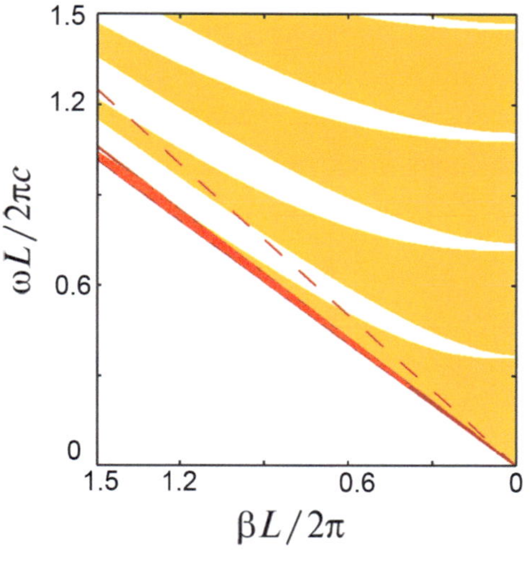

Figure 6. Band structure for TE polarization.

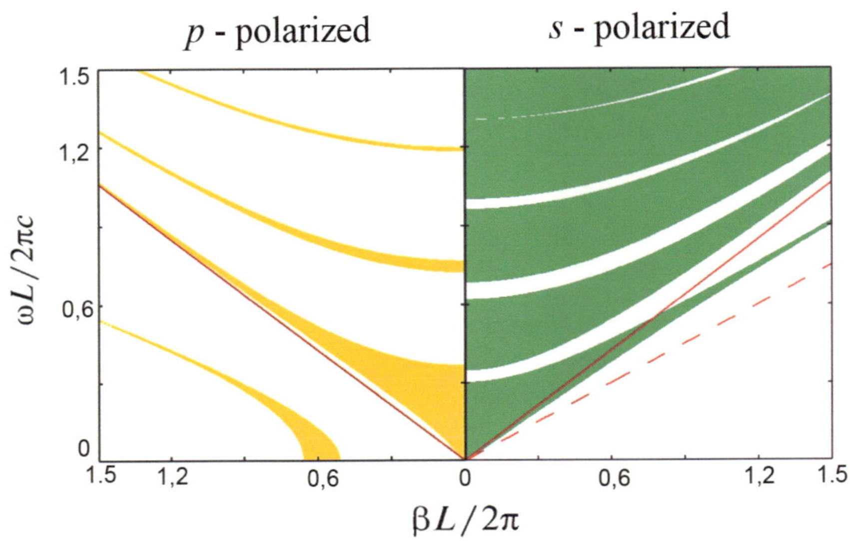

Figure 7. Dispersion diagrams of the magnetophotonic crystal containing plasma layers.

Figure 7 shows the dispersion diagrams for both polarizations in the case $\varepsilon_{\perp 2}<0$. Parameters of the problem were chosen as follows: $a/L=0.8$, $\mu_1=\mu_2=\mu_{\parallel 1}=\mu_{\parallel 2}=1$, $\varepsilon_1=2$, $\varepsilon_2=4$, $\varepsilon_{a2}=6.7$, and $\varepsilon_{\perp 2}=-7.2$. The solid line $k=\beta(\varepsilon_{\perp 1}\mu_{\parallel 1})^{-1/2}$ separates the areas of the existence of bulk waves and surface waves for *p*-polarization. One area relates to the modified Zenneck-Sommerfeld waves, and the other one relates to the gyrotropic surface wave which also exists for positive values of the effective permittivity, as shown above. We note that in the transmission bands of TE waves in dispersion diagrams, there are two regions in which surface waves propagate.

The case of MPC with a magnetodielectric and ferrite layer is analogous to that considered earlier by virtue of the permutation duality principle. Let us now consider the following case, when both layers on the structure period are ferrite with different material parameters. In this case, modes of bulk wave's existence in the transmission bands can be observed when two conditions $\xi_1^2=k^2\mu_{\perp 1}\varepsilon_{\parallel 1}-\beta^2>0$ and $\xi_2^2=k^2\mu_{2\perp}\varepsilon_{\parallel 2}-\beta^2>0$ are fulfilled. Regime of surface waves is realized when opposite conditions ($\xi_1^2<0$, $\xi_2^2<0$) and also the additional condition are fulfilled:

$$\frac{\mu_{\perp 1}}{|\mu_{\perp 1}|}\frac{\mu_{\perp 2}}{|\mu_{\perp 2}|}\left[\left|\frac{\xi_1}{\xi_2}\right|\left|\frac{\mu_{\perp 2}}{\mu_{\perp 1}}\right|+\left|\frac{\xi_2}{\xi_1}\right|\left|\frac{\mu_{\perp 1}}{\mu_{\perp 2}}\right|-\frac{\beta^2}{|\xi_1\xi_2|}\left|\frac{\mu_{\perp 2}}{\mu_{\perp 1}}\right|\left(\frac{\mu_{a1}}{\mu_1}-\frac{\mu_{\perp 1}}{\mu_{\perp 2}}\frac{\mu_{a2}}{\mu_2}\right)^2\right]<0 \qquad (24)$$

Note that this condition is equally suitable for both positive and negative values of the effective magnetic permeability of ferrite $\mu_{\perp j}$. Here, as in the case of the plasma semiconductor layer, the existence of gyrotropic surface waves is possible [36].

In **Figure 8**, we represent the dispersion diagrams of TM waves for the considered above MPC with two ferrite layers at the period of the structure. In the calculation, the following task parameters were chosen: $a/L=0.5$, $\mu_1=\mu_{\parallel 1}=2$, $\mu_2=\mu_{\parallel 2}=3$, $\varepsilon_1=2$, $\varepsilon_2=4$, $\mu_{a2}=2.9, \mu_{\perp 2}=0.197 \mu_{a1}=0.1$, and $\mu_{\perp 1}=1.995$.

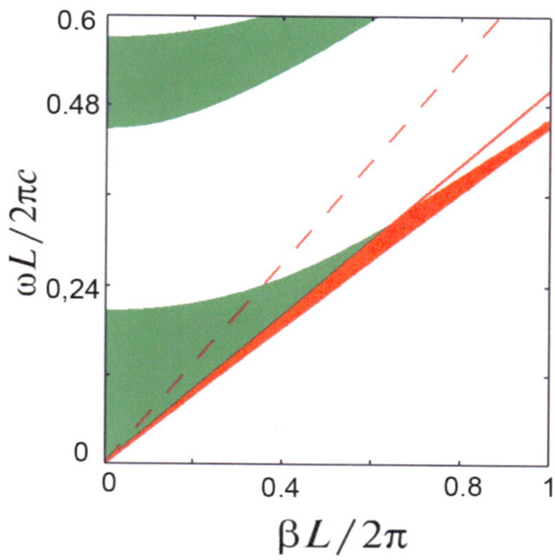

Figure 8. Band structure of the ferrite magnetophotonic crystal.

The solid line in the figure, below which the solutions of the dispersion equation are in the regime of surface waves, is determined by the equation $k=\beta(\mu_{\perp 1}\varepsilon_{\|1})^{-1/2}$. Analysis of the dispersion equation solutions and the conditions for the existence of surface waves show that for the chosen values of the material parameters in the ferrite MPC, there exists a range of parameters on the dispersion diagram that corresponds to the regime of propagation of the surface wave at the boundaries of the layers (in this figure it is marked as more shaded area in transmission zone). We emphasize that the regime of the surface wave is realized with positive values of the both magnetic permeabilities of the ferrite layers.

Figure 9 illustrates the case of the existence of a modified Zenneck-Sommerfeld wave for the case when the effective magnetic permeability of one of the layers is negative. Here, the dispersion diagrams are calculated with the following parameters of the problem: $a/L=0.85$, $\mu_1=\mu_{\|1}=2$, $\mu_2=\mu_{\|2}=3$, $\varepsilon_1=1.5$, $\varepsilon_2=1.8$, $\mu_{a2}=7.9$, $\mu_{\perp 2}=-17.8$, $\mu_{a1}=0.7$, and $\mu_{\perp 1}=1.755$.

Figure 10 illustrates the evolution of dispersion diagrams at change of ferrite effective magnetic permeability $\mu_{\perp 2}(\mu_{a2}=5.6;5.8;6.1)$ for $a/L=0.83$, $\mu_1=\mu_{\|1}=2$, $\mu_2=\mu_{\|2}=3$, $\varepsilon_1=4$, $\varepsilon_2=2$, and $\mu_{a1}=0.4$. A darker shade shows the transmission bands in which the conditions of existence of surface waves are fulfilled. The change of the bias magnetic field value leads to a significant change of the parameters of these areas. An increase in the value of the effective magnetic permeability $\mu_{\perp 2}$ owing to the magnitude μ_{a2} results in a change in the width and location of the transmission band of the modified Zenneck-Sommerfeld surface wave.

Figure 11 shows dispersion diagrams for the modified Zenneck-Sommerfeld surface wave of MPC with two ferrite layers on the structure period at change of width of the layer b ($b/L=0.3;0.5;0.7;0.8$) with $\mu_{\perp 2}<0$ and for the following parameters of the problem: $\mu_1=\mu_{\|1}=2$, $\mu_2=\mu_{\|2}=3$, $\varepsilon_1=4$, $\varepsilon_2=2$, $\mu_{a1}=0.4$, $\mu_{\perp 1}=1.92$, $\mu_{a1}=5.6$, and $\mu_{\perp 2}=-7.45$.

Increase of the second-layer width b leads to an expansion of existence area of surface waves with a simultaneous shift of the bandwidth toward the value $\beta=0$ (**Figure 11a** and **b**). The

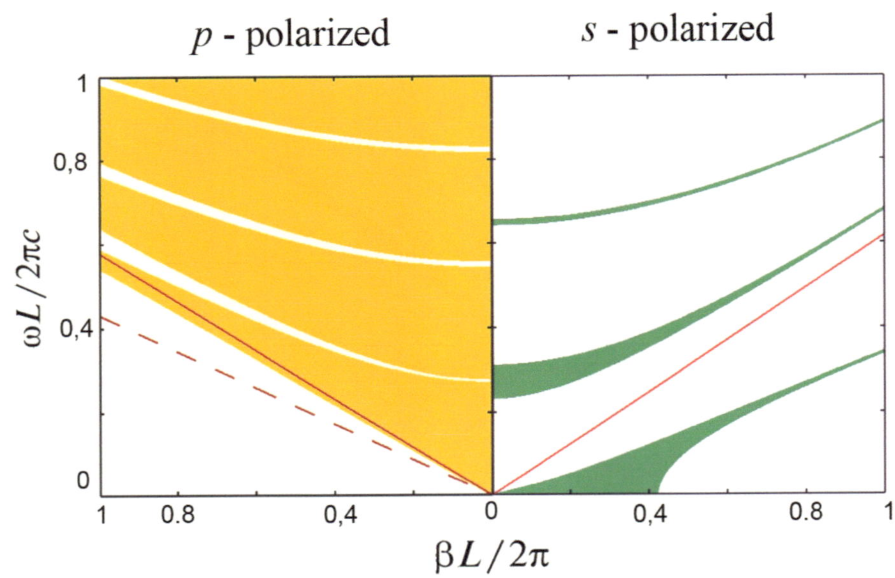

Figure 9. Dispersion diagrams of the ferrite magnetophotonic crystal for both polarizations.

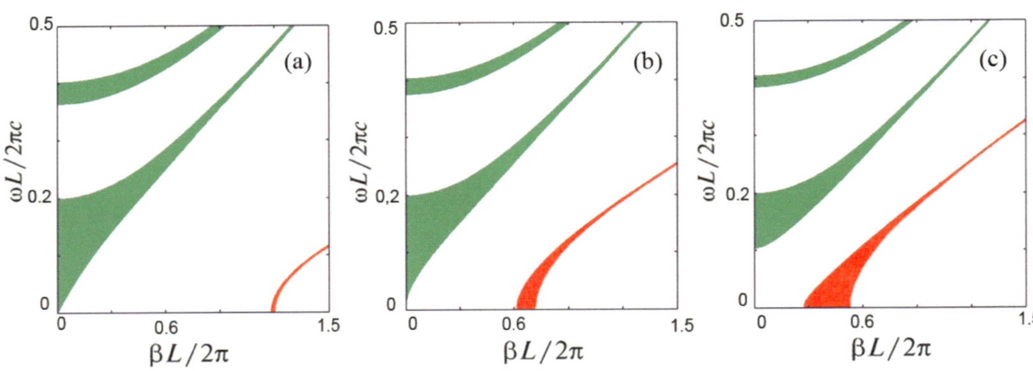

Figure 10. Dispersion diagrams for different values of bias magnetic field: (a) $\mu_{a2}=5.6$; (b) $\mu_{a2}=5.8$; (c) $\mu_{a2}=6.1$.

largest existence area of surface waves is realized in dispersion diagram for approximately equal values of the layer thicknesses. Further increase of the parameter b is accompanied by a displacement of this area to the opposite direction (**Figure 11c** and **d**). Thus the surface wave modes in a ferrite MPC are determined both by the material parameters of the system and by the width of the layers on the period of the structure.

Let us move on the dispersion diagrams for two bigyrotropic layers on the structure period. Taking into account the complete symmetry of the dispersion for TE and TM waves, the case of polarization indifference can be realized in this case. We will show this by example.

Figure 12 shows the dispersion diagrams for TE and TM waves for MPC consisting of two bigyrotropic layers on the period of the structure. **Figure 12a** corresponds to the following

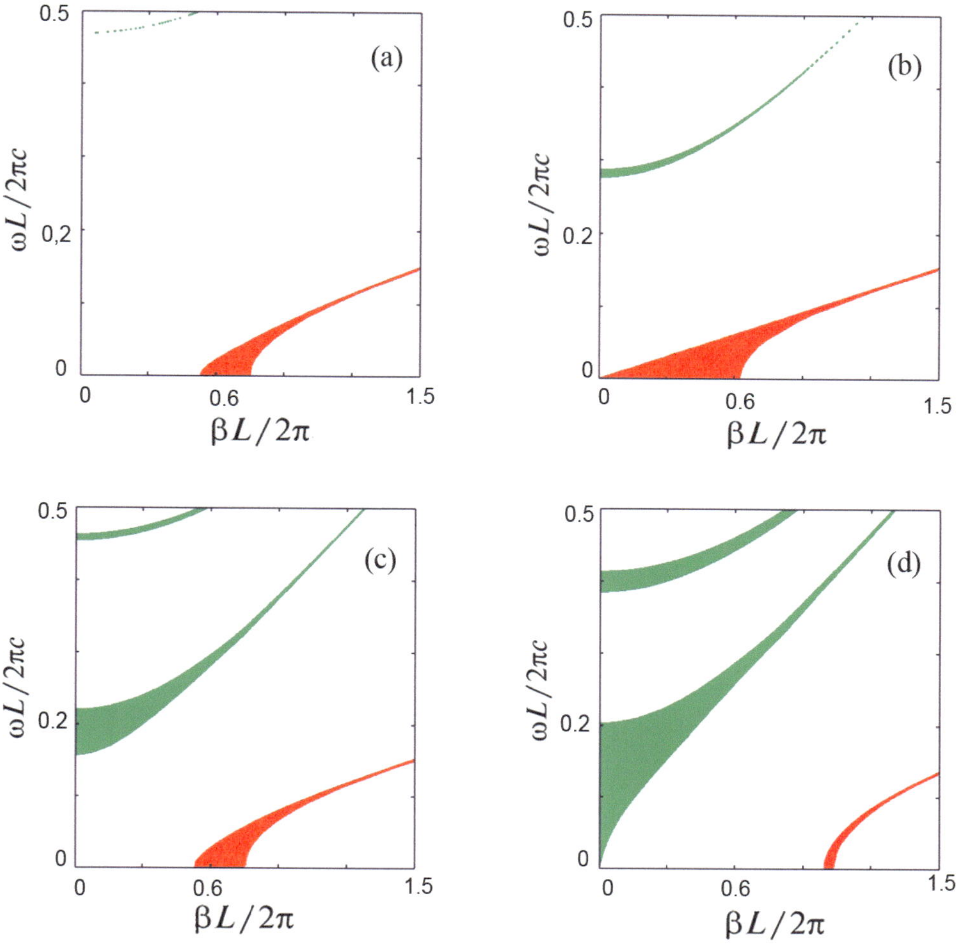

Figure 11. Dispersion diagrams for different values of second layer thickness: (a) $b/L = 0.3$; (b) $b/L = 0.5$; (c) $b/L = 0.7$; (d) $b/L = 0.8$.

values of the parameters: $a/L = 0.85$, $\mu_j = \mu_{\parallel j} = 2$, $\varepsilon_j = \varepsilon_{\parallel j} = 2$, $\varepsilon_{a1} = \mu_{a1} = 0.5$, $\varepsilon_{a2} = \mu_{a2} = 5.2$, $\mu_{\perp 1} = 1.875$, and $\mu_{\perp 2} = -11.52$. In **Figure 12b**, only the values of the effective permittivity and permeability of the second layer differ: $\varepsilon_{a2} = \mu_{a2} = 6.8$.

The solid lines in the figures distinguish the area of fast (upper part of figures) and slow wave (the lower part of figures). The complete identity of the dispersion diagrams for the transmission bands of both surface and bulk waves follows from the figures and formulas (18) and (23). By changing the bias magnetic field, it is possible to control the width and location of transmission bands for both polarizations.

Dispersion diagrams in **Figure 13** correspond to the case when only the surface wave transmission bands for both polarizations are realized. The parameters of the problem were chosen as follows: $a/L = 0.28$, $\mu_1 = \mu_{\parallel 1} = 1.5$, $\mu_2 = \mu_{\parallel 2} = 2$, $\varepsilon_1 = \varepsilon_{\parallel 1} = 1.2$, $\varepsilon_2 = \varepsilon_{\parallel 2} = 1.8$, $\varepsilon_{a1} = 4.95$, $\varepsilon_{a2} = 1.7$, $\mu_{a1} = 0.7$, and $\mu_{a2} = 5$. The polarization sensitivity of the MPC is realized in this case. Only the surface waves with certain polarization can propagate through the periodic structure for defined values of parameters k and β.

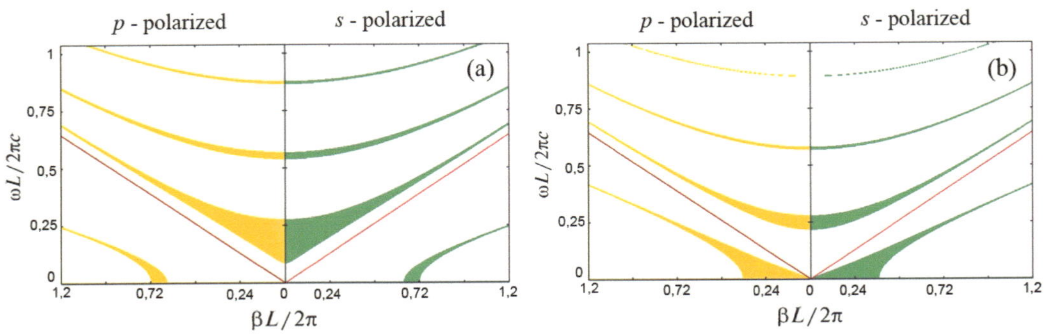

Figure 12. Dispersion diagrams of the bigyrotropic magnetophotonic crystal: (a) $\varepsilon_{a2}=\mu_{a2}=5.2$; (b) $\varepsilon_{a2}=\mu_{a2}=6.8$.

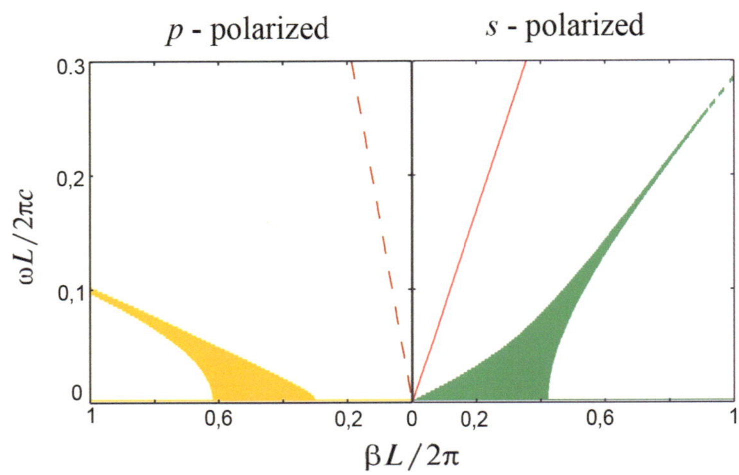

Figure 13. Transmission bands for surface waves in the bigyrotropic magnetophotonic crystal.

Therefore, we have considered the main features of the propagation of TE and TM waves in various magnetophotonic gyrotropic crystals. Important application of this one-dimensional PC theory is the problem of the electromagnetic waves scattering by a structures with limited number of periods.

4. Scattering of a plane wave by a MPC

In this section, the scattering of the plane wave on gyromagnetic MPC with N periods is considered. When a p-polarized plane wave is scattered on MPC with a limited number of periods, the problem is divided into three stages. At the first stage, the problem of scattering of a plane wave on the first gyromagnetic layer of MPC is solved. In the second stage, the coupling between the field coefficients of the first and last layers of the MPC is used in the problem for the MPC modes. And, finally, in the third stage, the problem of wave transmission from the last layer of the structure to the surrounding area is considered. Following

[37], we write out the final expression for the reflection and transmission coefficients for gyrotropic MPC:

$$R = \frac{[(n_{21}M_{11} + n_{22}M_{21})k_{11} + (n_{21}M_{12} + n_{22}M_{22})k_{21}]}{[(n_{11}M_{11} + n_{12}M_{21})k_{11} + (n_{11}M_{12} + n_{12}M_{22})k_{21}]} \tag{25}$$

$$T = \frac{1}{[(n_{11}M_{11} + n_{12}M_{21})k_{11} + (n_{11}M_{12} + n_{12}M_{22})k_{21}]} \tag{26}$$

Here, matrix M is the Nth power of an ABCD matrix. Other notations correspond to Ref. [37]. **Figure 14a** shows the dependences of the transmission coefficient modulus on the normalized frequency in the case of normal wave incidence ($\beta = 0$) on MPC with 20 periods in the regime of bulk waves. Calculation parameters are the same as in **Figure 7**.

There are three transmission bands in the frequency range under consideration. Each of these bands contains resonances which observed with respect to the frequency of the complete transmission of the wave (**Figure 14b** and **c**). Frequency resonances correspond to different modes of the periodic structure. The number of modes is determined by the number of periods of the structure ($N-1$).

Figure 15 depicts the frequency dependences of the transmission coefficient modulus in the regime of the surface waves. In this case the incident angle of wave is greater than the angle of total internal reflection. We can see one transmission band in this case. Inset in **Figure 15** shows enlarged frequency dependence within this band. Complete propagation is observed for each resonant frequency of modes of the limited periodic structure.

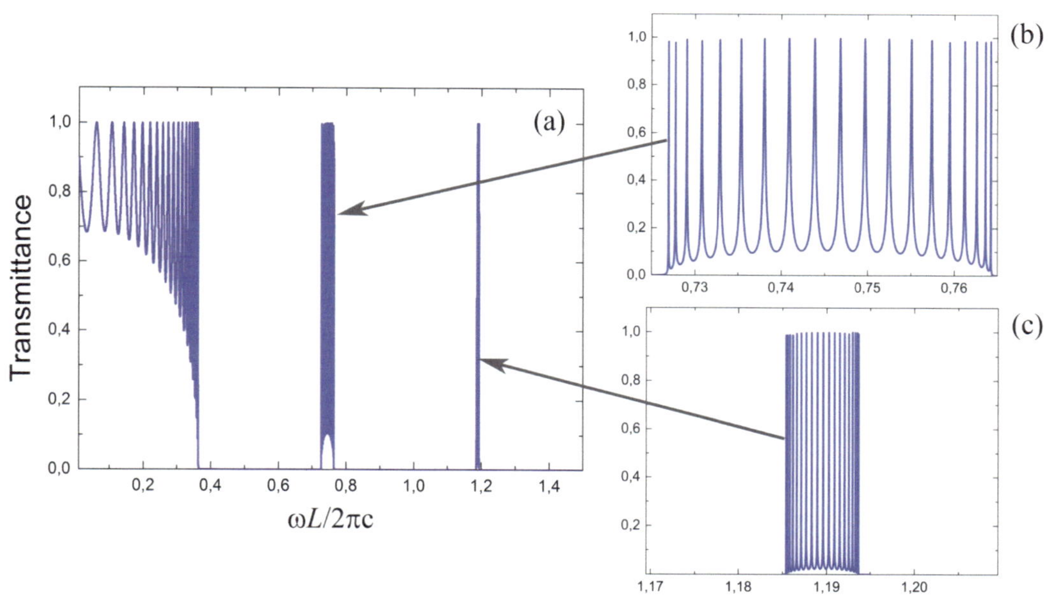

Figure 14. The transmittance vs frequencies for the case of normal incidence of wave: (a) spectral characteristic; (b) fine structure of spectral characteristic in second transmission band; (c) fine structure of spectral characteristic in third transmission band.

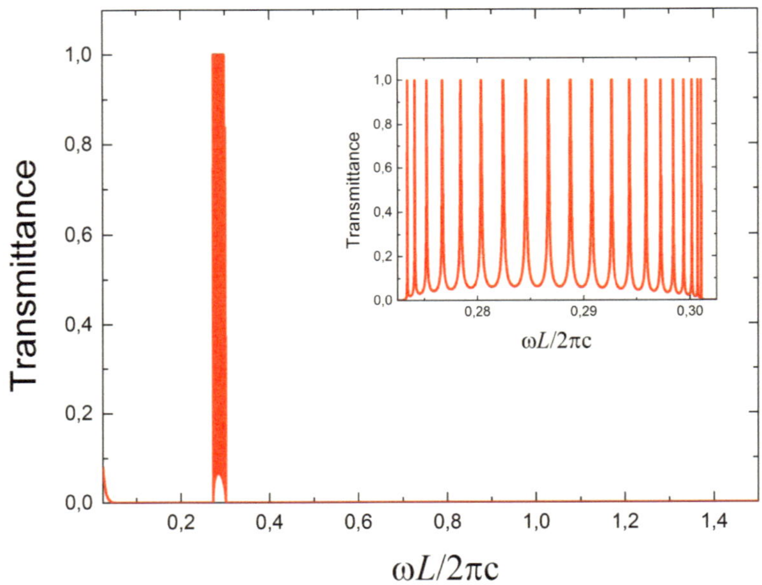

Figure 15. The transmittance vs. frequencies for the case of surface wave mode.

5. Conclusions

The electrodynamic problem is solved for the proper TE and TM waves of a MPC with two gyrotropic layers. The elements of the transmission matrix, the dispersion equation, and its solution are obtained analytically. An analysis of the dispersion properties of TE and TM waves for MPC is carried out, and features of the existence of fast and slow waves are revealed. Different regimes of gyrotropic surface waves are found. The conditions for the existence of surface waves are established for positive and negative values of the permittivity and permeability. Analytic expressions for the reflection and transmission coefficients for a limited MPC are obtained, and their analysis is performed for the regime of bulk and surface waves. Complete transmission of the wave through this structure is realized at resonant frequencies that correspond to different spatial distributions of the mode field in limited MPC.

Author details

Alexander A. Shmat'ko[1], Viktoria N. Mizernik[2], Eugene N. Odarenko[3]* and Viktor T. Lysytsya[1]

*Address all correspondence to: e.n.odarenko@gmail.com

1 V. N. Karazin Kharkiv National University, Kharkiv, Ukraine

2 V. N. Karazin Kharkiv National University, Scientific Physical-Technologic Centre of MES and NAS of Ukraine, Kharkiv, Ukraine

3 V. N. Karazin Kharkiv National University,Kharkiv National University of Radio Electronics, Kharkiv, Ukraine

References

[1] Yablonovitch E. Inhibited spontaneous emission in solid-state physics and electronics. Physical Review Letters. 1987;**58**(20):2059-2062. DOI: 10.1103/PhysRevLett.58.2059

[2] Sakoda K. Optical Properties of Photonic Crystals. 2nd ed. Berlin Heidelberg: Springer-Verlag; 2005. 258 p. DOI: 10.1007/b138376

[3] Lourtioz J-M, Benisty H, Berger V, Gerard J-M, Maystre D, Tchelnokov A. Photonic Crystals. Towards Nanoscale Photonic Devices. 2nd ed. Berlin Heidelberg: Springer-Verlag; 2008. 514 p. DOI: 10.1007/978-3-540-78347-3

[4] Massaro A, editor. Photonic Crystals - Innovative Systems, Lasers and Waveguides. Rijeka, Croatia: InTech; 2012. 358 p. DOI: 10.5772/2632

[5] Busch K, John S. Liquid-crystal photonic-band-gap materials: The tunable electromagnetic vacuum. Physical Review Letters. 1999;**83**(5):967-970. DOI: 10.1103/PhysRevLett.83.967

[6] Inoue M, Levy M, Baryshev AV, editors. Magnetophotonics from Theory to Applications. Berlin Heidelberg: Springer-Verlag; 2013. 228 p. DOI: 10.1007/978-3-642-35509-7

[7] Inoue M, Isamoto K, Yamamoto T, Fujii T. Magneto-optical faraday effect of discontinuous magnetic media with a one-dimensional array structure. Journal of Applied Physics. 1996;**79**(3):1611-1624. DOI: 10.1063/1.361005

[8] Inoue M, Arai K, Fujii T, Abe M. One-dimensional magnetophotonic crystals. Journal of Applied Physics. 1999;**85**(8):5768-5770. DOI: 10.1063/1.370120

[9] Lyubchanskii IL, Dadoenkova NN, Lyubchanskii MI, Shapovalov EA, Rashing T. Magnetic photonic crystals. Journal of Physics D: Applied Physics. 2003;**36**(18):R277-R287. DOI: 10.1088/0022-3727/36/18/R01

[10] Inoue M, Fujikawa R, Baryshev A, Khanikaev A, Lim PB, Uchida H, et al. Magnetophotonic crystals. Journal of Physics D: Applied Physics. 2006;**39**(8):R151-R161. DOI: 10.1088/0022-3727/39/8/R01

[11] Chernovtsev SV, Belozorov DP, Tarapov SI. Magnetically controllable 1D magnetophotonic crystal in millimetre wavelength band. Journal of Physics D: Applied Physics. 2007;**40**(2): 295-299. DOI: 10.1088/0022-3727/40/2/001

[12] Shramkova OV. Transmission properties of ferrite-semiconductor periodic structure. Progress In Electromagnetics Research M. 2009;**7**:71-85. DOI: 10.2528/PIERM09041305

[13] Abdi-Ghaleh R, Namdar A. Circular polarization bandpass filters based on one-dimensional magnetophotonic crystals. Journal of Modern Optics. 2013;**60**(19):1619-1626. DOI: 10.1080/09500340.2013.850540

[14] Aplet LJ, Carson JW. A faraday effect optical isolator. Applied Optics. 1964;**3**(4):544-545. DOI: 10.1364/AO.3.000544

[15] Kato H, Matsushita T, Takayama A, Egawa K, Nishimura M, Inoue M. Properties of one-dimensional Magnetophotonic crystals for use in optical isolator devices. IEEE Transactions on Magnetics. 2002;**38**(5):3246-3248. DOI: 10.1109/TMAG.2002.802511

[16] Chung KH, Kato T, Mito S, Takagi H, Inoue M. Fabrication and characteristics of one-dimensional magnetophotonic crystals for magneto-optic spatial light phase modulators. Journal of Applied Physics. 2010;**107**(9):09A930. DOI: 10.1063/1.3353020

[17] Vasiliev M, Kotov VA, Alameh KE, Belotelov VI, Zvezdin AK. Novel magnetic photonic crystal structures for magnetic field sensors and visualizers. IEEE Transactions on Magnetics. 2008;**44**(3):323-328. DOI: 10.1109/TMAG.2007.914675

[18] Romodina MN, Soboleva IV, Musorin AI, Nakamura Y, Inoue M, Fedyanin AA. Bloch-surface-wave-induced Fano resonance in magnetophotonic crystals. Physical Review B. 2017;**96**(8):081401(R). DOI: 10.1103/PhysRevB.96.081401

[19] Ardakani AG, Firoozi FB. Highly tunable bistability using an external magnetic field in photonic crystals containing graphene and magnetooptical layers. Journal of Applied Physics. 2017;**121**(2):023105. DOI: 10.1063/1.4973897

[20] Mehdian H, Mohammadzahery Z, Hasanbeigi A. Magneto-optical properties of one-dimensional conjugated photonic crystal heterojunctions containing plasma layers. Applied Optics. 2015;**54**(26):7949-7956. DOI: 10.1364/AO.54.007949

[21] Figotin A, Vitebskiy I. Electromagnetic unidirectionality in magnetic photonic crystals. Physical Review B. 2003;**67**(16):165210. DOI: 10.1103/PhysRevB.67.165210

[22] Chen Y, Wang X, Yong Z, Zhang Y, Chen Z, He L, et al. Experimental investigation of photonic band gap in one-dimensional photonic crystals with metamaterials. Physics Letters A. 2012;**376**(16):1396-1400. DOI: 10.1016/j.physleta.2012.01.044

[23] Aylo R, Nehmetallah G, Li H, Banerjee PP. Multilayer periodic and random metamaterial structures: Analysis and applications. IEEE Access. 2014;**2**:437-450. DOI: 10.1109/ACCESS.2014.2321661

[24] Abelès F. Recherches sur la propagation des ondes électromagnétiques sinusoïdales dans les milieux stratifiés. Annales de Physique. 1950;**12**(5):596-640 & 706-782. DOI: 10.1051/anphys/195012050596

[25] Yeh P, Yariv A, Hong C-S. Electromagnetic propagation in periodic stratified media. I. General theory. Journal of the Optical Society of America. 1977;**67**(4):423-438. DOI: 10.1364/JOSA.67.000423

[26] Gurevich AG. Ferrites at Microwave Frequencies. New York: Consultants Bureau; 1963 329 p

[27] Floquet G. Sur les équations différentielles linéaires à coefficients périodiques. Annales scientifiques de l'École Normale Supérieure. 1883;**12**:47-88

[28] Bloch F. Über die Quantenmechanik der Elektronen in Kristallgittern. Zeitschrift für Physik A. 1929;**52**(7–8):555-600. DOI: 10.1007/BF01339455

[29] Rayleigh L. On the maintenance of vibrations by forces of double frequency, and on the propagation of waves through a medium endowed with a periodic structure. Philosophical Magazine S. 5. 1887;**24**(147):145-159

[30] Brillouin L. Wave Propagation in Periodic Structures. New York, London: McGraw-Hill Book Company; 1946 247 p

[31] Bass FG, Bulgakov AA. Kinetic and Electrodynamic Phenomena in Classical and Quantum Semiconductor Superlattices. New York: Nova Science Publishers; 1997 498 p

[32] Sommerfeld A. Ueber die Fortpflanzung elektrodynamischer Wellen längs eines Drahtes. Annals of Physics. 1899;**303**(2):233-290. DOI: 10.1002/andp.18993030202

[33] Zenneck J. Über die Fortpflanzung ebener elektromagnetischer Wellen längs einer ebenen Leiterfläche und ihre Beziehung zur drahtlosen Telegraphie. Annals of Physics. 1907;**328**(10):846-866. DOI: 10.1002/andp.19073281003

[34] Li J, Zhou L, Chan CT, Sheng P. Photonic band gap from a stack of positive and negative index materials. Physical Review Letters. 2003;**90**(8):083901. DOI: 10.1103/PhysRevLett.90.083901

[35] Shmat'ko AA, Mizernik VN, Odarenko EN, Yampol'skii VA, Rokhmanova TN, Galenko A. Yu. Dispersion properties of a one-dimensional anisotropic magnetophotonic crystal with a gyrotropic layer. In: Advanced optoelectronics and lasers (CAOL), IEEE 7th International Conference on; 12–15 Sept.; Odessa, Ukraine. 2016. p. 123-125. DOI: 10.1109/CAOL.2016.7851399

[36] Mizernik V. N, Shmat'ko A. A. Surface Magnon-Polariton modes in the open transmission resonator partially filled ferrite. In: 20th Int. Crimean Conference "Microwave and Telecommunication Technology"; 13–17 Sept.; Sevastopol, Ukraine. 2010. p. 643-644. DOI: 10.1109/CRMICO.2010.5632685

[37] Shmat'ko A.A., Mizernik V.N., Odarenko E.N., Rokhmanova T.N. Bragg reflection and transmission of light by one-dimensional Gyrotropic Magnetophotonic crystal. In: Advanced Information and Communication Technologies (AICT-2017) 2nd International Conference on; 4–7 July; Lviv, Ukraine. 2017. p. 1-5

Chapter 4

Structure-Induced Ultratransparency in Photonic Crystals

Jie Luo and Yun Lai

Additional information is available at the end of the chapter

http://dx.doi.org/10.5772/intechopen.71274

Abstract

This chapter presents the recent progress on structure-induced ultratransparency in both one- and two-dimensional photonic crystals (PhCs). Ultratransparent PhCs not only have the omnidirectional impedance matching with the background medium, but also have the ability of forming aberration-free virtual images. In certain frequency regimes, such ultratransparent PhCs are the most transparent solid materials on earth. The ultratransparency effect has many applications such as perfectly transparent lens, transformation optics (TO) devices, microwave transparent devices, solar cell packaging, etc. Here, we demonstrate that the ultratransparent PhCs with "shifted" elliptical equal frequency contour (EFC) not only provide a low-loss and feasible platform for transformation optics devices at optical frequencies, but also enable new degrees of freedoms for phase manipulation beyond the local medium framework. In addition, microwave transparent devices can be realized by using such ultratransparent PhCs.

Keywords: ultratransparency, photonic crystals, impedance matching, spatial dispersion, transformation optics

1. Introduction

Photonic crystals (PhCs), as periodic arrangement of dielectrics, affect the motion of photons and electromagnetic (EM) waves in much the same way that semiconductor crystals affect the propagation of electrons, providing a new mechanism to control and manipulate the flow of light at wavelength scale [1–5]. The key property of the PhCs is the photonic band gap induced from the periodic modulation of photons and EM waves, which can block wave propagation in certain or all directions. As photonic band-gap materials, PhCs play vital roles in light confinement and optical manipulation, promising many important applications, such as omnidirectional reflectors [6, 7], waveguides [8, 9], fibers [10], high-Q nanocavities and laser [11, 12], and angular filters [13].

INTECH
open science | open minds

© 2018 The Author(s). Licensee InTech. This chapter is distributed under the terms of the Creative Commons Attribution License (http://creativecommons.org/licenses/by/3.0), which permits unrestricted use, distribution, and reproduction in any medium, provided the original work is properly cited.

However, little attention has been paid to the effect of enhancing transparency. Although there are pass bands in PhCs, they are usually reflective and, therefore, not transparent enough. As we know, transparent media are the foundation of almost all optical instruments, such as optical lens. However, perfect transparency has never been realized in natural transparent solid materials such as glass because of the impedance mismatch with free space. On the other hand, in the past decades, artificial EM materials like metamaterials [14–20] have been proposed to realize unusual EM properties beyond natural materials. However, most of the researches were focused on the realization of abnormal refractive behaviors such as negative refraction. Transparency over a large range of incident angle or a large frequency spectrum is theoretically possible, but the experimental realization is very difficult as complex and unusual parameters are required.

The photonic band-gap effect is actually induced by the periodic modulation of the reflections on the surfaces of dielectrics. That is, the periodic modulation strengthens the reflections on the surfaces of dielectrics to form a complete band gap at particular frequencies. Then, a natural question is: Is it possible to rearrange the periodic modulation of the reflections on the surfaces of dielectrics to make them cancel each other for all incident angles, so that omnidirectional impedance matching and omnidirectional perfect transmission can be realized?

In this chapter, we show the opposite effect of the band-gap effect in PhCs, i.e., the structure-induced ultratransparency effect [21, 22]. Ultratransparent PhCs not only have the omnidirectional impedance matching with the background medium, but also have the ability of forming aberration-free virtual images. The equal frequency contours (EFCs) of such ultratransparent PhCs are designed to be elliptical and "shifted" in the k-space and thus contain strong spatial dispersions and provide more possibilities for omnidirectional impedance matching. Interestingly, the combination of perfect transparency and elliptical EFCs satisfies the essential requirement of ideal transformation optics (TO) devices [23–27]. Therefore, such ultratransparent PhCs not only provide a low-loss and feasible platform for TO devices at optical frequencies, but also enable new freedom for phase manipulation beyond the local medium framework. Moreover, such ultratransparent PhCs have shown enormous potential applications in the designs of microwave transparent wall, nonreflection lens, omnidirectional polarizer, and so on.

2. Ultratransparency effect: the opposite of the band-gap effect

2.1. Definition of ultratransparency: omnidirectional impedance matching and aberration-free virtual image

It is well known that the band-gap effect induced by the periodic distribution of dielectrics can block the propagation of EM waves in certain or all directions, as illustrated in **Figure 1(a)**. The forbiddance of wave propagation is the result of lacking propagation modes within the PhCs, which can be seen from the EFCs in **Figure 1(a)**. The circle in the left denotes the EFC of free space, and there is no dispersion of the PhCs within the band gap. Interestingly, we would like to rearrange the periodic array of dielectrics to obtain the omnidirectional impedance matching effect, which allows near 100% transmission of light at all incident angles, as illustrated in **Figure 1(b)**.

http://dx.doi.org/10.5772/intechopen.71274

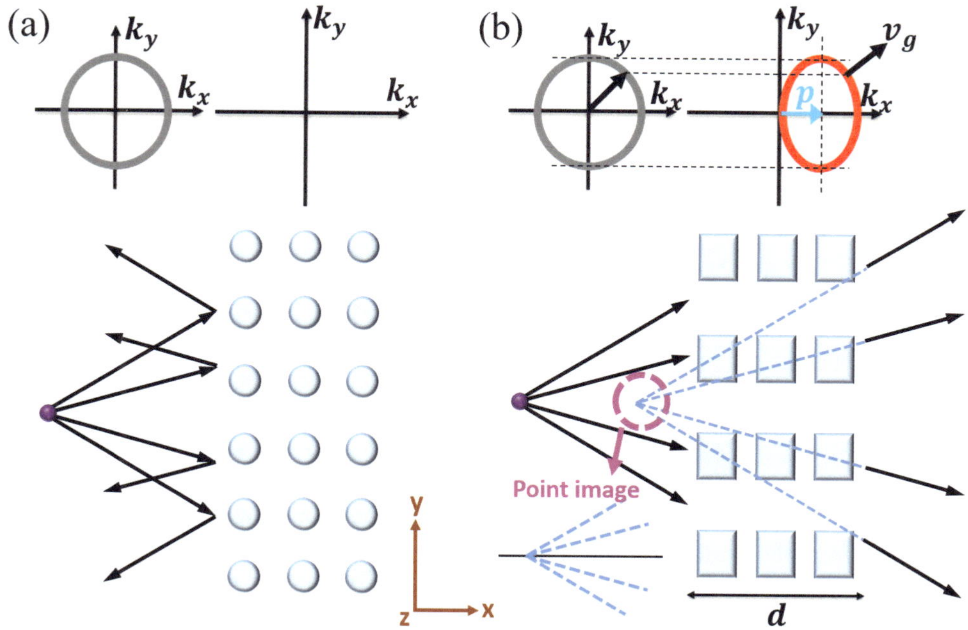

Figure 1. (a) Total reflection by a PhC slab with complete photonic band gap. (b) Aberration-free virtual image formation through an ultratransparent medium without any reflection due to omnidirectional impedance matching. The arrows and dashed lines in (a) and (b) represent the light rays from a point source, and the back-tracing lines, respectively. The inset graphs show the corresponding EFCs.

Moreover, we expect such media to have the ability of forming aberration-free virtual images, which is absent in normal transparent media like glass. By using ray optics, it can be easily shown that transmitted rays from a point source behind a dielectric slab would form a "blurred" area of virtual image rather than a point image. For demonstration, we placed a point source on the left side of a glass slab (with a thickness of d and a refractive index of 1.46) at a distance of d, as shown in **Figure 2(a)**. The virtual image is formed by the back tracing lines (thin lines) of the transmitted waves in the right side (thick lines), showing evident aberrations. In **Figure 2(b)**, we present the zoom in "blurred" image in the region marked by dashed lines in **Figure 2(a)**, showing the spatial distribution of the formed virtual image. This implies that the position of the virtual image will change if we observe from different positions, as illustrated in **Figure 2(c)**.

Actually, such aberrations of the virtual image originate from the mismatch between the EFCs of free space and dielectrics, i.e., their EFCs do not have the same height in the transverse direction (i.e., the k_y direction in **Figure 1**) [21]. By using ray optics, it can be demonstrated that a circle or elliptical EFC having the same height of the EFC of free space in the k_y direction enables the formation of aberration-free virtual images [21], i.e., the EFC has the form of

$$\frac{(k_x - p)^2}{q} + k_y^2 = k_0^2, \tag{1}$$

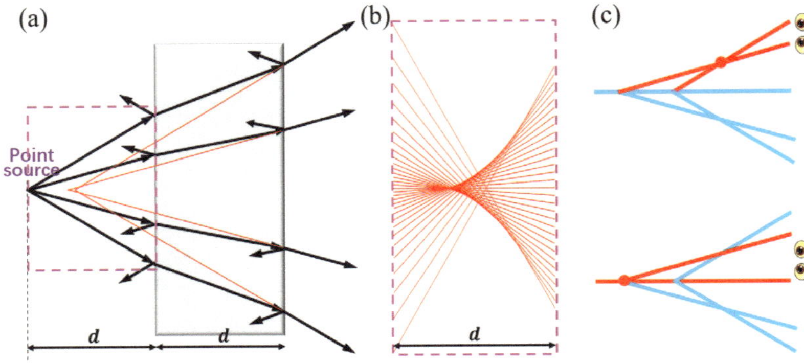

Figure 2. The formation of "blurred" virtual image by a glass slab. (a) The existence of general reflection and aberration in virtual images behind a glass slab. (b) The zoom in "blurred" image in the region marked by dashed lines in (a). (c) The position of the virtual image for the observation from different positions.

where k_0 is the wave number in air, $k_{x(y)}$ is the $x(y)$ component of the wave vector, p denotes the displacement from the Brillouin zone center, and q determines the ratio of the k_y and k_x axes of the ellipse, as shown in **Figure 1(b)**. We note that the "shift" of p in the k_x direction does not affect the formation of virtual images, because such a "shift" does not change the refractive behavior at all in this case.

Supposing that the rearrangement of the periodic array of dielectrics not only makes the whole structure impedance matched to free space for all incident angles, but also creates the unique EFC described by Eq. (1); thus, both omnidirectional 100% perfect transmission and aberration-free virtual imaging are enabled simultaneously. Apparently, such a level of transparency is superior to that of normal transparent media like dielectrics and is thus hereby denoted as ultratransparency.

2.2. Ultratransparency based on local and nonlocal media

According to Fresnel equations, reflection of light on the surface of dielectrics is inevitable, except at a single-incident angle referred to as the Brewster angle under transverse magnetic (TM) polarization, as demonstrated in **Figure 3(a)**. Here, we extend the impedance matching from one particular angle (i.e., Brewster angle) to all incident angles in a nonlocal or spatial dispersive medium, whose effective permittivity $\varepsilon(k_y)$ and permeability $\mu(k_y)$ are dependent on wave vectors.

To begin with, we assume that the nonlocal medium exhibits an irregular-shaped EFC shown in **Figure 3(b)**. An incident wave of transverse electric (TE) polarization with electric fields polarized in the z direction is considered. For the TE polarization, the dispersion relation of the nonlocal medium can be expressed in a general form as,

$$\frac{k_x^2}{\mu_y(k_y)} + \frac{k_y^2}{\mu_x(k_y)} = \varepsilon_z(k_y)k_0^2, \tag{2}$$

which determines the relationship between the components of wave vectors k_x and k_y.

http://dx.doi.org/10.5772/intechopen.71274

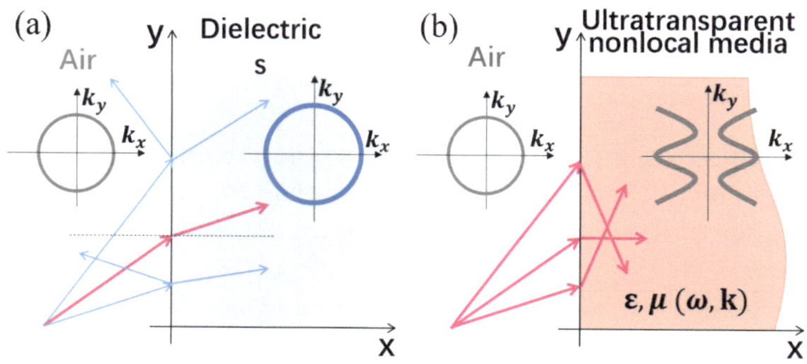

Figure 3. (a) There exist reflected waves at the interface of free space and dielectrics except for the Brewster angle for TM polarization. (b) Nonreflection and total transmission on the surface of ultratransparent nonlocal media for all incident angles. The thin circle, thick circle, and curves in the inset graphs denote the EFCs of air, dielectrics, and the nonlocal media, respectively.

In this case, the wave impedance of free space can be derived as $Z_0 \equiv \frac{E_z}{H_y} = -\frac{\mu_0}{k_{x,0}}\omega$ for TE polarization [28–30]. E_z and H_y are the z-component of electric fields and the y-component of magnetic fields, respectively. $k_{x,0}$, μ_0, and ω are the x-component of wave vector in free space, the permeability of vacuum, and angular frequency of the EM wave, respectively. Similarly, the wave impedance of the nonlocal medium can be written as $Z = -\frac{\mu_0\mu_y(k_y)}{k_x}\omega$. Then, the impedance matching of the nonlocal medium and free space requires $Z = Z_0$. By considering dispersion relations, the impedance matching condition is obtained as,

$$\frac{\mu_x(k_y)\mu_y(k_y)}{\mu_x(k_y)\varepsilon_z(k_y)k_0^2 - k_y^2} = \frac{1}{k_0^2 - k_y^2} \tag{3}$$

If Eq. (3) can be satisfied for all $|k_y| < k_0$, then omnidirectional impedance matching can be achieved, leading to near 100% transmission of light for all incident angles.

An obvious local medium solution of Eq. (3) is that $\mu_x\mu_y = 1$ and $\mu_x\varepsilon_z = 1$, which correspond to elliptical EFCs centered at the Brillouin zone center, are consistent with transformation optics theory [23–27]. However, this is not the only possible solution for spatially dispersive media. Now, we consider medium with the EFC described by Eq. (1), i.e., $k_x = p \pm \sqrt{q\left(k_0^2 - k_y^2\right)}$. By substituting this relation into Eq. (3) and the spatial dispersion, we obtain the analytical forms of spatially dispersive parameters as:

$$\mu_y^2(k_y) = \frac{\left[p \pm \sqrt{q\left(k_0^2 - k_y^2\right)}\right]^2}{k_0^2 - k_y^2} \text{ and } \varepsilon_z(k_y)k_0^2 - \frac{k_y^2}{\mu_x(k_y)} = \pm\left[p \pm \sqrt{q\left(k_0^2 - k_y^2\right)}\right]\sqrt{k_0^2 - k_y^2} \tag{4}$$

Eq. (4) shows that when EM parameters possess spatial dispersions, there exist infinite solutions of $\mu_y(k_y)$, $\varepsilon_z(k_y)$, and $\mu_x(k_y)$ for the satisfaction of Eq. (4) for all $|k_y| < k_0$, indicating infinite solutions for omnidirectional impedance matching.

3. Structure-induced ultratransparency in two-dimensional PhCs

3.1. Nonlocal effective medium theory

For the design of ultratransparent PhCs, we first propose a nonlocal effective medium theory for the homogenization of PhCs. Here, we consider uniform plane wave incidence. In this situation, the validity of the nonlocal effective medium theory lies in the satisfaction of the following four premises: (1) single-mode approximation [28], i.e., only one eigen-mode is excited; (2) the amplitudes of fields at the incident boundary are almost constant; (3) the phases of fields at the incident boundary obey the trigonometric functions; and (4) the electric and magnetic fields are in phase at the incident boundary. Although these premises are seemingly stringent, it turns out that most eigen-modes of the first few bands (e.g., monopolar and dipolar bands) can indeed satisfy these requirements (**Figure 4**).

With the assumption of the abovementioned premises, eigen-fields of TE polarization at the boundary $x = 0$ can be approximately expressed as $\mathbf{F} = \mathbf{F}_0 e^{i(k_x x + k_y y)} e^{-i\omega t}$ with $\mathbf{F}_0 = E_0\hat{z}, H_0\hat{y}$. E_0 and H_0 are the amplitudes of eigen-electric and eigen-magnetic fields, respectively, which are almost independent of y. This indicates that the eigen-fields are the boundary and have the same formula as those in a uniform medium. Thus, the surface impedance of the PhCs can be calculated in a similar formula as [28–30],

$$Z_x = \frac{\langle E_z \rangle_{x=0}}{\langle H_y \rangle_{x=0}} \text{ and } Z_y = \frac{\langle E_z \rangle_{x=0}}{\langle H_x \rangle_{x=0}} \tag{5}$$

where $\langle F \rangle_{x=0}$ denotes the average of eigen-field F along the unit cell boundary $x = 0$. E_z, H_x and H_y are the z-component of eigen-electric fields, x- and y-components of eigen-magnetic fields, respectively. Based on Eq. (5), the surface impedance of a PhC can be calculated as long as the eigen-fields of the PhC are solved.

On the other hand, the PhC satisfying previous premises generally can be described as a uniform medium with effective relative permittivity $\begin{pmatrix} \varepsilon_{x,eff} & & \\ & \varepsilon_{y,eff} & \\ & & \varepsilon_{z,eff} \end{pmatrix}$, relative

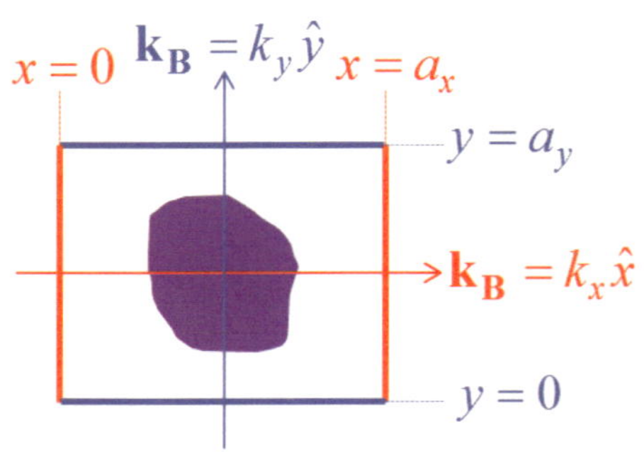

Figure 4. Schematic graph of the unit cell for the nonlocal effective medium theory.

permeability $\begin{pmatrix} \mu_{x,eff} & & \\ & \mu_{y,eff} & \\ & & \mu_{z,eff} \end{pmatrix}$, and dispersion $\frac{k_x^2}{\mu_{y,eff}} + \frac{k_y^2}{\mu_{x,eff}} = \varepsilon_{z,eff} k_0^2$. Thus, the impedance of the PhC can be rewritten as the function of effective parameters:

$$Z_x = -\frac{\mu_0 \mu_{y,eff}}{k_x}\omega \text{ and } Z_y = \frac{\mu_0 \mu_{x,eff}}{k_y}\omega \tag{6}$$

Compared with Eqs. (5) and (6), the effective parameters can be derived as,

$$\varepsilon_{z,eff} = \frac{1}{\varepsilon_0 \omega} \frac{k_y \langle H_x \rangle_{x=0} - k_x \langle H_y \rangle_{x=0}}{\langle E_z \rangle_{x=0}},$$

$$\mu_{x,eff} = \frac{k_y}{\mu_0 \omega} \frac{\langle E_z \rangle_{x=0}}{\langle H_x \rangle_{x=0}} \text{ and } \mu_{y,eff} = -\frac{k_x}{\mu_0 \omega} \frac{\langle E_z \rangle_{x=0}}{\langle H_y \rangle_{x=0}} \tag{7}$$

With Eq. (7), the effective parameters of the PhC can be obtained by analyzing the eigen-fields with Bloch wave vector $\mathbf{k_B} = k_x \widehat{x} + k_y \widehat{y}$. We note that the Bloch wave vector is restricted in the first Brillouin zone, and the choice of the eigen-fields of $-\mathbf{k_B}$ or $+\mathbf{k_B}$ depends on the direction of group velocity. In addition, we find that this method is still valid even for the eigen-modes far away from the Brillouin center, but the effective parameters are generally k-dependent, i.e., nonlocal or spatially dispersive.

3.2. Two-dimensional ultratransparent PhCs

PhCs contain strong spatial dispersions and thus provide the perfect candidate for realization of ultratransparency effect. Here, we demonstrate a type of PhCs composed of a rectangular array of dielectric rods in free space, with the unit cell shown in **Figure 5(a)**. Under TE polarization, the band structure is presented in **Figure 5(b)**, and the EFC of the third band is plotted in the reduced first Brillouin zone in **Figure 5(c)**. The working frequency is chosen as $fa/c = 0.3183$, where f, a, and c are the frequency, the lattice constant, and the speed of light in free space, respectively. We see that the corresponding EFC (the right dashed lines in **Figure 5(c)**) is indeed a "shifted" ellipse that can be described by Eq. (1) with $p \approx \pi/a$. And the dispersion can be approximately written as,

$$\left(\frac{k_x - \pi/a_1}{0.4541 a_2/0.5097 a_1}\right)^2 + k_y^2 = \left(\frac{0.5097\pi}{a_2}\right)^2 \tag{8}$$

Figure 5(d) shows the impedance difference of the PhC and free space of the third band, i.e., $\left|\frac{Z-Z_0}{Z+Z_0}\right|$, where Z is the impedance of the PhC obtained from Eq. (5) and Z_0 is the impedance of free space. From **Figure 5(d)**, it is seen that the impedance difference is very small for a very large range of k_y at the working frequency, indicating impedance matching in a large range of incidence angles.

Moreover, in **Figure 6(a)**, we present the effective parameters $\varepsilon_{z,eff}$ (square dots), $\mu_{x,eff}$ (circular dots), and $\mu_{y,eff}$ (triangular dots) of the PhC based on Eq. (7), showing the k-dependence.

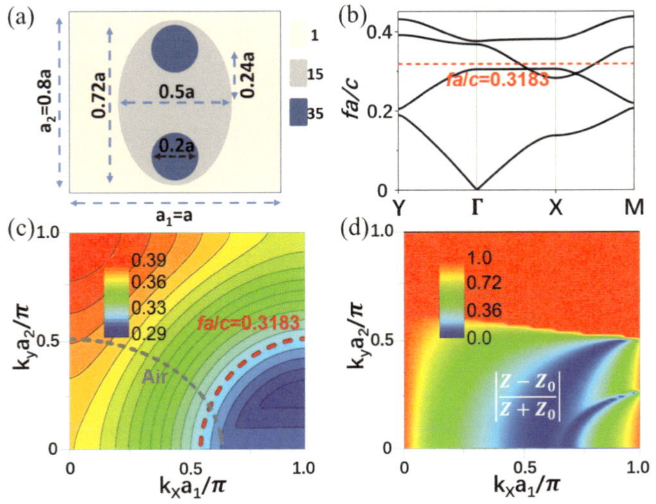

Figure 5. (a) Illustration of the unit cell of the ultratransparent PhC. (b) The band structure of the PhC. The dashed line denotes the working frequency $fa/c = 0.3183$. (c) The EFC of the third band. The left and right dashed lines denote the EFCs of air and the PhC at the frequency $fa/c = 0.3183$. (d) The impedance difference of the PhC and free space of the third band. Reprinted (figure) with permission from Ref. [21]. Copyright (2016) by the American Physical Society.

Interestingly, we find $\mu_{x,eff} \approx 1$, while $\varepsilon_{z,eff}$ and $\mu_{y,eff}$ are both dependent on k_y. The result of $\mu_{x,eff} \approx 1$ can be physically understood. Since the resonances are localized in the dielectrics and far away from the incident boundary, the modes propagating in the y direction are weakly modulated by the periodicity. As a consequence, $\mu_{x,eff}$ has the same value as the relative permeability of the background medium.

Now, by substituting the condition $\mu_{x,eff} \approx 1$ and the dispersion Eq. (8) into Eq. (4), the analytical form of the effective parameters $\varepsilon_{z,eff}$ and $\mu_{y,eff}$ can be solved as,

$$\varepsilon_{z,eff}(k_y) = \pm\frac{\pi}{a_1 k_0^2}\left[1 - 0.4541\sqrt{1-\left(\frac{k_y a_2/\pi}{0.5097}\right)^2}\right]\sqrt{k_0^2-k_y^2}+\frac{k_y^2}{k_0^2}$$
$$\text{and } \mu_{y,eff}(k_y) = \pm\frac{\pi}{a_1}\left[1 - 0.4541\sqrt{1-\left(\frac{k_y a_2/\pi}{0.5097}\right)^2}\right]\Big/\sqrt{k_0^2-k_y^2} \tag{9}$$

The choice of $\pm k_x$ depends on the direction of group velocity. Here, we choose $-k_x$, because hereby the band branch in the region $-\pi/a_1 < k_x < 0$ in the first Brillouin zone is excited, as determined by the direction of group velocity. In **Figure 6(a)**, $\varepsilon_{z,eff}$ and $\mu_{y,eff}$ based on Eq. (9) (dashed lines) show perfect match with those obtained from Eq. (7) for all $k_y < k_0$, demonstrating the omnidirectional impedance matching effect.

For further verification, the transmittance through such a PhC slab consisting of N (= 4, 5, 6, 15) layers of unit cells in the x direction is numerically calculated, as presented in **Figure 6(b)**. It is shown that the transmittance is near unity (>99%) for nearly all incident angles ($\theta < 89°$) and is

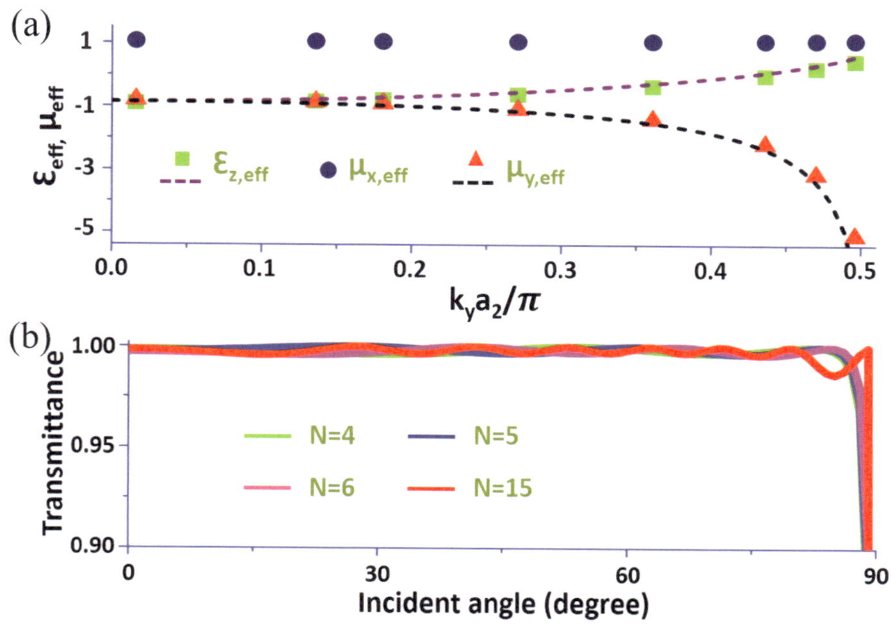

Figure 6. (a) The effective parameters $\varepsilon_{z,eff}$, $\mu_{x,eff}$, and $\mu_{y,eff}$ retrieved from the eigen-fields based on Eq. (7) (symbols) and predicted by the ultratransparency condition based on Eq. (9) (dashed lines). (b) Calculated transmittance through a PhC slab with N (= 4, 5, 6, 15) layers of unit cells as the function of incident angles. Reprinted (figure) with permission from Ref. [21]. Copyright (2016) by the American Physical Society.

almost irrespective of the layer number N, indicating that the total transmission is a result of impedance matching instead of tunneling effects.

3.3. Microwave experimental verification

In fact, for transparency in a relatively smaller range of incident angles, the design process is much easier and the effect can exist in much simpler structures. In the following, we demonstrate a simple ultratransparent PhC, which is verified by proof-of-principle microwave experiments. The PhC consists of rectangular alumina ($\varepsilon = 8.5$) bars in a square lattice, as illustrated by the inset in **Figure 7(a)**. The lattice constant is set to be a = 12 mm, and the length (W_1) and width (W_2) are 9.6 mm and 4.8 mm, respectively, corresponding to 0.8a and 0.4a. We consider TE polarization with electric fields polarized along the z direction, and the band structure is presented in **Figure 7(a)**. The normalized frequency is chosen to be $fa/c = 0.4723$ (marked by a dashed line), corresponding to a working frequency around 11.8 GHz. In **Figure 7(b)**, we can see that the shape of EFC of $fa/c = 0.4723$ is nearly a part of "shifted" ellipse. And the dispersion can be approximately written as,

$$\left(\frac{k_x - \pi/a_1}{0.4541a_2/0.5097a_1}\right)^2 + k_y^2 = \left(\frac{0.5097\pi}{a_2}\right)^2 \tag{10}$$

In addition, the impedance difference between the PhC and the free space is calculated by using Eq. (5), as shown in **Figure 7(c)**. Clearly, the impedance difference is very small on the

EFC of $fa/c = 0.4723$ for a large range of k_y, demonstrating the wide-angle impedance matching effect.

Moreover, the effective parameters obtained from Eq. (7) are presented in **Figure 8(a)** by solid lines with symbols, showing $\mu_{x,eff} \approx 1$ for $k_y < 0.818\pi/a$ as the result of weakly modulated

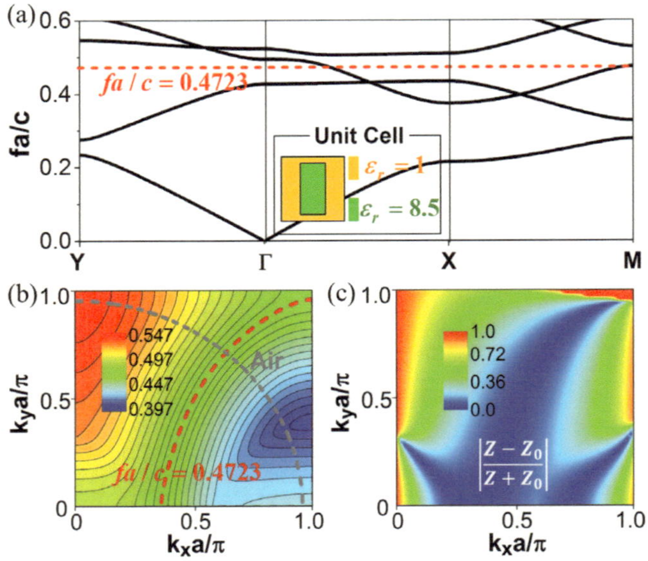

Figure 7. (a) Band diagram of the experimental PhC for TE polarization. The dashed line denotes the working frequency $fa/c = 0.4723$. The inset is the illustration of the unit cell of the PhC. (b) EFCs in the frequency range $0.397 \le fa/c \le 0.547$. The left and right dashed lines denote the EFCs of air and the PhC at the frequency $fa/c = 0.4723$. (c) The impedance difference of the PhC and the free space in the frequency range $0.397 \le fa/c \le 0.547$.

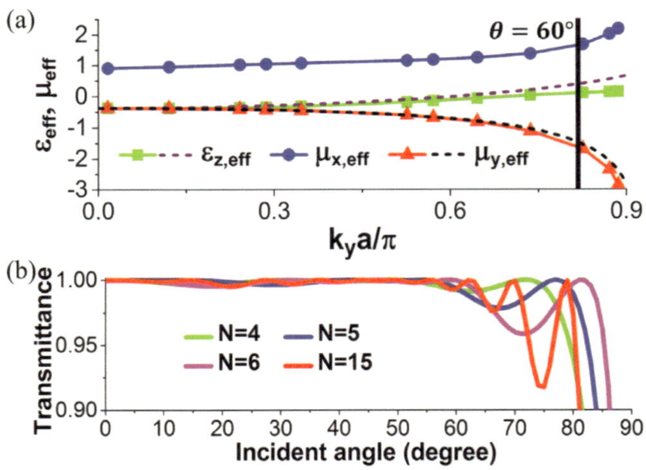

Figure 8. (a) Effective parameters $\varepsilon_{z,eff}$, $\mu_{x,eff}$, and $\mu_{y,eff}$ retrieved from the eigen-fields based on Eq. (7) (solid lines with symbols) and predicted by the ultratransparency condition based on Eq. (11) (dashed lines). (b) Transmittance through a N (= 4, 5, 6, 15)-layered PhC slab as the function of incident angles.

modes propagating in the y direction. By inserting the condition $\mu_{x,eff} \approx 1$ and the dispersion Eq. (10) into Eq. (4), the analytical expressions of $\varepsilon_{z,eff}$ and $\mu_{y,eff}$ can be derived as,

$$\begin{cases} \varepsilon_{z,eff}(k_y) = \pm \dfrac{\pi}{ak_0^2}\left[1.25 - 0.8945\sqrt{1 - \left(\dfrac{k_y a/\pi}{1.0415}\right)^2}\right]\sqrt{k_0^2 - k_y^2} + \dfrac{k_y^2}{k_0^2} \\ \mu_{y.eff}(k_y) = \pm \dfrac{\pi}{a}\left[1.25 - 0.8945\sqrt{1 - \left(\dfrac{k_y a/\pi}{1.0415}\right)^2}\right] / \sqrt{k_0^2 - k_y^2} \end{cases} \tag{11}$$

In **Figure 8(a)**, the dashed lines denote $\varepsilon_{z,eff}$ and $\mu_{y,eff}$ based on Eq. (11) (negative sign is chosen), showing excellent coincidence with those from Eq. (7) for $k_y < 0.818\pi/a$ (i.e., $\theta < 60°$).

Furthermore, we calculate the transmittance through the PhC slab with N (= 4, 5, 6, 15) unit cells, as shown in **Figure 8(b)**. It is clearly seen that there exists near-unity transmittance for all

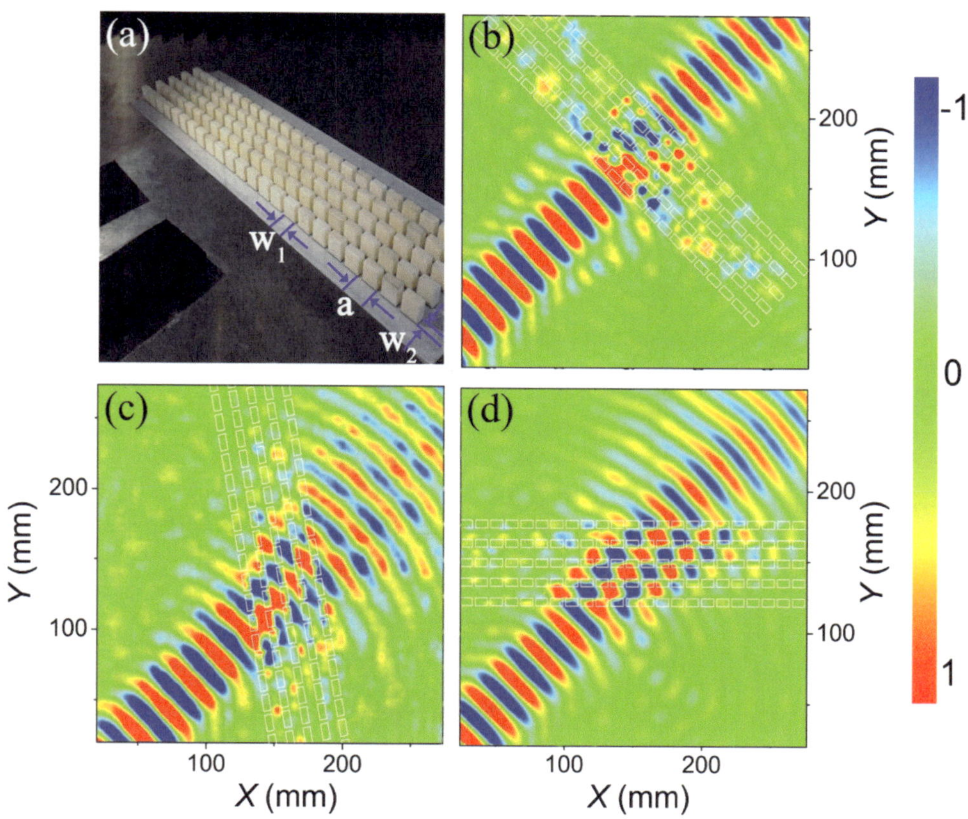

Figure 9. (a) Photo of the PhC composed of alumina bars (white) placed inside the microwave field mapper. The measured electric-field distributions for microwave beams with incident angles of (b) 0°, (c) 30°, and (d) 45°. Reprinted (figure) with permission from Ref. [21]. Copyright (2016) by the American Physical Society.

incident angles of $\theta < 60^\circ$ (i.e., $k_y < 0.818\pi/a$) irrespective of N, demonstrating the wide-angle impedance matching effect.

Next, we show microwave experimental results to verify the above theory. A 23 × 5 array of such a PhC is assembled in the xy plane inside a parallel-plate waveguide composed of two flat aluminum plates, as shown in **Figure 9(a)**. The separation between the two aluminum plates is 10.5 mm, slightly larger than the height of the bars (10 mm), but smaller than half of the wavelength of interest (25.4 mm for 11.8 GHz) to make sure that the whole experimental chamber can only support transverse EM modes. A microwave beam with a finite width was launched through a waveguide made of absorbing materials (in black in **Figure 9(a)**). The lower metal plate along with the PhC array was mounted on a translational stage. The electric field was measured via an antenna fixed in a hole in the upper metal plate (not shown here). Both the emitting and probing antennas are connected to our Agilent E5071C network analyzer to acquire the transmitted magnitude and phase of microwave signals. Such a setup allows us to measure the spatial distributions in the xy plane for a series of incident angles from 0° to 60° at the working frequency of 11.8 GHz.

The measured electric fields for 0°, 30°, and 45° incident angles are displayed in **Figure 9(b), (c)**, and **(d)**, respectively. Clearly, the reflection is barely noticeable, indicating impedance matching for all these incident angles. In **Figure 10**, the measured transmittance (triangular dots) coincides with simulation results (solid lines) quite well, both showing great enhancement compared with that through an alumina slab with the same thickness (dashed lines). Although the ultratransparency effect is hereby only verified at the microwave frequency regime, the principle can be extended to optical frequency regime by using PhCs composed of silicon or other dielectrics.

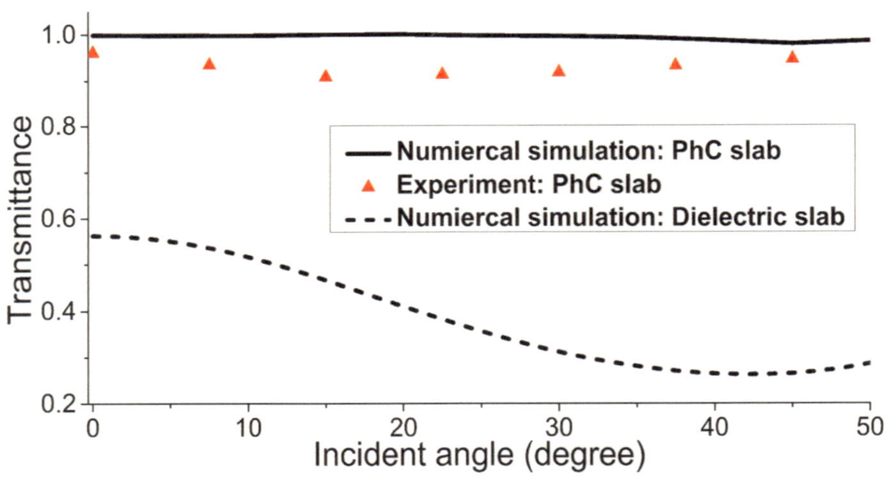

Figure 10. Transmittance through the PhC slab in simulations (solid lines) and experiments (triangular dots) and an alumina slab having the same thickness (dashed lines) as the function of incident angles. Reprinted (figure) with permission from Ref. [21]. Copyright (2016) by the American Physical Society.

4. Structure-induced ultratransparency in one-dimensional PhCs

In the above, we have shown the structure-induced ultratransparency in two-dimensional PhCs. In the following, we demonstrate the structure-induced ultratransparency in one-dimensional PhCs [22].

The one-dimensional ultratransparent PhC we studied is composed of two dielectric materials A and B stacked along the x direction (**Figure 11(c)**). As we know, neither the dielectric material A nor B is perfectly transparent due to impedance mismatch. As a result, a large amount of EM waves are always reflected at the surface of dielectrics, as illustrated in **Figure 11 (a)** and **(b)**. Interestingly, when we periodically stack the two dielectric slabs with appropriate filling ratio and lattice constant, the reflection waves on the dielectrics can cancel each other; thus, wide-angle and even omnidirectional nonreflection can be obtained, as illustrated in **Figure 11(c)**.

Figure 12(a) presents the band structure of the PhC, whose unit cell is constructed in a symmetric form, i.e., ABA structure, as shown by the inset in **Figure 12(a)**. The relative permittivity and filling ratio of the material A (B) are 2 (6) and 0.6 (0.4), respectively. The dashed line denotes the normalized frequency $fa/c = 0.397$, which is chosen as the working frequency.

In **Figure 12(b)** and **(d)**, the EFCs at the frequency $fa/c = 0.397$ for the TE and TM polarizations are plotted, respectively. It is seen that the EFCs can be approximately regarded as a part of an ellipse with the center located at the X point. In addition, the impedance difference between the PhC and air is shown in **Figure 12(c)** (for the TE polarization) and **Figure 12(e)** (for the TM polarization). It is seen that the impedance difference is very small for a large range of k_y in the

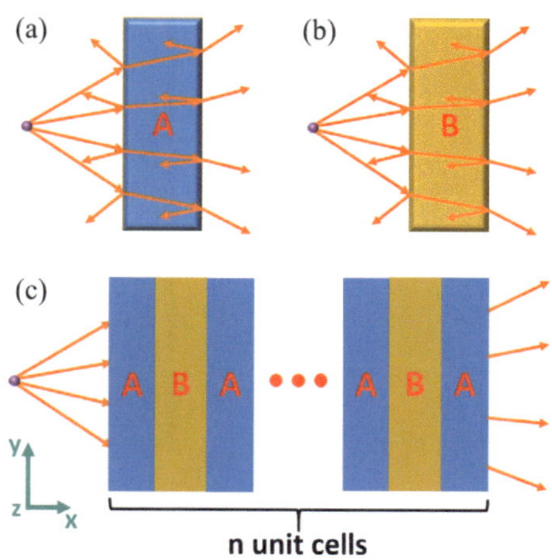

Figure 11. (a) and (b) show general reflection on the slab made of materials A and B, respectively. (c) Elimination of general reflection by a one-dimensional ultratransparent PhC composed of ABA unit cells. The orange arrows represent the rays of light emitted by a point source. Reprinted (figure) with permission from Ref. [22]. Copyright (2016) by the Optical Society.

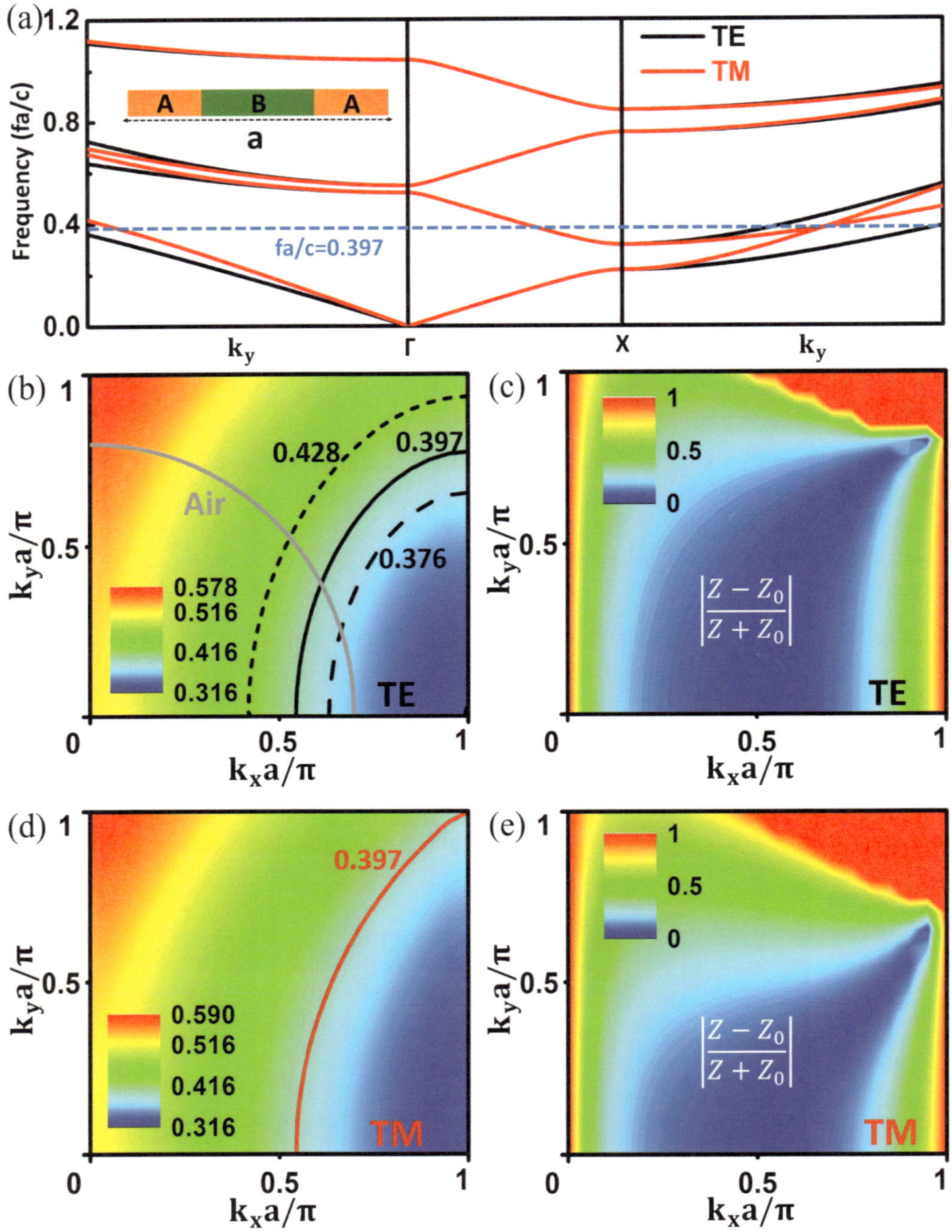

Figure 12. (a) Band structures of the PhC for TE and TM polarizations. The dashed lines denote the working frequency $fa/c = 0.397$. The inset denotes the illustration of the unit cell. (b, d) the EFCs of the PhC and (c, e) the impedance difference between PhC and free space in the second band for (b, c) TE and (d, e) TM polarizations. Reprinted (figure) with permission from Ref. [22]. Copyright (2016) by the Optical Society.

reciprocal space for both polarizations, demonstrating a wide-angle and polarization-insensitive impedance matching effect. Moreover, from **Figure 12(b)** and **(c)**, it is also seen that the impedance difference is small in the frequency range from $fa/c = 0.376$ to $fa/c = 0.428$, indicating a broadband impedance matching effect. Therefore, a one-dimensional PhC exhibiting a broadband, wide-angle, and polarization-insensitive impedance matching effect is realized.

For further verification, the transmittance through the PhC slab composed of N unit cells is calculated. The numerical setup for the transmission computation is shown in **Figure 13(a)**. A TE-polarized plane wave is incident from air in the left side. The upper and lower boundaries are set as periodic boundary condition. From the electric-field distribution in **Figure 13(a)**, we see that almost all the waves can propagate through the PhC slab with 6 unit cells under the incident angle of 45°. In **Figure 13(b)**, the transmittance is plotted as the functions of the incident angle and the number of unit cells (N = 4, 5, 6, 15) for the frequency $fa/c = 0.397$, showing N-independent almost perfect transmission for all incident angles of $\theta < 70°$.

Moreover, in **Figure 14(a)** and **(b)**, the transmittance through a PhC slab ($N = 10$) as the functions of the incident angle and normalized frequency for TE and TM polarizations is presented, which clearly demonstrates the broadband, wide-angle, and polarization-insensitive transparency of the PhC.

Although a wide-angle (0–70°) rather than omnidirectional impedance matching effect is obtained in such a one-dimensional PhC, we may still view such a PhC as an ultratransparent PhC. Compared with the two-dimensional ultratransparent PhCs, the one-dimensional ultratransparent PhCs have the advantages of broadband and polarization-insensitive impedance matching.

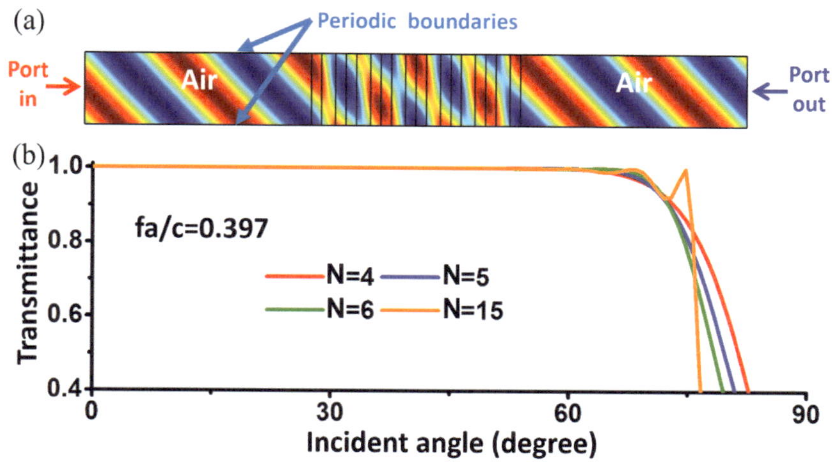

Figure 13. (a) The numerical setup for the transmission computation. The color denotes the distribution of electric fields under $\theta = 45°$ incident angle. (b) Transmittance as the function of the incident angle when EM waves propagate through the PhC slab composed of N (= 4, 5, 6, 15) unit cells. Reprinted (figure) with permission from Ref. [22]. Copyright (2016) by the Optical Society.

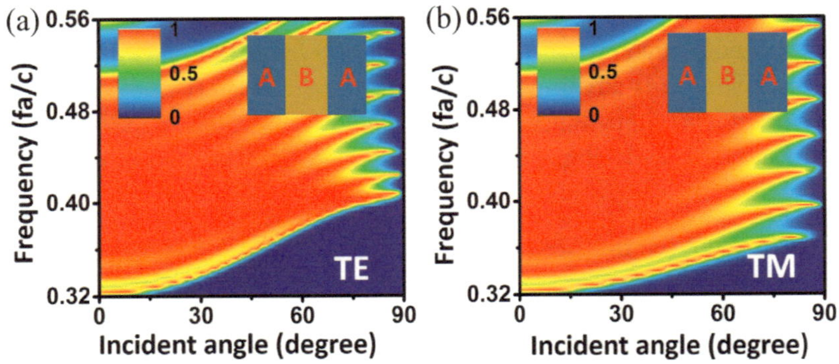

Figure 14. Transmittance as the functions of the incident angle and the frequency for (a) TE- and (b) TM-polarized waves propagating through the PhC slab with 10 unit cells. Reprinted (figure) with permission from Ref. [22]. Copyright (2016) by the Optical Society.

In addition, the easy design and fabrication of the one-dimensional PhCs may lead to more practical applications.

5. Applications

5.1. For transformation optics

In the above, we have demonstrated the ultratransparency in both one- and two-dimensional PhCs. In the following, we show some applications of such ultratransparent PhCs. It is interesting to note that the omnidirectional perfect transparency and elliptical EFCs of the ultratransparent media are essential for ideal TO devices. The theory of TO [23–27] promises many novel and interesting applications, such as invisibility cloaks [23, 25, 31, 32], concentrators [33], illusion optics devices [34–36], and simulations of cosmic phenomena [37, 38]. Generally, the TO devices are realized by using metamaterials [14–20], which require complicated designs of electric and magnetic resonances, hindering the realization and applications in practice. In fact, most of the previous TO experiments were realized by using the so-called reduced parameters, which maintain the refractive behavior, but sacrifice the impedance matching as well as the perfect transparency [25, 39–45]. Moreover, at optical frequencies, the inherent loss in metallic components of metamaterials makes the realization of perfect transparency as well as the ideal nonreflecting TO devices extremely difficult [46, 47], if not impossible. Interestingly, we find that the ultratransparent PhCs provide a low-loss and feasible platform for TO devices at optical frequencies.

To begin with, we consider a TO medium obtained by stretching the coordinate along the x direction in background medium of air. For TE polarization with electric fields polarized in the z direction, the parameters of the TO medium have the relationship [48]:

$$\varepsilon_z = 1/\mu_x = \mu_y = 1/s \tag{12}$$

where s is the stretching ratio. ε_z, μ_x and μ_y are the z component of relative permittivity tensor, x and y components of relative permeability tensor, respectively.

Considering Eq. (12), the dispersion of the TO medium, i.e., $\frac{k_x^2}{\mu_y} + \frac{k_y^2}{\mu_x} = \varepsilon_z k_0^2$, can be rewritten as

$$\frac{k_x^2}{1/s^2} + k_y^2 = k_0^2 \tag{13}$$

which has the similar form as that of Eq. (1). The EFC of the TO medium is an ellipse having the same height as the EFC of air in the k_y direction (**Figure 15(a)**), as that of the ultratransparent media. When the stretching ratio is larger than unity, i.e., $s > 1$, the EFC of the TO medium (the right ellipse) seems slimmer than the EFC of air (the left circle) in **Figure 15(a),** and the coordinate mesh is looser in the x direction in the TO medium region, as illustrated in **Figure 15(a)**.

The only difference of the EFCs between the TO medium and the ultratransparent medium is that there may exist a "shift" of p in the k_x direction (Eq. (1)) for the ultratransparent medium (**Figure 15(b)**). Interestingly, such a "shift" does not affect the refractive behaviors, but enables new freedom for phase manipulation beyond the local medium framework.

For demonstration, we show a specific example in **Figure 16**. The ultratransparent PhC is one-dimensional and composed of components I and II. The unit cell is constructed in a symmetric way with a lattice constant of a, as illustrated by the inset in **Figure 16(a)**. The relative permittivity and thickness of the component I (II) are 2.132 (5.522) and $0.3a$ ($0.4a$), respectively. In **Figure 16(a)**, the EFC of the PhC at the normalized frequency $fa/c = 0.402$ for TE polarization is plotted as the solid lines, which can be viewed as a part of an elliptical located at the X point. Compared with the EFC of the TO medium with $\left\{\varepsilon_z, \mu_x, \mu_y\right\} = \{0.5, 2, 0.5\}$ (dashed lines in **Figure 16(a)**), we see that the EFC of the PhC has the same height in the k_y direction, which satisfies the requirement proposed above.

Moreover, simulations of wave propagation through the TO medium slab with a thickness of $5a$ (upper inset) and a PhC slab with 5 unit cells (lower inset) are performed, as shown in **Figure 16(b)**.

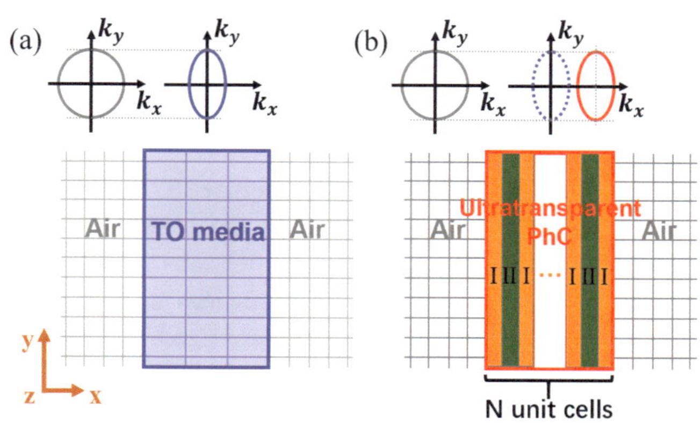

Figure 15. (a) Upper inset: the EFCs of air (the left circle) and TO media obtained by stretching air in the x direction (the right ellipse). Lower inset: scheme of the TO medium slab (with loose mesh) in the background medium of air (with dense mesh). (b) Upper: the EFC of the ultratransparent PhC (the right ellipse with solid lines), which has a "shift" in the k_x direction compared with the EFC of the TO media (dashed lines). Lower inset: the ultratransparent PhC is used to replace the TO medium slab in (a).

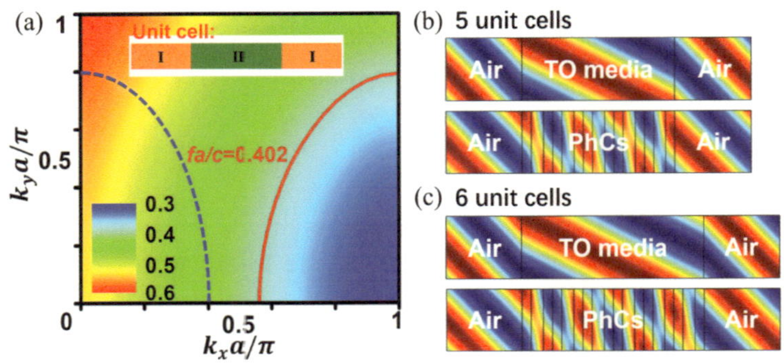

Figure 16. (a) The EFCs of the PhC for TE polarization. The solid line and dashed line denote the EFCs of the PhC and the TO medium at the normalized frequency $fa/c = 0.402$, respectively. (b, c) The snapshot of electric fields for wave propagating through a TO medium slab (upper inset) and a PhC slab (lower inset) under an incident angle of 45°. The thickness of the TO medium slab is (b) 5*a*, (c) 6*a*, and the number of unit cells of the PhC slab is (b) 5, (c) 6.

Simulation results show perfect transmission under the incident angle of 45°. Interestingly, we notice a π phase difference in the transmission waves, which actually is caused by the "shift" of the PhC's EFC and odd number of unit cells. On the other hand, if the total number is even, there will be no phase difference, which is confirmed in the simulation results in **Figure 16(c)**. In **Figure 16(c)**, the thickness of the TO medium slab is 6*a* (upper inset), and the number of unit cells of the PhC slab is 6 (lower inset). Comparing the transmission waves, we see identical phases under the incident angle of 45°. The results demonstrate that the ultratransparent PhC can approximately work as the TO medium with $\left\{\varepsilon_z, \mu_x, \mu_y\right\} = \{0.5, 2, 0.5\}$.

Next, we show an example of TO device by using one-dimensional ultratransparent PhCs. The design process is shown in **Figure 17(a)**, in which the original shell of a concentrator [33] is discretized into four layers and each layer is further replaced by a corresponding ultratransparent PhC. **Figure 17(b)** shows the parameters of the discretized layers of TO media and the ideal profile. The corresponding four types of ultratransparent PhCs are of the same lattice constant *a* and of 4, 2, 2, and 1 units for the A, B, C, and D layers, respectively. The EFCs of PhCs and the discretized layers of TO media are shown in **Figure 17(c)**. It can be seen that the EFCs of PhCs have almost the same shapes with their corresponding layers, but are "shifted" by π/a in the k-space.

The detailed parameters of the PhCs are presented in **Figure 18**. The insets present the illustrations of unit cells, relative permittivities, and thicknesses of each component of the four different PhCs. Moreover, the transmittance through PhC slabs with 10 unit cells is plotted as the function of incident angles, as shown by the solid lines in **Figure 18(a–d)**. During the calculation, the background media are chosen as the discretized TO media (**Figure 17(b)**) with the parameters $\left\{\mu_x, \mu_y, \varepsilon_z\right\}$ being $\{1.52, 0.658, 0.793\}$, $\{1.64, 0.61, 0.855\}$, $\{1.84, 0.543, 0.96\}$, and $\{2.045, 0.489, 1.067\}$, respectively. It is clearly seen that the four PhCs all have near-unity transmittance in a wide-angle range ($0 - 70°$).

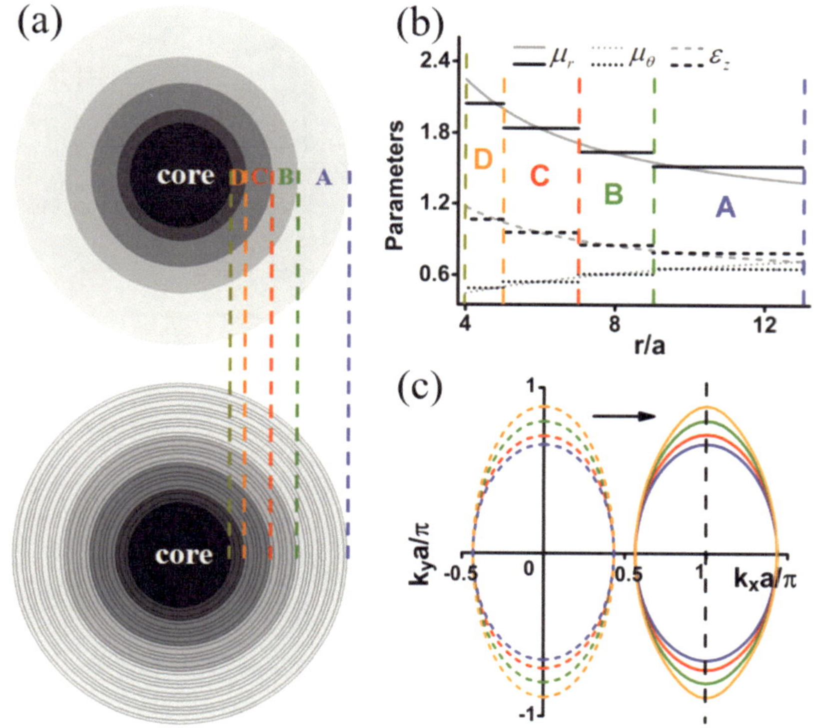

Figure 17. (a) Illustration of the design process from a concentrator composed of discretized layers of TO media (upper) to one composed of ultratransparent PhCs (lower). (b) Parameters of the discretized layers of TO media and the ideal profile. (c) EFCs of the discretized layers of TO media (dashed lines) and the corresponding PhCs (solid lines). Reprinted (figure) with permission from Ref. [21]. Copyright (2016) by the American Physical Society.

Moreover, numerical simulations are performed to demonstrate the functionality of the concentrator. **Figure 19(a)** and **(b)** corresponds to the concentrator composed of the original discretized TO media and the ultratransparent PhCs, respectively. It is seen that under an incident beam of Gaussian wave from the lower left, both concentrators exhibit good concentration effects in the core areas and induce almost no scattering of waves. Interestingly, the waves inside the core areas exhibit a distinct phase difference of π. This discrepancy is a result of the new freedom introduced by the "shift" of EFCs in the k-space, i.e., spatial dispersion.

Therefore, we have demonstrated that ultratransparent media can work as the TO media to realize TO devices. Such ultratransparent media not only provide a low-loss and feasible platform for TO devices at optical frequencies, but also enable new freedom for phase manipulation beyond the local medium framework.

5.2. For microwave transparency

In the microwave regime, the ultratransparent media are also very useful and may have many applications in the design of radome, transparent wall, and so on. Here, we show an example

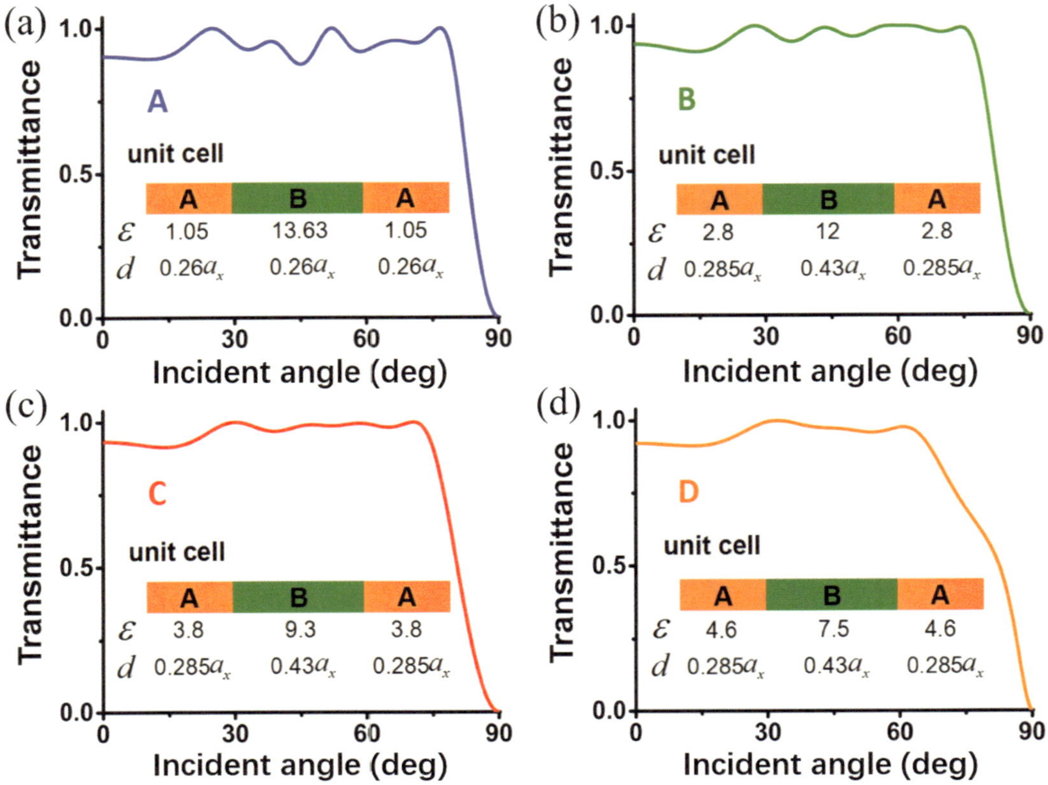

Figure 18. Transmittance through the PhC slabs with 10 unit cells as the function of incident angles. The PhCs in (a), (b), (c), and (d) correspond to the PhCs in regions A, B, C, and D, respectively. The insets are the structure, the relative permittivities, and the thicknesses of the components of the PhCs.

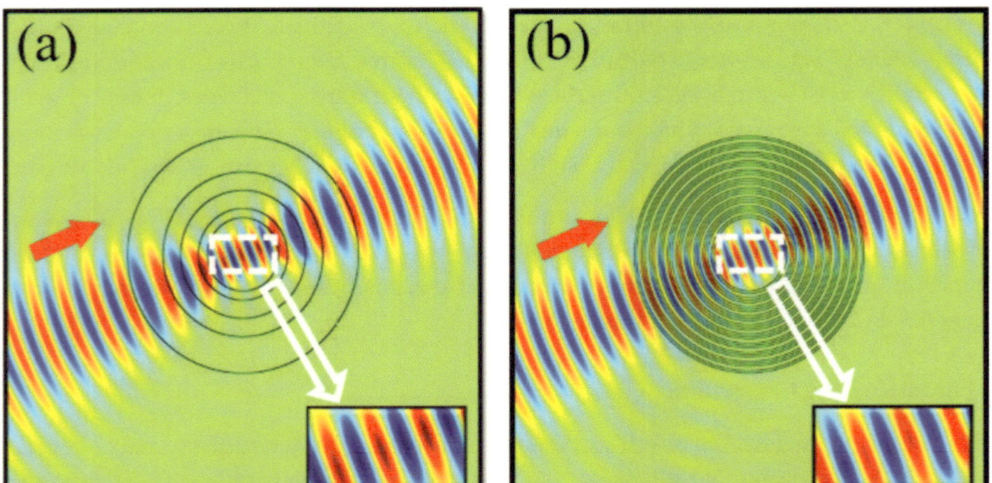

Figure 19. Electric field distributions in the concentrators made of (a) discretized TO media and (b) ultratransparent PhCs. Reprinted (figure) with permission from Ref. [21]. Copyright (2016) by the American Physical Society.

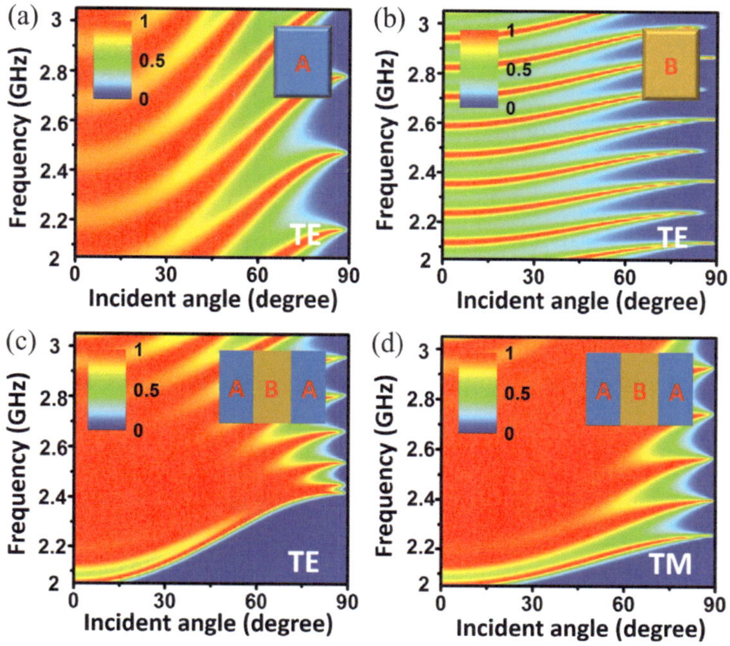

Figure 20. Transmittance with respect to the incident angle and the frequency for EM waves propagating through a slab made of (a) material A for the TE polarization, (b) material B for the TE polarization, (c, d) 10 unit cells of ABA for the (c) TE and (d) TM polarizations. Reprinted (figure) with permission from Ref. [22]. Copyright (2016) by the Optical Society.

of microwave transparent wall which allows the WiFi and 4G signals to pass through freely, and thus may find applications in architectural designs.

The microwave transparent wall is composed of one-dimensional ultratransparent PhCs with ABA unit cells. Materials A and B are chosen as polypropylene ($\varepsilon_A = 2.3$) and concrete ($\varepsilon_B = 9$). As we know, a wall (with a thickness of 42.5 cm) composed of only material A or B would lead to Fabry-Pérot resonance-induced oscillation in the transmission spectrum, as shown by the **Figure 20(a)** and **(b)**. Interestingly, if we utilize the composite structure with ABA unit cells, broadband, wide-angle, and polarization-insensitive transparency can be obtained. Specifically, the lattice constant a of the unit cell is set to be 4.25 cm, and the filling ratio of the material A (B) is 0.6 (0.4). In **Figure 20(c)** and **(d)**, the transmittance through a wall of 10 unit cells is plotted as the function of the incident angle and the working frequency for TE and TM polarizations, respectively. Obviously, almost 100% transmittance is obtained in a broad frequency band (from 2.3 GHz to 2.7 GHz) and a wide-angle range (for almost all $\theta < 60^\circ$) for both polarizations. We note that the frequency range covers the working frequency of Wi-Fi signal (with 2.4 GHz) and 4G signal (2.3 GHz~2.7 GHz). Thus, such microwave transparent walls may have applications in architectural designs.

6. Conclusions and outlook

In this chapter, we introduced the recent results of the structure-induced ultratransparency effect in both one- and two-dimensional PhCs, which allow near 100% transmission of light for all incident angles and create aberration-free virtual images. The ultratransparency effect is well explained by nonlocal effective medium theory for PhCs and verified by both simulations and proof-of-principle microwave experiments. The design principle lies in systematic tuning of the microstructures of the PhCs based on the retrieved nonlocal effective parameters.

With the ultratransparent media, many applications can be expected such as the perfectly transparent optical lens, ideal TO devices, microwave transparent devices, and solar cell packaging. Interestingly, the ultratransparent media with "shifted" elliptical EFC not only provides a low-loss and feasible platform for TO devices at optical frequencies, but also enables new degrees of freedoms for phase manipulation beyond the local medium framework. In addition, microwave transparent walls allowing the WiFi and 4G signals to pass through freely can also be realized.

Although the ultratransparency effect is mainly demonstrated for TE polarization here, the principle is general and can be extended to TM polarization, or even both polarizations. Polarization-independent ultratransparency has wide and important applications. On the other hand, polarization-dependent ultratransparent media could also have some special applications. For instance, if the PhC is ultratransparent for TE polarization, while the working frequency falls in an omnidirectional band gap for TM polarization, such a PhC would work as an omnidirectional polarizer.

The concept and theory of ultratransparency give a guideline for pursuing solid materials with the ultimate transparency, i.e., broadband, omnidirectional, and polarization-insensitive total transparency. In the future, ultratransparent solid materials may be optimized to exhibit an unprecedented level of transparency and find vital applications in various fields.

Acknowledgements

This chapter is supported by National Natural Science Foundation of China (No. 11374224, 11574226, 11704271), Natural Science Foundation of Jiangsu Province (No. BK20170326), Natural Science Foundation for Colleges and Universities in Jiangsu Province of China (No. 17KJB140019), Jiangsu Planned Projects for Postdoctoral Research Funds (1701181B) and a Project Funded by the Priority Academic Program Development of Jiangsu Higher Education Institutions (PAPD).

Author details

Jie Luo and Yun Lai*

*Address all correspondence to: laiyun@suda.edu.cn

College of Physics, Optoelectronics and Energy and Collaborative Innovation Center of Suzhou Nano Science and Technology, Soochow University, Suzhou, China

References

[1] Yablonovitch E. Inhibited spontaneous emission in solid-state physics and electronics. Physical Review Letters. 1987;**58**:2059-2062

[2] John S. Strong localization of photons in certain disordered dielectric superlattices. Physical Review Letters. 1987;**58**:2486-2489

[3] Joannopoulos JD, Villeneuve PR, Fan S. Photonic crystals: Putting a new twist on light. Nature. 1997;**386**:143-149

[4] Joannopoulos JD, Johnson SG, Winn JN, Meade RD. Photonic Crystals: Molding the Flow of Light. 2nd ed. Princeton, USA: Princeton University Press; 2008:63-121

[5] Sakoda K. Optical Properties of Photonic Crystals. New York, USA: Springer; 2005:99-250

[6] Ho K, Chan C, Soukoulis C. Existence of a photonic gap in periodic dielectric structures. Physical Review Letters. 1990;**65**:3152-3155

[7] Fink Y, Winn JN, Fan S, Chen C, Michel J, Joannopoulos JD, Thomas EL. A dielectric omnidirectional reflector. Science. 1998;**282**:1679-1682

[8] Mekis A, Chen JC, Kurland I, Fan SH, Villeneuve PR, Joannopoulos JD. High transmission through sharp bends in photonic crystal waveguides. Physical Review Letters. 1996;**77**: 3787-3790

[9] Lin S, Chow E, Hietala V, Villeneuve PR, Joannopoulos JD. Experimental demonstration of guiding and bending of electromagnetic waves in a photonic crystal. Science. 1998;**282**: 274-276

[10] Russell P. Photonic crystal fibers. Science. 2003;**299**:358-362

[11] Painter O, Lee RK, Scherer A, Yariv A, O'Brien JD, Dapkus IKPD. Two-dimensional photonic band-gap defect mode laser. Science. 1999;**284**:1819-1821

[12] Yamamoto T, Pashkin YA, Astafiev O, Nakamura Y, Tsai JS. High-Q photonic nanocavity in a two-dimensional photonic crystal. Nature. 2003;**425**:941-944

[13] Shen Y, Ye D, Celanovic I, Johnson SG, Joannopoulos JD, Soljacic M. Optical broadband angular selectivity. Science. 2014;**343**:1499-1501

[14] Pendry JB, Holden AJ, Stewart WJ, Youngs II. Extremely low frequency plasmons in metallic mesostructures. Physical Review Letters. 1996;**76**:4773-4776

[15] Pendry JB, Holden AJ, Robbins DJ, Stewart WJ. Magnetism from conductors and enhanced nonlinear phenomena. IEEE Transactions on Microwave Theory and Techniques. 1999;**47**: 2075-2084

[16] Smith DR, Padilla WJ, Vier DC, Nemat-Nasser SC, Schultz S. Composite medium with simultaneously negative permeability and permittivity. Physical Review Letters. 2000;**84**: 4184-4187

[17] Shelby RA, Smith DR, Schultz S. Experimental verification of a negative index of refraction. Science. 2001;**292**:77-79

[18] Pendry JB. Negative refraction makes a perfect lens. Physical Review Letters. 2000;**85**: 3966-3969

[19] Zheludev NI. The road ahead for metamaterials. Science. 2010;**328**:582-583

[20] Liu Y, Zhang X. Metamaterials: A new frontier of science and technology. Chemical Society Reviews. 2011;**40**:2494-2507

[21] Luo J, Yang Y, Yao Z, Lu W, Hou B, Hang ZH, Chan CT, Lai Y. Ultratransparent media and transformation optics with shifted spatial dispersions. Physical Review Letters. 2016;**117**:223901

[22] Yao Z, Luo J, Lai Y. Photonic crystals with broadband, wide-angle, and polarization-insensitive transparency. Optics Letters. 2016;**41**:5106-5109

[23] Pendry JB, Schurig D, Smith DR. Controlling electromagnetic fields. Science. 2006;**312**: 1780-1782

[24] Leonhardt U. Optical conformal mapping. Science. 2006;**312**:1777-1780

[25] Schurig D, Mock JJ, Justice BJ, Cummer SA, Pendry JB, Starr AF, Smith DR. Metamaterial electromagnetic cloak at microwave frequencies. Science. 2006;**314**:977-980

[26] Chen H, Chan CT, Sheng P. Transformation optics and metamaterials. Nature Materials. 2010;**9**:387-396

[27] Pendry JB, Aubry A, Smith DR, Maier SA. Transformation optics and subwavelength control of light. Science. 2012;**337**:549-552

[28] Śmigaj W, Gralak B. Validity of the effective-medium approximation of photonic crystals. Physical Review B. 2008;**77**:235445

[29] Lu Z, Prather DW. Calculation of effective permittivity, permeability, and surface impedance of negative-refraction photonic crystals. Optics Express. 2007;**15**:8340-8345

[30] Lawrence FJ, de Sterke CM, Botten LC, McPhedran RC, Dossou KB. Modeling photonic crystal interfaces and stacks: Impedance-based approaches. Advances in Optics and Photonics. 2013;**5**:385

[31] Li J, Pendry JB. Hiding under the carpet: A new strategy for cloaking. Physical Review Letters. 2008;**101**:203901

[32] Lai Y, Chen H, Zhang Z, Chan C. Complementary media invisibility cloak that cloaks objects at a distance outside the cloaking shell. Physical Review Letters. 2009;**102**:093901

[33] Rahm M, Schurig D, Roberts DA, Cummer SA, Smith DR, Pendry JB. Design of electromagnetic cloaks and concentrators using form-invariant coordinate transformations of Maxwell's equations. Photonics and Nanostructures. 2008;**6**:87-95

[34] Lai Y, Ng J, Chen H, Han D, Xiao J, Zhang Z, Chan C. Illusion optics: The optical transformation of an object into another object. Physical Review Letters. 2009;**102**:253902

[35] Pendry J. All smoke and metamaterials. Nature. 2009;**460**:79-580

[36] Lai Y, Ng J, Chen H, Zhang Z, Chan CT. Illusion optics. Frontiers of Physics China. 2010;**5**:308-318

[37] Genov DA, Zhang S, Zhang X. Mimicking celestial mechanics in metamaterials. Nature Physics. 2009;**5**:687-692

[38] Sheng C, Liu H, Wang Y, Zhu SN, Genov DA. Trapping light by mimicking gravitational lensing. Nature Photonics. 2013;**7**:902-906

[39] Cai W, Chettiar UK, Kildishev AV, Shalaev VM. Optical cloaking with metamaterials. Nature Photonics. 2007;**1**:224-227

[40] Liu R, Ji C, Mock JJ, Chin JY, Cui TJ, Smith DR. Broadband ground-plane cloak. Science. 2009;**323**:366-369

[41] Valentine J, Li J, Zentgraf T, Bartal G, Zhang X. An optical cloak made of dielectrics. Nature Materials. 2009;**8**:568-571

[42] Gabrielli LH, Cardenas J, Poitras CB, Lipson M. Silicon nanostructure cloak operating at optical frequencies. Nature Photonics. 2009;**3**:461-463

[43] Ergin T, Stenger N, Brenner P, Pendry JB, Wegener M. Three-dimensional invisibility cloak at optical wavelengths. Science. 2010;**328**:337-339

[44] Ma HF, Cui TJ. Three-dimensional broadband ground-plane cloak made of metamaterials. Nature Communications. 2010;**1**:21

[45] Xu S, Cheng X, Xi S, Zhang R, Moser HO, Shen Z, Xu Y, Huang Z, Zhang X, Yu F, Zhang B, Chen H. Experimental demonstration of a free-space cylindrical cloak without superluminal propagation. Physical Review Letters. 2012;**109**:223903

[46] Xiao S, Drachev VP, Kildishev AV, Ni X, Chettiar UK, Yuan HK, Shalaev VM. Loss-free and active optical negative-index metamaterials. Nature. 2010;**466**:735-738

[47] Hess O, Pendry JB, Maier SA, Oulton RF, Hamm JM, Tsakmakidis KL. Active nanoplasmonic metamaterials. Nature Materials. 2012;**11**:573-584

[48] Lai Y, Zheng H, Zhang Z, Chan CT. Manipulating sources using transformation optics with 'folded geometry'. Journal of Optics. 2011;**13**:024009

Chapter 5

Influence of Substrate Wettability on Colloidal Assembly

Junchao Liu, Jingxia Wang and Lei Jiang

Additional information is available at the end of the chapter

http://dx.doi.org/10.5772/intechopen.71991

Abstract

In this paper, we presented a detailed discussion about the influence of the substrate wettability on the colloidal assembly and the resultant functionality of the films. It covers the basic assembly principle for colloidal crystals, the basic understanding of the substrate wettability on colloidal assembly, and the detailed explanation of the influence by give a full examples of various assembly from the substrate with distinct wettability, such as superhydrophilic, hydrophilic, hydrophobic, superhydrophobic and hydrophilic-hydrophobic pattern substrate.

Keywords: wettability of substrate, colloidal crystals, assembly

1. Introduction

Colloidal crystals [1–4] has aroused wide research attention owing to its fascinating light manipulation properties and important application in sensing [5–18], detecting [19–22], catalytic [23, 24] and some special optic devices [25–32]. Colloidal crystals are generally fabricated from the well-ordered assembly of the monodispersed latex particles or infiltrating the functional materials into the template and subsequent template removal (the typical fabrication process for the inverse opals). Accordingly, colloidal assembly plays an important role on the resultant functional, unique properties and the resulted potential applications. In this paper, we presented a detailed discussion about the influence of the substrate wettability on the colloidal assembly and the resultant functionality of the films. It covers the basic assembly principle for colloidal crystals, the basic understanding of the substrate wettability on colloidal assembly, and the detailed explanation of the influence by giving a full examples of various assembly from the substrate with distinct wettability, such as superhydrophilic, hydrophilic, hydrophobic, superhydrophobic and hydrophilic-hydrophobic pattern substrate.

INTECH
open science | open minds

© 2018 The Author(s). Licensee InTech. This chapter is distributed under the terms of the Creative Commons Attribution License (http://creativecommons.org/licenses/by/3.0), which permits unrestricted use, distribution, and reproduction in any medium, provided the original work is properly cited. (cc) BY

2. Basic assembly principle for colloidal crystals

Colloidal assembly has become a well-known process since more than 5 decades' investigation of the colloidal assembly [33]. As shown in **Figure 1**, the colloidal assembly process includes the crystal nucleus by capillary force among latex particles, when the liquid front covers the half of the particles and crystal growth driven by the convective force owing to solvent evaporation and solvent reflux. A common mode for colloidal assembly includes vertical deposition approach. Where, the substrate is vertically placed in a colloidal solution, then a film of colloidal particles is formed at the interface between the substrate and the liquid surface with the solvent evaporation and the liquid surface drops. The formation of colloidal assembly is mainly originated from the driven force of capillary force between the drying colloidal particles at the meniscus of the solvent [34].

It is well known that the colloidal assembly is mainly depended on the assembly temperature, assembly humidity, and plenty of researches literatures took a careful investigation of the influence of various influencing factors on assembly behavior, assembly structure, and resultant property of colloidal crystals. Meantime, many assembly approach, such as spin coating, spray coating, inkjet printing, has been developed to improve the duplicated, assembly rate, controllable assembly, scalable assembly and high-quality assembly. Where, the vertical deposition approach has become a common method owing to its economic and easily duplicated in most laboratory (**Figure 1C**). In this following part, we would like to discuss the influence of the substrate wettability on colloidal assembly.

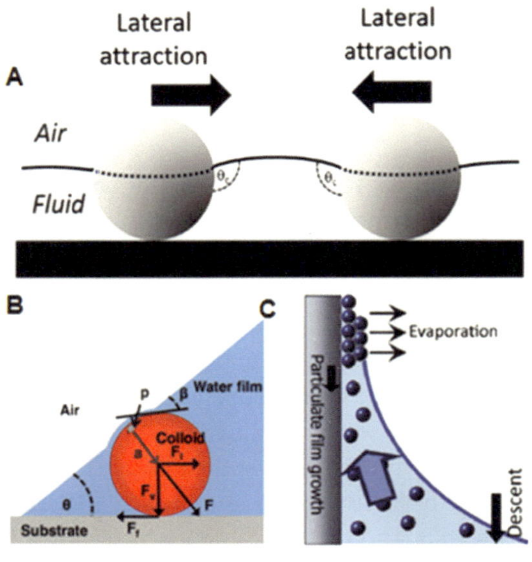

Figure 1. Mechanism of colloidal PCs assembly. (A) Colloidal particles assembly forced by the liquid flow as evaporation of the solvent; (B) acting on a colloid trapped at an air water substrate interface; (C) schematic illustration of the convective self-assembly technique in vertical deposition way [33, 34].

3. Basic understanding of the substrate's wettability on colloidal assembly

To understand the basic influence of the wettability on colloidal assembly, we present a basic concept of the substrate wettability and depict a detailed analysis about the effect of substrate wettability on the assembly process, assembly force and etc.

3.1. Basic concept of substrate wettability and the basic understanding of the wettability on the colloidal assembly

Wettability is a basic property of liquid on solid surface, it is determined by surface chemical composition and surface roughness, as shown in **Figure 2**. Generally, the wettability is evaluated by the water contact angle (CA), i.e., static CA. It is defined by integrated angle for the three-phase contact line of liquid, gas and solid after a droplet spread on the substrate (**Figure 2A**). Based on different statistic CA, distinct substrate is determined. The hydrophilic substrate with CA < 65°, indicating a strong attraction of liquid on solid surface, which results in droplet spreading. Especially superhydrophilic substrate is obtained when CA is around 0° (**Figure 2C**). Hydrophobic substrate with CA > 90°, meaning a limited attraction between solid and liquid surface, a less spread is observed for droplet on the substrate. Recently, the diving line is set as 65°, a new dividing line

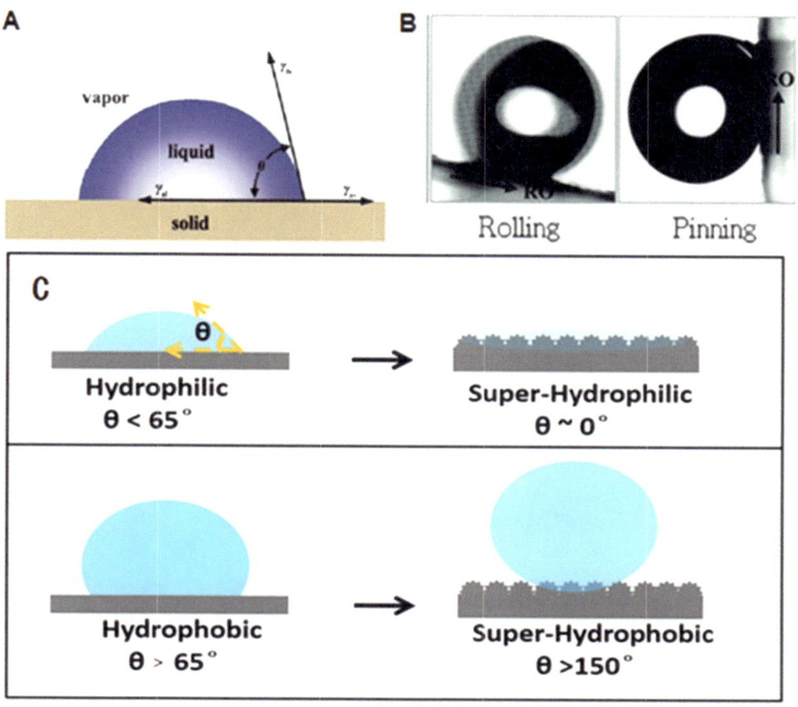

Figure 2. Wettability of interfaces. (A) Statistic contact angle, (B) sliding angle with different substrate adhesive, (C) schematic illustration of different wettability state [35].

between hydrophilic and hydrophobic, is defined by Prof. Lei Jiang. Superhydrophobic substrate with water CA > 150°, is a special substrate, it is difficult for the droplet to spread upon it. Such as the lotus surface. Besides statistic CA, sliding angle is also a judgment of wettability of substrate. As shown in **Figure 2B**, when a droplet and three phase contact line (TCL) can stay steadily on a solid substrate it indicates high-adhesive, in the contrast, droplet and TCL can easily move on a low-adhesive substrate. Both static CA and the sliding angle plays an important role on the colloidal assembly. The substrate with different water CA showed various potential applications.

3.2. Basic understanding of the influence of the substrate wettability on colloidal assembly from static and dynamic wettability

Here, we understand the influence of the substrate's wettability on colloidal assembly by qualitatively analyzing the relationship between the substrate wettability on the evaporation time/rate and the evaporation force when a droplet spreading on the substrate. As shown in **Figure 3**, for the hydrophilic substrate with lower CA, having a large spreading area. In this

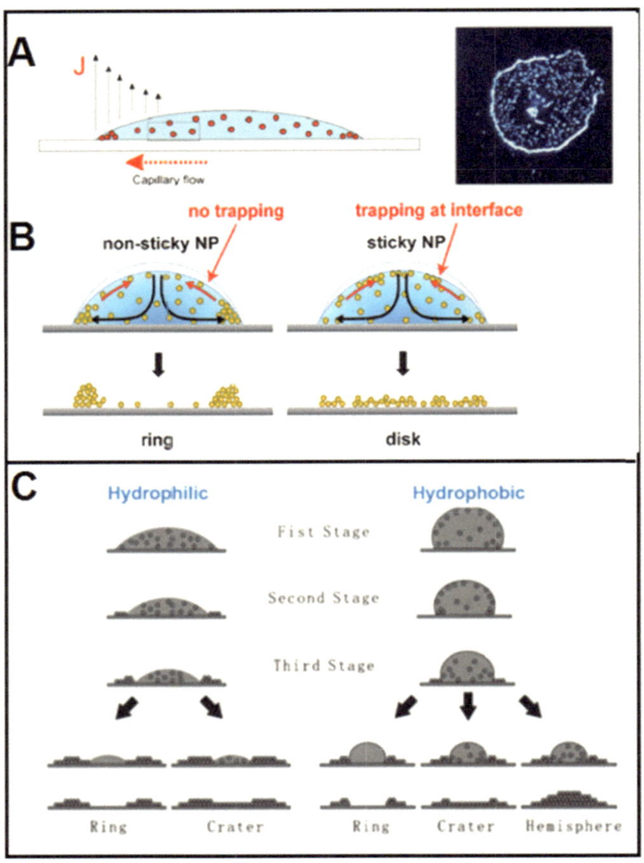

Figure 3. Formation of colloidal PCs as substrate is hydrophilic. (A) Illustration of coffee stain formation and a microscope image of a coffee stain formed by fluorescently labeled 5-μm particles; (B) on a hydrophilic substrate, the competition of Marangoni flow (red arrows) and the evaporation-driven capillary flow (black arrows) resulted the ring-shape or the disk shape. (C) Schematic depiction of the aggregation of monodisperse spheres on hydrophilic PMMA and hydrophobic FAS-coated glass substrates [36].

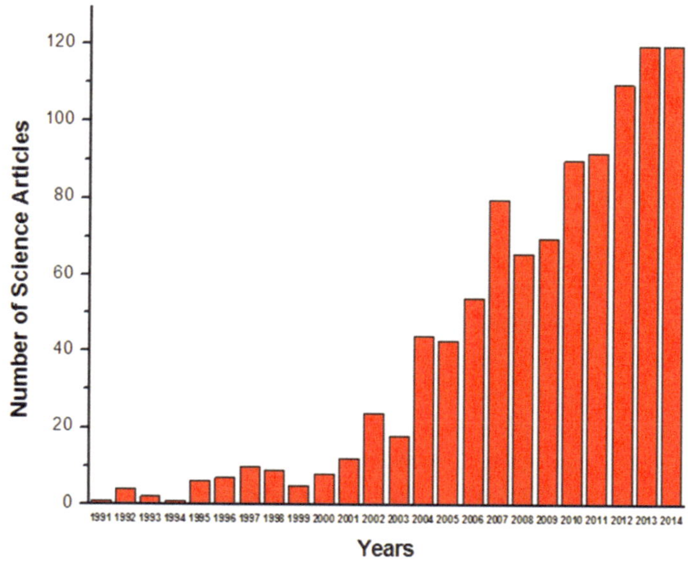

Figure 4. Number of science articles that searching result of "assembly" + "wettability".

case, the droplet keeps constant contact area and decreased water CA during evaporation process, as a result, a short time or fast evaporation rate is observed for the droplet evaporation. In contrast, when droplet is on the hydrophobic substrate, the droplet keeps constant CA but a decreased contact area during evaporation process. As a result, a longer evaporation time and slower evaporation rate occurs in the evaporation system. Accordingly, there would be a short assembly time on hydrophilic substrate, and a longer assembly time on hydrophobic substrate. On the other hand, substrate wettability will affects the pinning or sliding or TCL on the substrate, the sliding TCL will produce an addition driven force for the latex assembly, will contributing to distinct assembly structure and colloidal property as well.

As regarding of the research progress related to the influence of the wettability on colloidal assembly, there is an obvious increased paper published after Year 2004, as shown in **Figure 4**. Which presented the research papers number (related to colloidal assembly & wettability) ranging from 1991 to 2014, the data is searched from database of Web of Science. Clearly, less papers is related to the colloidal assembly &wettability prior to 2004, there are only 1–2 papers published in 1991. In comparison, a rapid growth of the paper numbers related to colloidal assembly/wettability occurred after the period. More than 100 papers published related to the topic in Year 2014. In the following part, we will list a full example to understand what the influence of the substrate's wettability on the colloidal assembly.

4. Full examples for the understanding of the influence of wettability on colloidal assembly

Wettability of substrate plays an important effect on the colloidal assembly by affecting its assembly time, assembly rate and assembly process. In this section, we summarized various examples to give a full description of the influence of the wettability on the colloidal assembly process and its resultant assembly structure. Wherein, the substrate with wettability, such as superhydrophilic, hydrophilic, hydrophobic, superhydrophobic, pattern-substrate is considered.

4.1. Superhydrophilic substrate

Superhydrophilic substrate with water CA of 0°, shows excellent spreading behavior for the colloidal suspension. In this case, the resultant colloidal assembly structure showed an uniform distribution. Therefore, superhydrophilic substrate is generally thought to be optimal substrate for the colloidal assembly at the earlier assembly literature. How to achieve a superhydrophilic substrate has ever to be an important technical issue for the well-ordered colloidal assembly [37]. Accordingly, many early assembled literatures are carried out on superhydrophilic substrate. Many outstanding and impressing assembly work is done on superhydrophilic substrate. Typically, many colloidal crystals are assembled on superhydrophilic substrate by vertical deposition, spin coating, spray coating.

4.2. Hydrophilic substrate

Hydrophilic substrate with water CA <65° have a special "coffee-ring" effect on colloidal assembly. Coffee ring is aroused owing to the faster evaporation rate in the exterior region of the droplet comparing that interior region, thus more latex transfer from the interior toward the exterior region, which resulted in the more deposit and assembly of the latex particles at the brim of the droplet, leaving less latex at the center of the droplet, forming a ring-shaped structure (left part of **Figure 3B** and **C**). Much work is developed to remove coffee ring effect by introducing temperature filed, surfactant. Of course, some researchers fabricated some interesting pattern by taking advantage of the coffee-ring effect. For example, Gu et al. [36] displayed different assemble modes of PC in hydrophilic and hydrophobic substrate in **Figure 3**. They used different concentration suspension dripple onto polymethyl methacrylate (PMMA) substrate and FAS-treated glass substrate in the hydrophilic substrate, ring and crater structure can be obtained, while on hydrophobic substrate, hemisphere structure also appear for the liquid shrink to a dot. The assembly process of colloidal particles on different wettability substrate can be described as follows. Furthermore, Gu et al. developed a ring-shape colloidal crystal by taking advantage of coffee-ring phenomenon for artificial eye-pupil structure.

4.3. Hydrophobic substrate

Hydrophobic substrate with water CA of higher than 65°, showed special evaporation behavior than that on the hydrophilic substrate. Particularly, TCL recedes owing to the lower adhesive force of particles on the substrate when droplet evaporating on hydrophobic substrate. Just because the receding TCL, the coffee ring effect can be effectively removed on hydrophobic substrate (right part in **Figure 3B** and **C**). In this case, a sphere-rich phase near the solvent/air interface was caused by evaporation of the solvent (first stage); the crystallization of spheres at the rim of drop occurred when the concentration exceeded a critical value (second stage); receding TCL aggregate the latex toward the center of the droplet (third stage). As a result, a final compact assembly structure can be obtained "similar to dome". The application of the hydrophobic substrate may produce some novel and functional assembly structure.

Liu et al. [38] fabricated controllable inkjet printing lines by adjusting the ink droplets' dynamic wettability on hydrophobic substrate (**Figure 5A** and **B**). Mixing the ink with water and ethylene glycol to adjust the ink droplet's surface tension and the nanoparticle concentration. Distinct dynamic wettability of ink droplets on the hydrophobic substrate could be achieved owing to the different surface tension and nanoparticle concentration. In the first case, the surface tension of droplet 1 and 2 were similar, the TCL of droplet was hardly pinned owing to no particle assembly at TCL. A spherical cap was obtained after coalescing and drying of ink, as shown in **Figure 5A$_1$**. In contrast, if the surface tension of droplet 1 was smaller than droplet 2, the TCL of droplet was pinned and a straight line could be obtained after coalescing and drying owing to nanoparticles assembled at the TCL of the droplet 1. Furthermore, a stronger pinning TCL and dumbbell shape was formed with more nanoparticles assembling at the TCL, as shown in **Figure 5A$_2$** and **A$_3$**. As a result, different line shapes of wave, straight footprint and wave with straight footprint were obtained, as shown in **Figure 5B**. Hydrophobic substrate was used for the removal of the coffee ring. Typically, Cui et al. [40] made a research about a drop of a colloidal suspension of latex spheres dropped onto a hydrophobic-silica pillar array (HSPA) with high contact hysteresis to remove coffee ring effect. In details, a drop of colloidal suspension with latex spheres presents a Wenzel state with high CA hysteresis on the surface of HSPA, which leads to the pinning of contact line (CL) during the solvent evaporation. Then more latex spheres will be deposited on the periphery of the drop to accelerate growth of the porous gel foot (means the aggregation of latex spheres at the edge of the droplet). Subsequently, the capillary

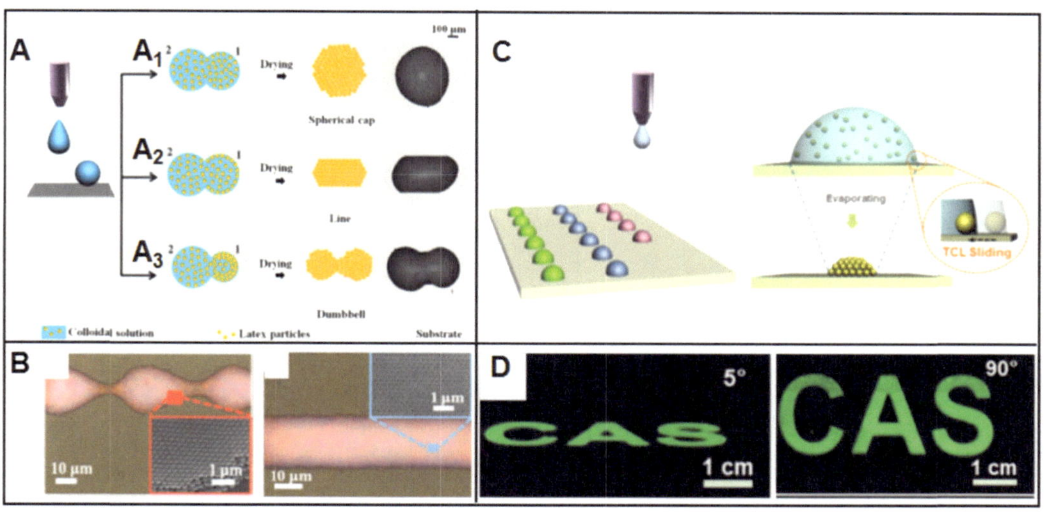

Figure 5. (A) Three typical coalescing cases of the neighboring ink droplets induced by different dynamic wettability of ink droplets on the substrates. (B) Optical microscope images of the as-printed PC lines with wave and straight footprints. Three typical straight PC lines (red, green, and blue) demonstrating good optical properties and the insets are the corresponding SEM images [38]. (C, D) fabrication of photonic crystal dome on hydrophobic substrate by inkjet printing [39].

flow formed based on a growing gel foot. Finally, a uniform colloidal deposition without a coffee ring structure is obtained owing to the existence of gel foot and capillary flow. Finally, "coffee ring" diminished as well as closed-packing PC structure can be obtained. Due to the fast shrinking of the TPCL. Furthermore, hydrophobic substrate was used for the fabricated of the PC dome array by inkjet printing by Kuang et al. [39] in **Figure 5C–D**. They inkjet printed latex suspension on the hydrophobic substrate, obtained dome-like PC sphere, which can effectively avoid the angle-dependent property of the stopband of PC, providing an important insight for the wide-view display applications of PCs. That is, the hydrophobic substrate provides an effective approach for the fabrication of colloidal crystals with excellent wide-angle property [41–46].

4.4. Superhydrophobic substrate

Superhydrophobic substrate is a special substrate with water CA > 150°, the rolling angle less than 5°, as well as low adhesive surface. The large water CA makes it possible for the spherical colloidal assembly. While the low-adhesive property of substrate contributed to the formation of crack-free colloidal assembly.

The application of superhydrophobic substrate is helpful for the fabrication of spherical PCs and crack-free colloidal PCs. For example, Velev et al. [47] reported an approach for colloidal assembly in droplets on superhydrophobic substrates (**Figure 6A**), which yields

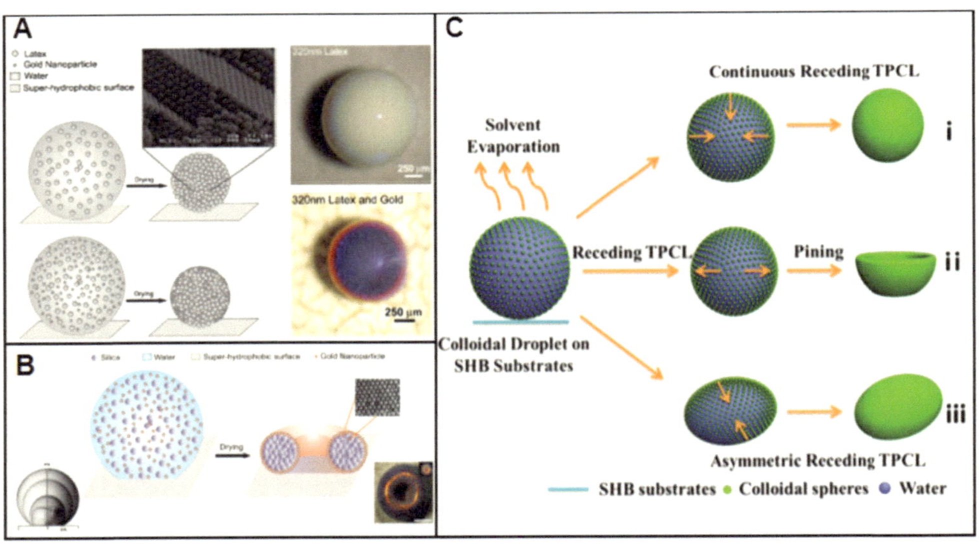

Figure 6. (A) Schematics of the process for making spherical colloidal assemblies on superhydrophobic. The inset displays the hexagonal close-packed structure of latex spheres inside an opal ball of 540 nm latex [47]. (B) Schematic of supra-particle formation by evaporating a droplet containing suspension of silica and gold nanoparticles. Top and angled view of "golden doughnut" supra-particles on the right-down inset. The inset on the left-down shows an overlay of experimental images for the side-view profile of a drying droplet with silica particles over time fabricated from a droplet containing 330-nm-diameter silica [48]. (C) Schematic illustrations of the microshape fabrication of PC particles on superhydrophobic substrates based on different dynamic behaviors of the TCL, (i) continuous receding, (ii) receding and then pinning, and (iii) asymmetric receding of the TCL [49].

better control over the final shape and creates supra-particles that are easily detached and ready to use. The resultant colloidal assembly is near-spherical and spheroidal supraballs. The process of sample are as follows: the droplet of colloidal latex or mixed with gold nanoparticles showed superhydrophobic and high CA hysteresis owing to the small pinning areas formed between low-density polyethylene and substrate. Thus the colloidal spheres were inclined to form a close-packed microsphere crystals with the evaporation of solvents due to the decrease of free volume between latex particles. As a result, the spherical colloidal PC with colored ringlike diffraction patterns was formed. Later, they developed this method and obtained shape-anisotropic ("doughnut") and composition-anisotropic ("patchy magnetic") supra-particles [48] in **Figure 6B**. At the initial stage, near-spherical shape was obtained due to a high contact angle. During the process of solvent evaporation, silica suspension droplets undergo shape transitions (concaving) leading the structure of the final assemblies to doughnut supra-particles. Furthermore, composition anisotropy is achieved by drying a droplet containing a mixed suspension of latex and magnetic nanoparticles among a magnetic field gradients. The magnetic nanoparticles assembled into single, bilateral, or trilateral, patched spherical supra-particles. Additionally, the shape of the microsphere can be tuned by Zhou et al. from microbeads to microwells to micro-ellipsoids via adjusting the dynamic behaviors of the three-phase contact line (TCL) during the evaporating process on superhydrophobic substrates [49] (**Figure 6C**). The assembly of PC structure is prepared by dispensing aqueous colloidal latex spheres droplet (acting as self-assembly template) onto superhydrophobic substrate and evaporates naturally. The high CA (≈151°) of colloidal latex droplet on the superhydrophobic substrate contributed to the formation of self-assemblies templates. Microbead PC assemblies was obtained by continuous homogeneous receding of the TCL of the aqueous colloidal droplet evaporating on superhydrophobic substrate (scheme C_i). But for dimpled microbead or microwell PC assemblies (scheme C_{ii}), the reason for its formation maybe the change in the moving directions of the colloidal spheres caused by pinning of the TCL during aqueous colloidal droplet evaporation. Furthermore, asymmetric receding of TCL benefits the interesting anisotropic PC assemblies (scheme C_{iii}). It worth to be noted that high-quality crack-free colloidal PC can be fabricated from superhydrophobic substrate. For instance, Huang et al. [50] fabricated a centimeter scale PCs on low-adhesive super-hydrophobic substrates (**Figure 7**). It shows the schematic illustration of colloidal PCs assembled on high F_{ad} substrate and low F_{ad} superhydrophobic substrate. The red dashed lines indicate the changing trend of the TCL during different drying conditions. The latex particles at first assemble on the surface of the suspension, and then shrink with further solvent evaporation. When the latex particles dry on a substrate with high F_{ad}, the pinned TCL and latex shrinkage causes tensile stress and crack formation. In contrast, large-scale crack-free colloidal PCs are achieved on a superhydrophobic substrate with low F_{ad} due to the timely release of tensile stress as the TCL recedes. The sample showed high quality and crack-free property, it is especially important, the sample showed an evident narrow stopband with full-width-at-half-maxima of the stopbands of just 12 nm due to the receding TCL during evaporation process, which releases the tensile stress induced by latex shrinkage. The work is of great significance for the creation of novel and high-quality PC optic devices.

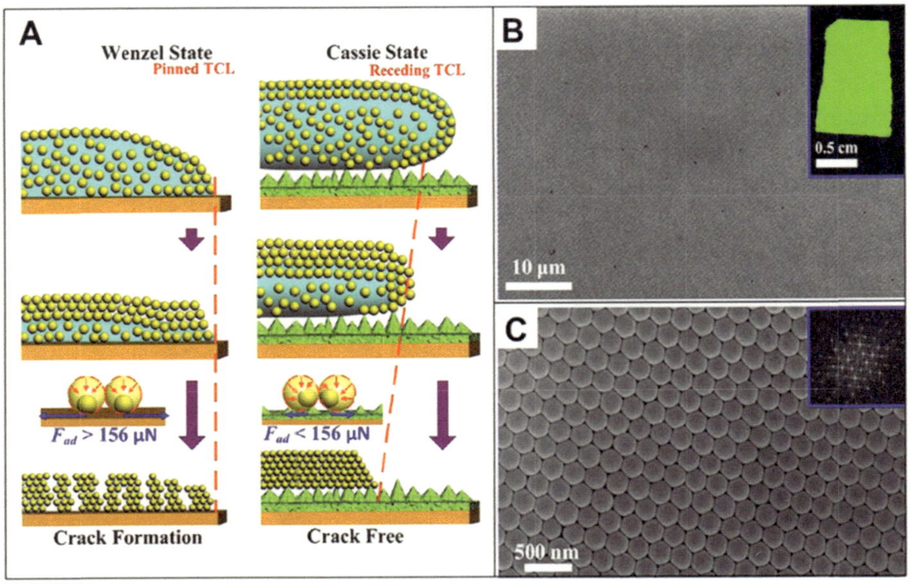

Figure 7. (A) Schematic illustration for the fabrication of crack-free PC on low-adhesive superhydrophobic substrate. (B, C) SEM images of the as-prepared colloidal PCs with diameter of 224 nm assembled on low-adhesive superhydrophobic substrate. The images demonstrate perfectly ordered latex arrangement and close-packed assembly structures of the PCs [50].

4.5. Pattern colloidal crystals from the combination of the hydrophilic/hydrophobic substrate

Wettability also can be used for design patterned PCs [51]. Colloidal crystals with different patterns in various colors may be obtained through tuning the wettability of assemble substrate, and different particles, Also, asymmetric or dissymmetric pattern can also be created in wettability gradient surface.

Figure 8A exhibited a detailed description about the fabrication of the substrate with modified wettability. For example, Young et al. demonstrated the structure of the PS colloidal crystals which were fabricated on the hydrophilic/hydrophobic Si wafers by a spin-coating technique (**Figure 8A)**. PS spheres organized as ordered close-packed face-centered cubic structure on the hydrophilic surface while they gathered without the crystal structure on the hydrophobic surface. Lee et al. [52] made site-selective assembly available and designed a surface with alternating wetting region and dewetting region in **Figure 8B**, then the particles only confined and assembled in the wet region. Therefore, wettability induced PC pattern can be obtained. The basic concept of site-selective assembly of colloidal particles on a wettability-patterned surface are shown in **Figure 8B**. The relative fraction of the wet region W to the dewet region D and droplet volume decided the wetting layer morphology including the CA. First, a generate patterned wettability was obtained by liquid crystal (LC) alignment layer irradiated via UV light through a photomask. Then a periodic array of circular W regions surrounded by intact D regions was obtained. When dropping an LC/prepolymer solution with colloidal particles on the as-prepared wettability patterned substrate, it could

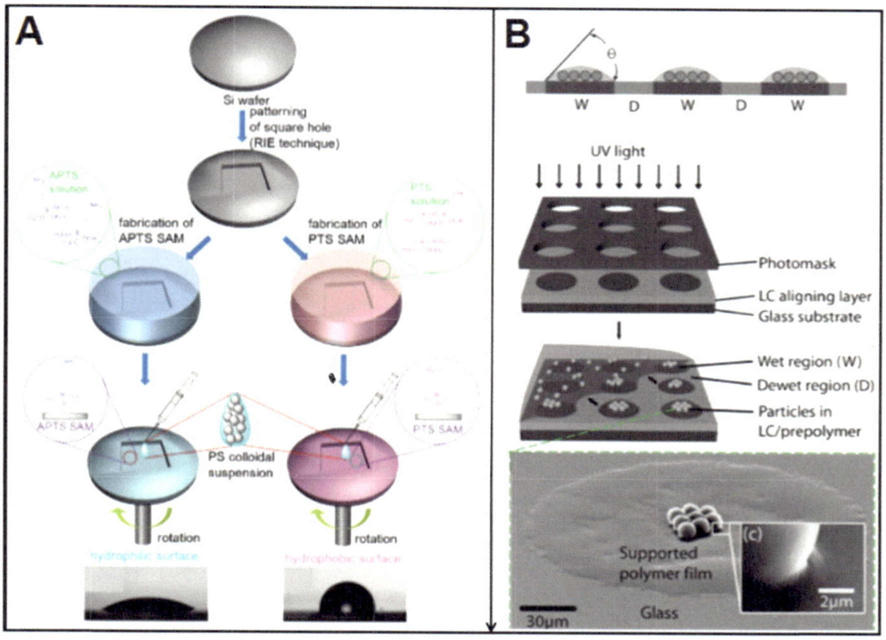

Figure 8. (A) a detailed description about the fabrication of the substrate with modified wettability. (B) schematic diagrams showing underlying concept of site-selective assembly of colloidal particles: Liquid crystal (LC)/prepolymer solution containing colloidal particles on wettability-patterned surface with wet regions (W) and dewet regions (D) and wettability-patterning process using UV exposure on LC alignment layer (top) and array of LC/prepolymer droplets produced only in wet regions where the colloidal particles are confined (bottom) [52].

be confined to the W regions. A highly close-packed structure of colloidal particles located in the W region was formed. Trough different value of H, two assembly cases of colloidal particles that in central or boundary regions could be formed.

Wu et al. [53] presented a strategy to fabricate controllable 3D structures and morphologies from one single droplet via inkjet printing (**Figure 9A** and **B**). The 3D morphology of microcolloidal crystal pattern was controlled by hydrophilic pattern induced asymmetric dewetting. First, using a hydrophobic silicon wafer with patterned hydrophilic pinning spots (green shading) as substrate. Then nanoparticles contained precisely designed droplets array was prepared by inkjet printing. Arrayed 3D microcolloidal crystals with controllable morphology were achieved owing to the hydrophilic pattern induced asymmetric dewetting. Many morphologies of quadrilateral, pentagon, hexagon and etc. could be obtained through different hydrophilic patterns. Wang et al. [54] fabricated a microring PC made of colloidal particles by taking advantage of a superhydrophilic flat transfer substrate and a superhydrophobic groove–structured silicon template (**Figure 9C**). The process of "sandwich assembly" was mainly used by a superhydrophobic groove structured template and a flat superhydrophilic transfer substrate. The as-prepared microrings showed homogeneous bring–green color owing to the fluorescent signal and favorable waveguide property. Yoo et al. [55] reported a flexible superhydrophobic PDMS cage formed by superhydrophobic patterns encompassing the unmodified region for aqueous

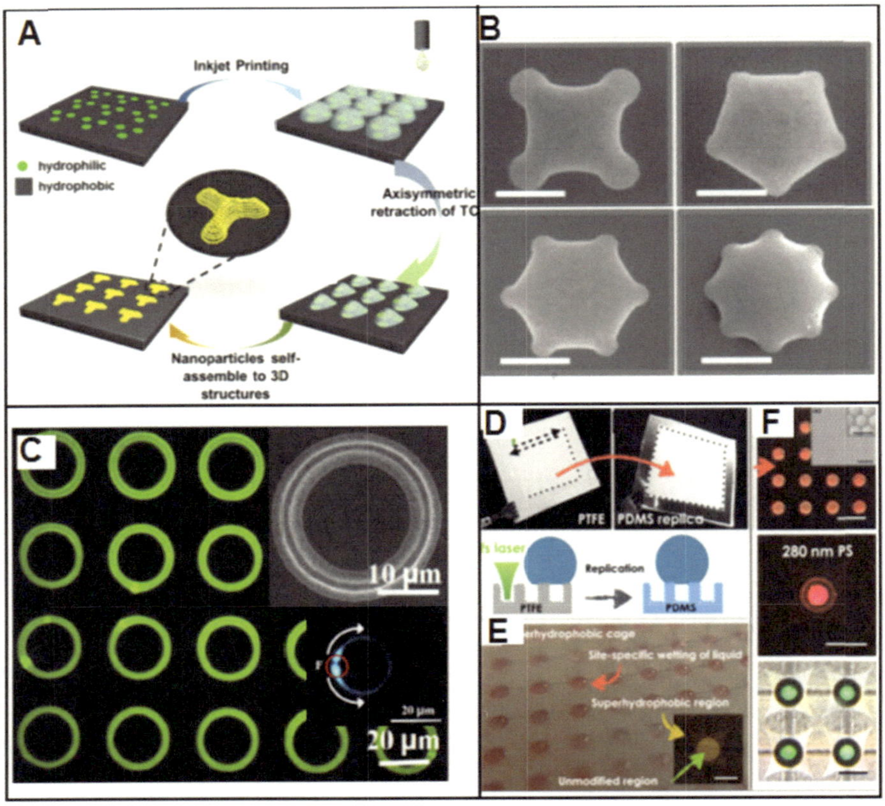

Figure 9. (A) Manipulating 3D morphology of microcolloidal crystal pattern through hydrophilic pattern induced asymmetric dewetting. (B) SEM images of various assembly units through designed hydrophilic pinning pattern [53]. (C) Fluorescence optical and SEM images of micro-ring PC and its waveguide property [54]. (D) Schematics of superhydrophobic surface production by FS laser direct writing. Photographs of superhydrophobic polytetrafluoroethylene (PTFE) film and superhydrophobic polydimethylsiloxane (PDMS) replica. (E) PDMS superhydrophobic cage. (F) Microscopic pictures of self-assembled structures on a superhydrophobic cage. The scale bars are 1 mm. Magnified reflection and transmission microscopic pictures of self-assembled structures consisting of 280 nm PS particles from colloidal droplets. Scale bars are 500 μm [55].

droplet positioning and trapping in **Figure 9D**. A PTFE film with nano-sized bumpy structures could be fabricated by femtosecond laser ablation under ambient conditions. Then a superhydrophobic PDMS surface was easily replicated from the superhydrophobic PTFE mold owing to the tribological properties of PTFE. As a result, a superhydrophobic cage structure on a PDMS replica surface was prepared by laser enabled direct writing. The optical image of one superhydrophobic cage structure was shown in **Figure 9D**. Subsequently, dropping a liquid with colloidal particles on the surface of superhydrophobic PDMS, colloidal particles self-assembly in this superhydrophobic cage during evaporation process, forming a self-assembled PC.

Choi et al. [44] employed a fast, high-through method to fabricate size-tunable micro/nanoparticle clusters via evaporative assembly in picoliter-scale droplets of particle suspension on hydrophobic substrate. Various morphologies that particle clusters with accurate positioning and alignment are demonstrated (shown in **Figure 10A** and **B**). the whole fabrication process includes: (i) Particle suspension menisci are extruded to the upfront end of the membrane by gravity. (ii) Contact of the head with the substrate is achieved. (iii) Surface tension of the substrate attracts a fraction of the suspension fluid. (iv) Picoliter-scale droplets are transferred to the substrate via pinch-off processes. (v) Rapid evaporative self-assembly of the particles forms 3-D clusters. In this case, the printing head was fabricated by the method of applying traditional microfabrication technology to SOI (Silicon-On-Insulator) substrates while a micro/nanoparticle suspension container was achieved by being wet-etched of backside. A microporous membrane with a 200 nm thickness was released on the head after the above whole process. After the suspension was loaded into the head, the meniscus of the droplet was completely extruded to the front of the head. Then, multiple picoliter-scale (2 − 20pL) droplet of particle suspension was transferred from the bulk suspension to the substrate by direct contact of the head with the substrate. This whole process could be achieved less than 5 s. The evaporative self-assembly process is controlled by gravity force and surface tension force of a contacting surface and controlled sizes and spacing of particle clusters.

Kim et al. [56] reported a novel and controllable patterning technique for 3D or 2D colloidal arrays of polymeric domes using photocurable emulsion droplets as templates (**Figure 10C–E**). The oil-in-water emulsion droplets will adhere selectively on the surface with a high interfacial affinity. The preparation process includes the following parts: first, a patterned glass substrate was prepared by microcontact printing with a hydrophobic ink of octadecyltrichlorosilane (OST). The OTS molecules binded covalently on the clean glass which from a poly(dimethylsiloxane) (PDMS) stamp. PDMS stamps had cylindrical posts of patterned arrays or characters by soft-lithography. Then, dropping oil-in-water emulsion in the surface of prepatterned glass substrate.

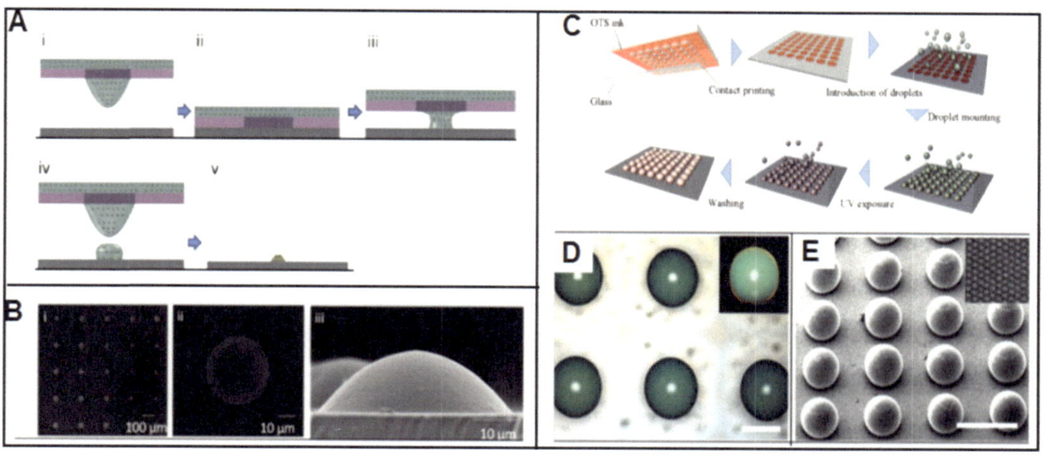

Figure 10. Schematics of printing procedure. (A) Serial processes for the fabrication of the PC sphere on hydrophobic substrate. (B) SEM views of zinc oxide nanoparticle (diameter ≈ 30 nm) clusters [44]. (C) Schematic of mounting and solidifying of droplets on a substrate pre-patterned with a hydrophobic moiety. (D) Optical microscopy images of patterned photonic domes of a 260-mm diameter with a 500-mm interval. They are composed of 170-nm silica particles (33 vol%) in an ETPTA matrix. (E) SEM image of the dome pattern; the inset shows the surface morphology of the dome. Scale bars, 500 μm [56].

The oil phase was ethoxylated trimethylolpropane triacrylate (ETPTA), which could be photocurabled. In water, a high CA of 158.4°for ETPTA on the bare glass was showed but it was inclined to spread over the OTS-coated glass with a low CA of 8°. As a result, the ETPTA drops could be selectively adsorbed on the patterned OTS dots. Finally, a pattern of hemispherical colloidal PC was prepared by using ETPTA droplets with SiO_2 particles. The as-prepared colloidal PC dome presented hexagonal arrangement of colloidal silica particles on its surface, which corresponded to the (111) plane of the FCC lattice.

Ultratrace detection is of enormous interest in early diagnosis, drugs testing, explosives detection, and ecopollution determination [57–60]. Fog collecting structure on Stenocara beetle's back gives a good example to fabricate a PC microchip with hydrophilic–hydrophobic micropattern by inkjet printing. It makes high-sensitive ultratrace detection of fluorescence analytes and fluorophore-based assays possible. For example, Hou et al. [20] prepared a PC microchip which was printed by the hydrophilic monodispersed poly(styrene-methylmethacrylate-acrylic acid) (poly(St-MMA-AA)) spheres on a hydrophobic polydimethylsiloxane (PDMS) substrate. All PC dots (**Figure 11A**) were about 200 mm in diameter. They were assembled from the monodispersed colloidal spheres with diameters of 180, 215, and 240 nm, named as PC_{180}, PC_{215}, and PC_{240}, respectively. The wettability between a PC dot and a PDMS substrate

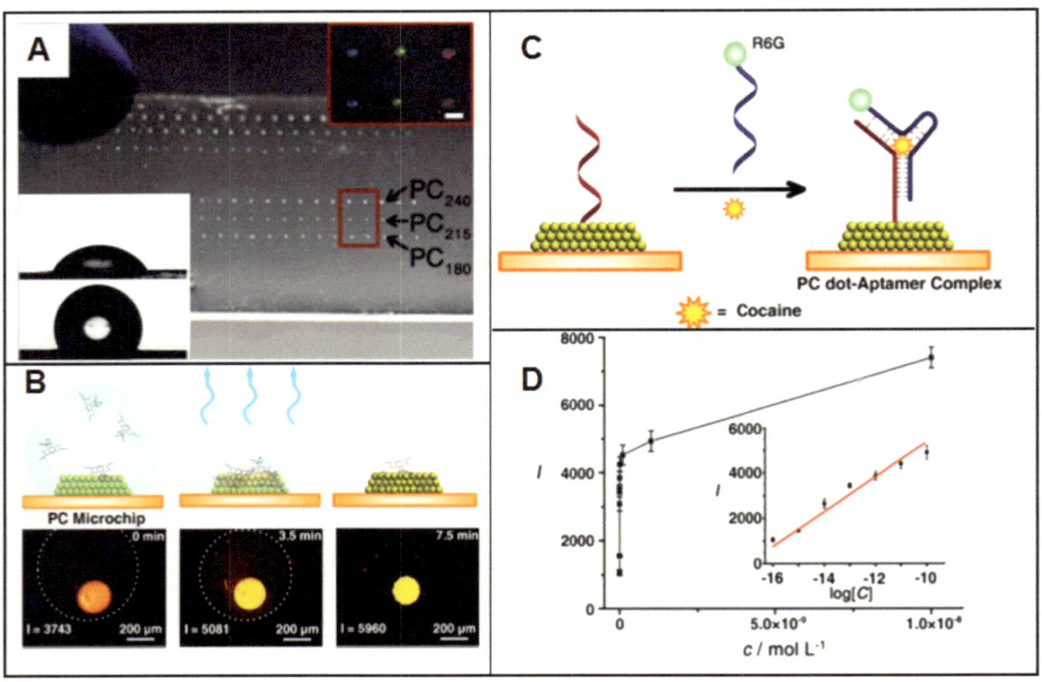

Figure 11. (A) Photography of the bio-inspired PC microchip with hydrophilic PC dots on hydrophobic substrate. The inset shows the magnified picture corresponding to PC dots of different stopbands. The scale bar is 200 mm. The CA of PDMS substrate is 115.0 ± 3.1° while CA of the PC dot was 46.4 ± 3.4°.(B) schematic representation of the cocaine detection mechanism and the fluorescence intensities at different cocaine concentrations. The inset shows a linear relationship between fluorescence intensity and the logarithm of the cocaine concentration (1×10^{-10} to 1×10^{-16} mol L^{-1}) [20].

is quite different. Because of hydrophilic-hydrophobic micropattern, the analyte in the highly diluted solution droplet can be concentrated onto the PC dot. This phenomenon is mainly due to the wettability difference between the hydrophilic PC dot and the hydrophobic PDMS substrate. The solution dewetted from the hydrophobic substrate and was concentrated to the hydrophilic PC dot with the water evaporation. In **Figure 11B**, the PC dot was functionalized with capture DNA. By enriching the cocaine molecules and the R6G-labeled target DNA from solution onto the PC dot, it was possible to detect cocaine. After enriching cocaine on the PC dot and specifically capturing by a DNA-functionalized PC dot and R6G-labeled target DNA, the high-efficient fluorescence detection could be realized (**Figure 11C** and **D**).

5. Conclusions

In conclusion, we presented a summarization about the influence of the substrate wettability on colloidal assembly. It is clear that the substrate with different wettability produces a distinct influence on the colloidal assembly process, the resultant assembly structure and the resulting functionality, it will bring about new insight for the creation of novel-type PC optic devices.

Author details

Junchao Liu, Jingxia Wang* and Lei Jiang

*Address all correspondence to: jingxiawang@mail.ipc.ac.cn

CAS, Key Laboratory of Bio-inspired Materials and Interface Sciences, Technical Institute of Physics and Chemistry Chinese Academy of Sciences, Beijing, China

References

[1] Yablonovitch E. Inhibited spontaneous emission in solid-state physics and electronics. Physical Review Letters. 1987;**58**:2059-2062. DOI: 10.1103/PhysRevLett.58.2059

[2] John S. Strong localization of photons in certain disordered dielectric superlattices. Physical Review Letters. 1987;**58**:2486-2489. DOI: 10.1103/PhysRevLett.58.2486

[3] Yablonovitch E. Photonic band-gap crystals. Journal of Physics: Condensed Matter. 1993;**5**:2443-2460. DOI: 10.1088/0953-8984/5/16/004

[4] Joannopoulos JD, Villeneuve PR, Fan SH. Photonic crystals: Putting a new twist on light. Nature. 1997;**386**:143-149. DOI: 10.1038/386143a0

[5] Fang Y, Ni YL, Leo SY, Taylor C, Basile V, Jiang P. Reconfigurable photonic crystals enabled by pressure-responsive shape-memory polymers. Nature Communications. 2015;**6**:7416. DOI: 10.1038/ncomms8416

[6] Liu JC, Wan L, Zhang MB, Jiang KJ, Song K, Wang JX, Ikeda T, Jiang L. Electrowetting-induced morphological evolution of metal-organic inverse opals toward a water-lithography approach. Advanced Functional Materials. 2017;**27**:1605221. DOI: 10.1002/adfm.201605221

[7] Burgess IB, Mishchenko L, Hatton BD, Kolle M, Loncar M, Aizenberg J. Encoding complex wettability patterns in chemically functionalized 3D photonic crystals. Journal of the American Chemical Society. 2011;**133**:12430-12432. DOI: 10.1021/ja2053013

[8] Teyssier J, Saenko SV, van der Marel D, Milinkovitch MC. Photonic crystals cause active colour change in chameleons. Nature Communications. 2015;**6**:6368. DOI: 10.1038/ncomms7368

[9] Qin M, Huang Y, Li YN, Su M, Chen BD, Sun H, Yong PY, Ye CQ, Li FY, Song YL. A rainbow structural-color Chip for multisaccharide recognition. Angewandte Chemie, International Edition. 2016;**55**:6911-6914. DOI: 10.1002/anie.201602582

[10] Sun XM, Zhang J, Lu X, Fang X, Peng HS. Mechanochromic photonic-crystal fibers based on continuous sheets of aligned carbon nanotubes. Angewandte Chemie, International Edition. 2015;**54**:3630-3634. DOI: 10.1002/anie.201412475

[11] Kuno T, Matsumura Y, Nakabayashi K, Atobe M. Electroresponsive structurally colored materials: A combination of structural and Electrochromic effects. Angewandte Chemie, International Edition. 2016;**55**:2503-2506. DOI: 10.1002/anie.201511191

[12] Yin SN, Yang SY, Wang CF, Chen S. Magnetic-directed assembly from Janus building blocks to multiplex molecular-analogue photonic crystal structures. Journal of the American Chemical Society. 2016;**138**:566-573. DOI: 10.1021/jacs.5b10039

[13] Ge JP, Yin YD. Responsive photonic crystals. Angewandte Chemie, International Edition. 2011;**50**:1492-1522. DOI: 10.1002/anie.200907091

[14] Li HL, Wang JX, Yang LM, Song YL. Superoleophilic and Superhydrophobic inverse opals for oil sensors. Advanced Functional Materials. 2008;**18**:3258-3264. DOI: 10.1002/adfm.200800507

[15] Ye XZ, Li Y, Dong JY, Xiao JY, Ma YR, Qi LM. Facile synthesis of ZnS Nanobowl arrays and their applications as 2D photonic crystal sensors. Journal of Materials Chemistry C. 2013;**1**:6112-6119. DOI: 10.1039/c3tc30118d

[16] Gur D, Palmer BA, Leshem B, Oron D, Fratzl P, Weiner S, Addadi L. The mechanism of color change in the neon tetra fish: A light-induced tunable photonic crystal Array. Angewandte Chemie, International Edition. 2015;**54**:12426-12430. DOI: 10.1002/anie.201502268

[17] Liu CH, Ding HB, ZQ W, Gao BB, FF F, Shang LR, ZZ G, Zhao YJ. Tunable structural color surfaces with visually self-reporting wettability. Advanced Functional Materials. 2016;**26**:7937-7942. DOI: 10.1002/adfm.201602935

[18] YN W, Li FT, Zhu W, Cui JC, Tao CA, Lin CX, Hannam PM, Li GT. Metal-organic frameworks with a three-dimensional ordered macroporous structure: Dynamic photonic materials. Angewandte Chemie, International Edition. 2011;**50**:12518-12522. DOI: 10.1002/anie.201104597

[19] Zhang YQ, Fu QQ, Ge JP. Photonic sensing of organic solvents through geometric study of dynamic reflection Spectrum. Nature Communications. 2015;**6**:7510. DOI: 10.1038/ncomms8510

[20] Hou J, Zhang HC, Yang Q, Li MZ, Song YL, Jiang L. Bio-inspired photonic-crystal microchip for fluorescent Ultratrace detection. Angewandte Chemie, International Edition. 2014;**53**:5791-5795. DOI: 10.1002/anie.201400686

[21] Vogel N, Belisle RA, Hatton B, Wong TS, Aizenberg J. Transparency and damage tolerance of Patternable Omniphobic lubricated surfaces based on inverse colloidal monolayers. Nature Communications. 2013;**4**:2176. DOI: 10.1038/ncomms3176

[22] Sano K, Kim YS, Ishida Y, Ebina Y, Sasaki T, Hikima T, Aida T. Photonic water dynamically responsive to external stimuli. Nature Communications. 2016;**7**:12559. DOI: 10.1038/ncomms12559

[23] Cubillas AM, Schmidt M, Euser TG, Taccardi N, Unterkofler S, Russell PS, Wasserscheid P, Etzold BJM. In situ heterogeneous catalysis monitoring in a hollow-Core photonic crystal fiber microflow reactor. Advanced Materials Interfaces. 2014;**1**:1300093. DOI: 10.1002/admi.201300093

[24] Mazingue T, Lomello TM, Hernandez RC, Passard M, Goujon L, Rousset JL, Morfin F, Bosselet F, Maulion G, Kribich R. Pellet photonic Innovant gas sensor using catalysis and integrated photonics. Sensors and Actuators B: Chemical. 2016;**222**:133-140. DOI: 10.1016/j.snb.2015.07.107

[25] Chen M, Zhang YP, Jia SY, Zhou L, Guan Y, Zhang YJ. Photonic crystals with a reversibly inducible and erasable defect state using external stimuli. Angewandte Chemie, International Edition. 2015;**54**:9257-9261. DOI: 10.1002/anie.201503004

[26] Chen K, Tuysuz H. Morphology-controlled synthesis of Organometal halide Perovskite inverse opals. Angewandte Chemie, International Edition. 2015;**54**:13806-13810. DOI: 10.1002/anie.201506367

[27] Fan J, Li YN, Bisoyi HK, Zola RS, Yang DK, Bunning TJ, Weitz DA, Li Q. Light-directing omnidirectional circularly polarized reflection from liquid-crystal droplets. Angewandte Chemie, International Edition. 2015;**54**:2160-2164. DOI: 10.1002/anie.201410788

[28] Gur D, Palmer BA, Weiner S, Addadi L. Light manipulation by guanine crystals in organisms: Biogenic Scatterers, mirrors, multilayer reflectors and photonic crystals. Advanced Functional Materials. 2017;**27**:1603514. DOI: 10.1002/adfm.201603514

[29] Lee SY, Choi JK, Jeong JR, Shin JH, Kim SH. Magnetoresponsive photonic microspheres with structural color gradient. Advanced Materials. 2017;**29**:1605450. DOI: 10.1002/adma.201605450

[30] Kim KH, Hwang MS, Kim HR, Choi JH, No YS, Park HG. Direct observation of exceptional points in coupled photonic-crystal lasers with asymmetric optical gains. Nature Communications. 2016;**7**:13893. DOI: 10.1038/ncomms13893

[31] Zhai Y, Ma YG, David SN, Zhao DL, Lou RN, Tan G, Yang RG, Yin XB. Scalable-manufactured randomized glass-polymer hybrid Metamaterial for daytime Radiative cooling. Science. 2017;**355**:1062-1066. DOI: 10.1126/science.aai7899

[32] Wu H, Kuang XH, Cui LY, Tian D, Wang MH, Luan GY, Wang JX, Jiang L. Single-material solvent-sensitive actuator from poly(ionic liquid) inverse opals based on gradient Gewetting. Chemical Communications. 2016;**52**:5924-5927. DOI: 10.1039/c6cc01442a

[33] Kralchevsky PA, Denkov ND. Capillary forces and structuring in layers of colloid particles. Current Opinion in Colloid & Interface Science. 2001;**6**:383-401. DOI: 10.1016/S1359-0294(01)00105-4

[34] Denkov ND, Velev OD, Kralchevsky PA, Ivanov IB, Yoshimura H, 2-Dimensional Crystallization NK. Nature. 1993;**361**:26-26. DOI: 10.1038/361026a0

[35] Tian Y, Jiang L. Wetting intrinsically robust hydrophobicity. Nature Materials. 2013;**12**:291-292. DOI: 10.1038/nmat3610

[36] Gu ZZ, Yu YH, Zhang H, Chen H, Lu Z, Fujishima A, Sato O. Self-assembly of Monodisperse spheres on substrates with different wettability. Applied Physics A: Materials Science & Processing. 2005;**81**:47-49. DOI: 10.1007/s00339-004-3020-4

[37] Sun ZJ, Bao B, Jiang JK, He M, Zhang XY, Song YL. Facile fabrication of a Superhydrophilic-superhydrophobic patterned surface by inkjet printing a sacrificial layer on a Superhydrophilic surface. RSC Advances. 2016;**6**:31470-31475. DOI: 10.1039/c6ra02170k

[38] Liu MJ, Wang JX, He M, Wang LB, Li FY, Jiang L, Song YL. Inkjet printing controllable footprint lines by regulating the dynamic wettability of coalescing ink droplets. ACS Applied Materials and Interfaces. 2014;**6**:13344-13348. DOI: 10.1021/am5042548

[39] Kuang MX, Wang JX, Bao B, Li FY, Wang LB, Jiang L, Song YL. Inkjet printing patterned photonic crystal domes for wide viewing-angle displays by controlling the sliding three phase contact line. Advanced Optical Materials. 2014;**2**:34-38. DOI: 10.1002/adom.201300369

[40] Cui LY, Zhang JH, Zhang XM, Li YF, Wang ZH, Gao HN, Wang TQ, Zhu SJ, HL Y, Yang B. Avoiding coffee ring structure based on hydrophobic silicon pillar arrays during single-drop evaporation. Soft Matter. 2012;**8**:10448-10456. DOI: 10.1039/c2sm26271a

[41] Kim SH, Lim JM, Jeong WC, Choi DG, Yang MY. Patterned colloidal photonic domes and balls derived from viscous Photocurable suspensions. Advanced Materials. 2008;**20**:3211-3217. DOI: 10.1002/adma.200800782

[42] Bao B, Li MZ, Li Y, Jiang KJ, ZK G, Zhang XY, Jiang L, Song YL. Patterning fluorescent quantum dot Nanocomposites by reactive inkjet printing. Small. 2015;**11**:1649-1654. DOI: 10.1002/smll.201403005

[43] Kawamura A, Kohri M, Yoshioka S, Taniguchi T, Kishikawa K. Structural color tuning: Mixing melanin-like particles with different diameters to create neutral colors. Langmuir. 2017;**33**:3824-3830. DOI: 10.1021/acs.langmuir.7b00707

[44] Choi S, Jamshidi A, Seok TJ, MC W, Zohdi TI. Fast, high-throughput creation of size-tunable micro/nanoparticle clusters via evaporative self-assembly in Picoliter-scale droplets of particle suspension. Langmuir. 2012;**28**:3102-3111. DOI: 10.1021/la204362s

[45] Ding HB, Zhu C, Tian L, Liu CH, Fu GB, Shang LR, Gu ZZ. Structural color patterns by Electrohydrodynamic jet printed photonic crystals. ACS Applied Materials and Interfaces. 2017;**9**:11933-11941. DOI: 10.1021/acsami.6b11409

[46] Qin M, Huang Y, Li YN, Su M, Chen BD, Sun H, Yang PY, Yang CQ, Li FY, Song YL. A rainbow structural-color Chip for multisaccharide recognition. Angewandte Chemie, International Edition. 2016;**55**:6911-6914. DOI: 10.1002/anie.201602582

[47] Rastogi V, Melle S, Calderón OG, García AA, Marquez M, Velev OD. Synthesis of light-diffracting assemblies from microspheres and nanoparticles in droplets on a Superhydrophobic surface. Advanced Materials. 2008;**20**:4263-4268. DOI: 10.1002/adma.200703008

[48] Rastogi V, García AA, Marquez M, Velev OD. Anisotropic particle synthesis inside droplet templates on Superhydrophobic surfaces. Macromolecular Rapid Communications. 2010;**31**:190-195. DOI: 10.1002/marc.200900587

[49] Zhou JM, Yang J, Gu ZD, Zhang GF, Wei Y, Yao X, Song YL, Jiang L. Controllable fabrication of noniridescent microshaped photonic crystal assemblies by dynamic three-phase contact line behaviors on Superhydrophobic substrates. ACS Applied Materials and Interfaces. 2015;**7**:22644-22651. DOI: 10.1021/acsami.5b07443

[50] Huang Y, Zhou JM, Su B, Shi L, Wang JX, Chen SR, Wang LB, Zi J, Song YL, Jiang L. Colloidal photonic crystals with narrow Stopbands assembled from low-adhesive Superhydrophobic substrates. Journal of the American Chemical Society. 2012;**134**:17053-17058. DOI: 10.1021/ja304751k

[51] Gu ZZ, Fujishima A, Sato O. Patterning of a colloidal crystal film on a modified hydrophilic and hydrophobic surface. Angewandte Chemie, International Edition. 2002;**4**:2068-2070. DOI: 10.1002/1521-3757

[52] Lee SW, Na YJ, Choi Y, Lee SD. Site-selective assembly and fixation of colloidal particles into two-dimensional Array on wettability-patterned surface. Japanese Journal of Applied Physics. 2007;**46**:1129-1131. DOI: 10.1143/JJAP.46.L1129

[53] Wu L, Dong ZC, Kuang MX, Li YN, Li YF, Jiang L, Song YL. Printing patterned fine 3D structures by manipulating the three phase contact line. Advanced Functional Materials. 2015;**25**:2237-2242. DOI: 10.1002/adfm.201404559

[54] Wang YZ, Wei C, Cong HL, Yang Q, Wu YC, Su B, Zhao YS, Wang JX, Jiang L. Hybrid top-down/bottom-up strategy using Superwettability for the fabrication of patterned colloidal assembly. ACS Applied Materials and Interfaces. 2016;**8**:4985-4993. DOI: 10.1021/acsami.5b11945

[55] Yoo JH, Kwon HJ, Paeng D, Yeo J, Elhadj S, Grigoropoulos CP. Facile fabrication of a superhydrophobic cage by laser direct writing for site-specific colloidal self-assembled photonic crystal. Nanotechnology. 2016;**27**:145604. DOI: 10.1088/0957-4484/27/14/145604

[56] Kim SH, Kim SH, Yang SM. Patterned polymeric domes with 3D and 2D embedded colloidal crystals using Photocurable emulsion droplets. Advanced Materials. 2009;**21**: 3771-3775. DOI: 10.1002/adma.200901243

[57] Arya SK, Bhansali S. Lung cancer and its early detection using biomarker-based biosensors. Chemical Reviews. 2011;**111**:6783-6809. DOI: 10.1021/cr100420s

[58] Apple FS, Smith SW, Pearce LA, Ler R, Murakami MM. Use of the centaur TnI-ultra assay for Ddetection of myocardial infarction and adverse events in patients presenting with symptoms suggestive of acute coronary syndrome. Clinical Chemistry. 2008;**54**:723-728. DOI: 10.1373/clinchem.2007.097162

[59] Steinfeld JI, Wormhoudt J. Explosives detection: A challenge for physical chemistry. Annual Review of Physical Chemistry. 1998;**49**:203-232. DOI: 10.1146/annurev.physchem.49.1.203

[60] Zhang HX, Hu JS, Yan CJ, Jiang L, Wan L. Functionalized carbon nanotubes as sensitive materials for electrochemical detection of ultra-trace 2,4,6-trinitrotoluene. Physical Chemistry Chemical Physics. 2006;**8**:3567-3572. DOI: 10.1039/b604587c

Section 2

Design and Application

Chapter 6

The Dawn of Photonic Crystals: An Avenue for Optical Computing

Renju Rajan, Padmanabhan Ramesh Babu and Krishnamoorthy Senthilnathan

Additional information is available at the end of the chapter

http://dx.doi.org/10.5772/intechopen.71253

Abstract

In this chapter, a new paradigm is developed for optical computation using photonic crystals. As photonic crystals are the most sophisticated optical materials to date, information processing using this structure is one of the most sought-after technologies in photonics. While the semiconductor industry is striving hard to increase the microprocessors' processing power, it is certain that the trend would not last forever as against Moore's prediction. At this juncture, photonics technologies have to compete with the upcoming quantum computing technology to emerge as a promising successor for semiconductor microprocessors. This chapter is devoted to the introduction of photonic crystals as the workhorse for an all-optical computational system with a myriad of logic gates, memory units, and networks which can be constructed using these structures.

Keywords: optical computing, optical logic gates, optical memory, nanocavity, optical Kerr effect

1. Introduction

Modern computers evolved out of the semiconductor technology which began with the invention of transistor in 1948 [1]. Compared to previous generation of computers which used vacuum tubes, transistors were smaller, reliable, and efficient. In a subsequent development, integrated circuit which incorporates several transistors into a single chip was invented in 1958 [2]. This was a real revolution for the semiconductor technology which has enabled scalability of the processing power with an increase in the number of transistors that are inducted into the chip. In addition, these chips consume less power, run faster, and are smaller than their transistor counterparts. With an ever-increasing demand for processing power, there is a corresponding increase in demand for fabricating chips with more and more transistors in them. This has resulted in miniaturization of the individual transistor components in the chip,

INTECH
open science | open minds

© 2018 The Author(s). Licensee InTech. This chapter is distributed under the terms of the Creative Commons Attribution License (http://creativecommons.org/licenses/by/3.0), which permits unrestricted use, distribution, and reproduction in any medium, provided the original work is properly cited.

which has now crossed into the nano regime [3]. This trend is supposed to end somewhere in the near future, due to the size of individual components reaching the size of individual atoms. This roadblock in semiconductor technology can be solved by embracing alternative technologies such as optical computing.

Ever since the advent of lasers, there have been deliberate efforts to develop an optical analog for computation. Optical computation was initially envisaged as a hybrid system consisting of electronic and optical components. This venture turned out to be unsuccessful due to the unfeasible conversion time required from one system to another [4]. Even today, this is a daunting task, provided the communication networks are all made of optical fibers. However, the processing part is currently done by semiconductor microprocessors. With the introduction of photonic crystals, there is again a renewed interest in an otherwise dropped plan of optical computation. Photonic crystals have got the potential to create an all-optical information processing system. This will have an overwhelming influence on the information processing capacity of the communication system as a whole. In this chapter, an overview of such an optical computational system which can be implemented using a photonic crystal is outlined. An introduction to computer architecture is given in Section 2. The basics of planar photonic crystals are described in Section 3. The implementation of logic gates using photonic crystals is discussed in Section 4. An optical memory unit which can be implemented using photonic crystals is delineated in Section 5. The chapter is concluded in Section 6.

2. Computer architecture

Computer architecture deals with the design which stipulates the working of a computer with its components such as microprocessor, memory, and input/output devices [5]. Ever since the introduction of von Neumann architecture in 1945 for the computer called EDVAC, it has been the de facto architecture for electronic computers for the subsequent generations to date [6]. Although the discussion in this section centers on this architecture, one cannot be sure if the future optical computers would continue to lean on this architecture. For the first time, the von Neumann architecture embarked on the concept of stored program for realizing a general purpose computer. This enabled the computer to perform different computational tasks based on the program stored in its memory. Other salient features of this architecture include the usage of binary digits and sequential execution of instructions from a given program. Major components of von Neumann architecture consist of central processing unit (CPU), memory unit, and input/output units. The general outline of von Neumann architecture is shown in **Figure 1**.

In a modern computer, microprocessor occupies the position of central processing unit by handling data processing and system control [7]. The data processing part is done by arithmetic/logic unit and register array of the microprocessor, whereas the system control is done by the control unit. Primitive computers had these units separately, but with the advancement in circuit integration technology, it is possible to incorporate them onto a single chip. A microprocessor takes data from input devices such as keyboard and mouse, processes those data

http://dx.doi.org/10.5772/intechopen.71253

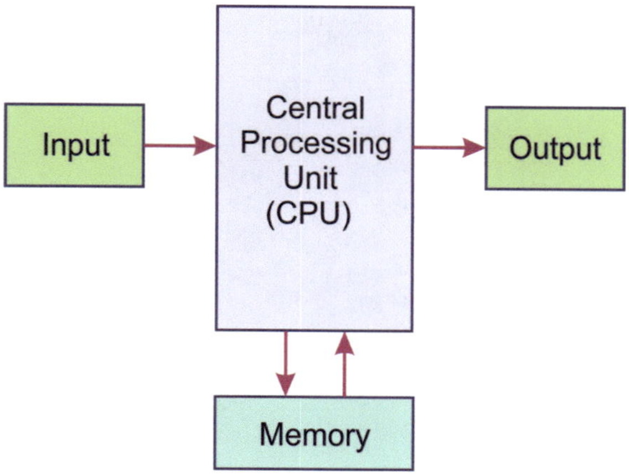

Figure 1. Outline of von Neumann architecture for computer.

according to the instructions stored in the memory, and sends the processed data into an output device such as monitor [8]. In this process, the data and control signals are transferred between microprocessor, memory, and input/output devices through a communication network known as the system bus. A typical microprocessor operation consists of fetch-execute cycles. A microprocessor fetches an instruction from the memory and executes that instruction on the input data. Afterward, it goes to the next instruction in the memory, fetches, and executes it. This process goes on until the last instruction before finishing a given task according to the program stored in the memory.

3. Planar photonic crystals

Photonic crystals (PhCs) are artificial periodic structures made of dielectric materials [9]. They are generally classified into three categories based on the dimensionality of the structure—1D, 2D, and 3D. Two dimensional structures such as planar photonic crystals, which are useful in integrated photonic circuits, are considered here for discussion. In these structures, air holes are arranged periodically across the plane of the structure such that the spacing between them is less than the wavelength of light propagating through them [10]. This causes light to reflect away from the air hole structure when trying to pass through them. On the other hand, light can propagate through a channel made within this structure with width more than the wavelength of light [11]. In some ways, this is analogous to cutting a road for transportation. This type of light localization feasible within these structures has made them suitable for integrated photonics circuits. Similar to optical fibers used in communication, these structures also exhibit very low loss for the passage of light as against the case of their electronic counterparts [12]. In this way, this technology is an energy efficient alternative for integrated circuits with minimal heat loss. A schematic representation of PhC slab with channel for light passage is shown in **Figure 2**.

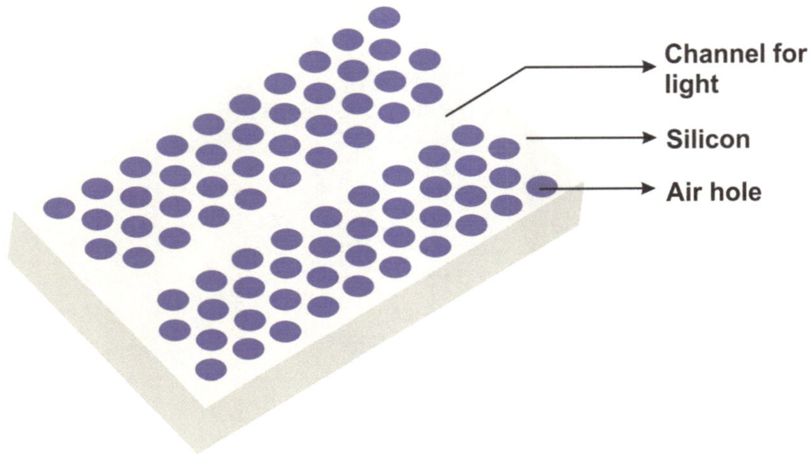

Figure 2. Schematic representation of photonic crystal slab with channel for light passage.

3.1. Structure of photonic crystal slabs

Photonic crystal slab is a planar structure made of silicon or other compound semiconductor material with periodically varying refractive index [13]. Toward this goal, air holes are etched in these structures such that a photonic band gap (PBG) is formed for certain wavelengths at which light cannot propagate through this structure. For these wavelengths, when some defects are created within the structure by disturbing the periodicity, light can be channelized through them with little propagation loss [14]. This is the working principle of a PhC slab. The PBG formation in a periodically varying refractive index structure is analogous to band gap formation for electrons in a crystal structure with Bragg diffraction from multiple ion lattice sites [15]. In planar photonic crystals, the Bragg diffraction is due to the periodic variation in refractive index by air holes. The PBG formed in a planar photonic crystal is not three dimensional. Rather, the band formed in it is two dimensional. In this way, there is light confinement for a defect in the horizontal plane of the PhC slab, whereas the vertical directions remain unconfined [16]. This can give rise to loss along the vertical directions. To do away with this loss, total internal reflection with the underlying silica or air layer is sought such that the light is confined within the structure [17]. Accordingly, for a silicon PhC slab, two types of structures are possible, one with silica and another with air as the underlying medium, as shown in **Figure 3**.

3.2. Nanocavities in photonic crystal slabs

Similar to a line defect which creates a channel waveguide, a point defect can create a nanocavity in the PhC slab structure [18]. Due to the requirement of tight light confinement in the vertical directions, an air-cladding photonic crystal structure having high refractive index contrast with the slab is preferred to a silica cladding for realizing a nanocavity [19]. A nanocavity can be created in a PhC slab by various means. It can be realized by creating a point defect, a line defect, or a width-modulated line defect in a photonic crystal [20]. A fourth method is also in practice wherein a double heterostructure results in a nanocavity. Each one of these

http://dx.doi.org/10.5772/intechopen.71253

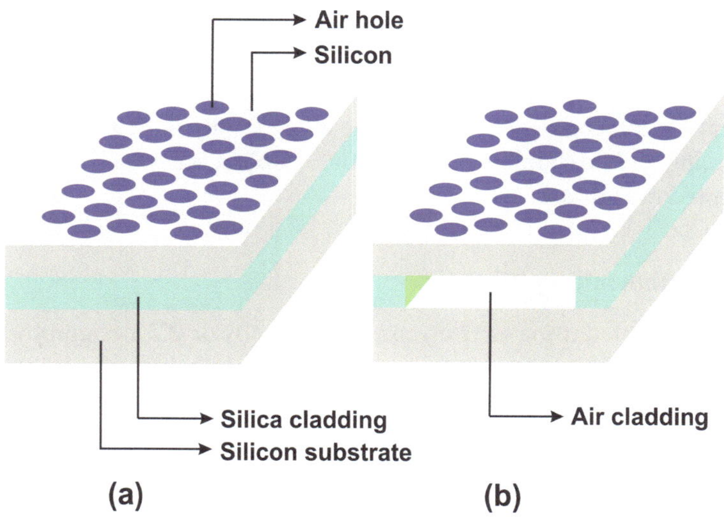

Figure 3. Schematic diagrams of two classes of photonic crystal slab structures, (a) with silica as the cladding layer and (b) with air as the cladding layer.

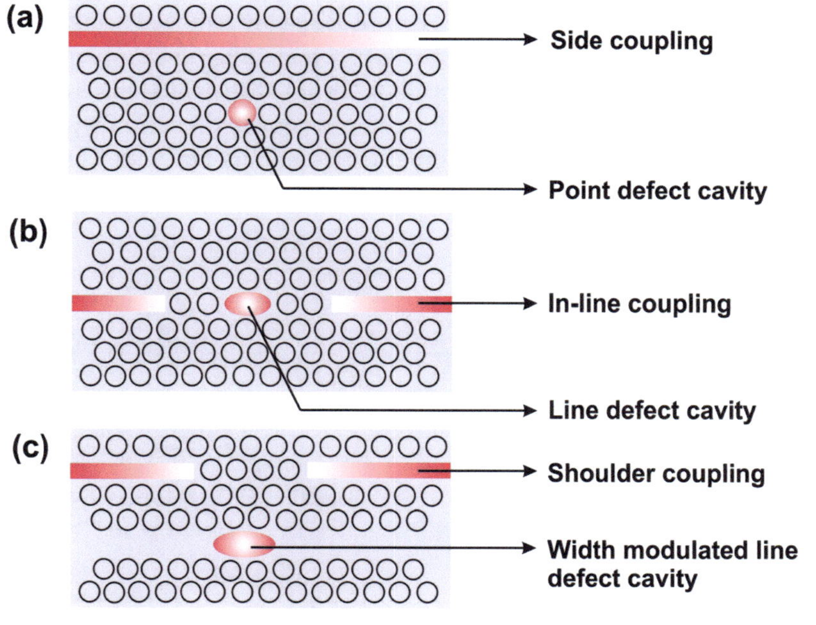

Figure 4. Schematic representation of nanocavities along with various coupling schemes: (a) a point-defect nanocavity with side coupling with the waveguide, (b) a line-defect nanocavity with in-line coupling with the waveguide, and (c) a width-modulated line defect nanocavity with shoulder coupling with the waveguide.

nanocavities is illustrated in **Figure 4**. For a point-defect nanocavity, a slight shift of the air holes from their regular positions away from the cavity region is found to produce significant increase in the Q factor of the cavity. Similarly, for a line defect cavity, a shift in the end holes away from the cavity region offers an appreciable increase in the Q factor. Among these configurations, width-modulated line defect cavity and double-heterostructure cavity offer the highest Q factors and are the preferred configurations for nanocavity-based applications. There are various coupling schemes for coupling a nanocavity with a channel waveguide. Of these, side coupling, in-line coupling, and shoulder coupling schemes which are commonly in use are represented in **Figure 4**.

3.3. Fabrication of photonic crystal slabs

PhC slabs can be fabricated using electron beam lithography or UV lithography technique [21]. While electron beam (e-beam) lithography is suitable for fabrication on a laboratory scale, UV lithography suits to the requirements for mass production. In both schemes, there are a series of steps to be followed for realizing a PhC slab with an air cladding. For silicon PhC slabs, the readily available SOI (silicon-on-insulator) wafer used in semiconductor industry is used as a substrate for realizing PhC slabs [22, 23]. On the other hand, for compound semiconductors, the substrates have to be custom-made in the laboratory [24]. The process steps involved in the fabrication of silicon PhC slabs are considered here. Initially, the SOI wafer is coated with an e-beam resist such as polymethyl methacrylate (PMMA). The pattern that has to be imprinted is transferred from a CAD onto the surface of the resist using an e-beam. The resist is sensitive to e-beam, and the exposed parts of the resist are subsequently removed using chemicals in the development stage. This process is analogous to photographic film development in analog photography. The resulting resist which contains holes at lattice points serves as a template for the subsequent etching process on SOI. After etching is done in the silicon layer, the resist and the underneath silica layer are removed to form the air-cladding PhC slab. These process steps involved in the fabrication of silicon PhC slabs are illustrated in **Figure 5**.

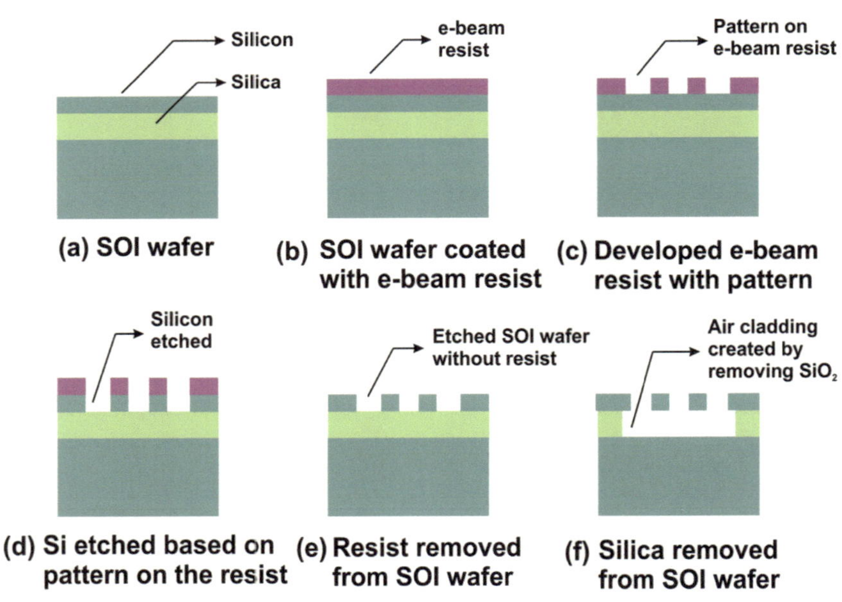

Figure 5. Series of process involved in the fabrication of silicon photonic crystal slabs.

4. Logic gates using photonic crystals

Many of the phenomena in optics can very well be described within the framework of wave theory of light based on Maxwell's equations, without resorting to the complexity of quantum mechanics. Even today, this is the preeminent theory which stands out against the test of time [25]. Nonlinear optics which is an offshoot of classical electrodynamics is also a branch of optics. Even though nonlinearity is routine in electronic devices such as diodes and transistors, it was not so in optics until the invention of laser. Laser, due to its high intensity, can have electric field strength comparable to bond strength between atoms, resulting in nonlinear response from the optical materials through which it propagates [26]. The nonlinear response from the medium can be modeled by taking into account of the Taylor series expansion of the electric field of the laser light. This can be explicitly represented in terms of polarization of the medium, wherein higher order terms in series describe the nonlinear polarization [27],

$$P(E) = \varepsilon_0\left(\chi_1 E + \chi_2 E^2 + \chi_3 E^3 + \cdots\right), \tag{1}$$

where first term denotes linear polarization, and higher order terms represent nonlinear polarization. Here, E is electric field strength, ε_0 is permittivity in free space, and $\chi_1, \chi_2, \chi_3, \ldots$ denote first-order, second-order, and third-order electric susceptibilities, and so on. Eq. (1) can be used for describing various nonlinear optical processes which can occur in a system. Accordingly, first-order term in Eq. (1) is responsible for linear polarization, second-order term is responsible for second harmonic generation, and third-order term is responsible for third harmonic generation, optical Kerr effect, and so on. Optical Kerr effect is responsible for change in refractive index corresponding to intensity of the light source [28]. Intensity dependent refractive index is expressed as,

$$n(I) = n_0 + n_2 I, \tag{2}$$

where n_0 is the linear refractive index and n_2 is the nonlinear refractive index coefficient. The value of n_2 can be positive or negative depending on the type of material. Thus, for a material with positive n_2, the overall refractive index increases and vice versa. The intensity dependent variation in refractive index can be used for making optical logic gates. But, in this approach, the nonlinearity of the medium turns out to be crucial as it relates to the operation of the logic gates. Hence, materials with high nonlinearity are preferred to those with weak nonlinearities. Since nonlinearity of silicon is very low when compared to other compound semiconductors, it may not be the suitable candidate for realizing such all-optical logic gates [29]. Further, with the materials of high nonlinearity, deploying ultrashort pulses down to femtosecond range results in faster switching [30]. In this way, nonlinear response of the medium can be put to good use for realizing logic gates in PhC slabs.

4.1. NOT gate in a photonic crystal

Logic gates are fundamental building blocks of a digital computer [31]. Basic logic operations such as NOT, AND, OR, as well as their derivatives can be generated using these digital circuits. Truth table of a logic circuit represents the output of the circuit under various conditions of input. The truth table and circuit symbol of a NOT gate are shown in **Figure 6(a)**. It is possible to carry

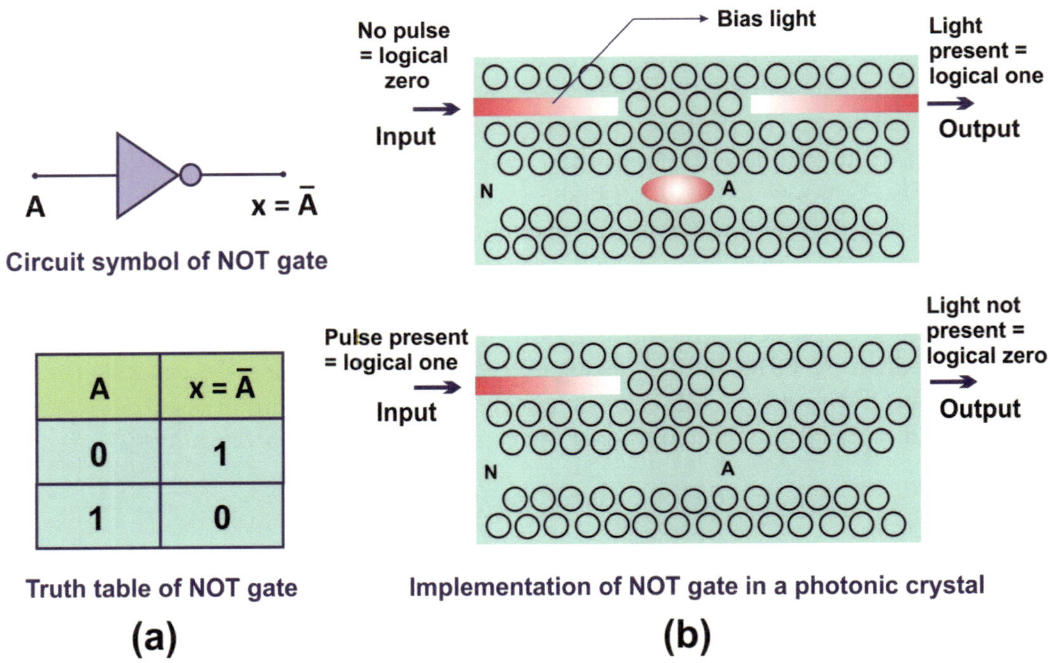

Figure 6. (a) Circuit symbol and truth table of a NOT gate and (b) implementation of NOT operation using a photonic crystal slab.

out the inverting operation done by the NOT gate using a PhC slab by exploiting the optical nonlinearity. A nonlinear medium with negative value for n_2 can be used for achieving this goal as shown in **Figure 6(b)**. Here, the coupling between the waveguide and nanocavity is such that when there is no optical pulse at the input waveguide, the bias light would emerge at the output waveguide. On the other hand, when an optical pulse is passed through the input waveguide, refractive index of the medium gets reduced due to negative nonlinearity of the medium. As a result, the resonant wavelength of the resonator differs considerably from the input wavelength of the bias light, resulting in decoupling of the bias light from the nanocavity and the output waveguide. In this way, the requirement of a NOT gate can be satisfied using this design, wherein the presence of an optical pulse at the input end gives no light at the output, and the absence of an optical pulse gives light at the output end of the waveguide. Here, the presence of a pulse at the input waveguide denotes a logic one, and the absence of a pulse denotes a logical zero. The same can also be implemented using materials with positive value for n_2.

It would be useful to adopt some naming conventions before dwelling further into the topic. In this regard, it is useful to note that there are two types of nonlinear media, one with positive Kerr nonlinearity (n_2 is positive) and the other with negative Kerr nonlinearity (n_2 is negative). Here, these two nonlinearities are represented in the sketch of PhC slab by **P** and **N**, respectively. In the sketches, a lighter shade is used to represent a material with positive Kerr nonlinearity, whereas the one with a darker shade represents negative Kerr nonlinearity. Moreover, based on the resonance wavelength of the nanocavity, three types of cavities are possible. A cavity having resonance at the same wavelength as that of the input light is denoted by **A** in the sketch, whereas a cavity with resonance wavelength lower than the input

http://dx.doi.org/10.5772/intechopen.71253

light is denoted by **L,** and a cavity whose resonance wavelength is set higher than the input light is denoted by **H** in the sketch. When an incoming radiation satisfies the resonance condition of the cavity [32],

$$2nd = p\lambda, \tag{3}$$

it is allowed to pass through the cavity. In other cases, the cavity would block the passage of light through it. In Eq. (3), d is cavity length, λ is wavelength of light, p is an integer, and n is the refractive index of the medium within the cavity given by Eq. (2). From this equation, it is clear that the resonance wavelength of the cavity is decided by the cavity length and refractive index of the medium. A cavity can be designed such that a slight deviation from the resonance cavity length can be compensated by the variation in refractive index due to the optical Kerr effect. In this way, when the refractive index of the medium gets altered by the input pulse, the resonance condition of the cavity sets in, allowing the passage of light through the cavity. When there is no input light pulse, resonance condition of the cavity is disturbed, resulting in blockage of the light. This is the working principle of a cavity-based optical switching circuit. In total, two types of nonlinearities along with three types of cavity configurations result in six possible combinations. Of these, four can be put to good use for creating logic gates.

4.2. OR gate in a photonic crystal

For an OR gate, the output is a logical one even if either of the inputs is a logical one. It gives a logical zero only when all of the inputs are at logical zero. For two inputs A and B, the OR logical operation is expressed by A + B. The circuit symbol and truth table of an OR gate are given in **Figure 7(a)**. This logical operation can be implemented in a PhC slab by using a combination of three nanocavities as shown in **Figure 7(b)**. Here, the PhC slab is made of

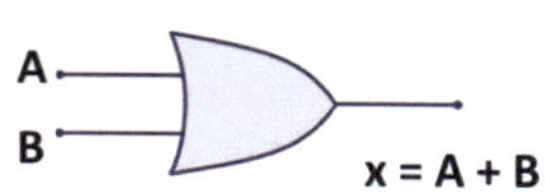

Circuit symbol of OR gate

A	B	x = A + B
0	0	0
0	1	1
1	0	1
1	1	1

Truth table of OR gate

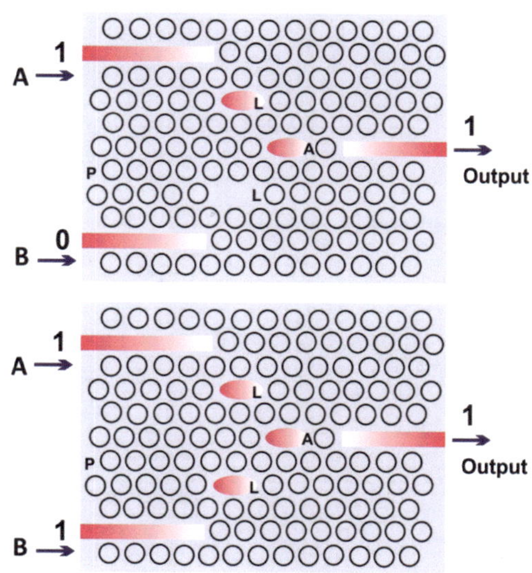

Implementation of OR gate in a photonic crystal

(a) (b)

Figure 7. (a) Circuit symbol and truth table of OR gate and (b) implementation of OR gate operation in a photonic crystal slab.

material with positive Kerr nonlinearity. But, it can also be implemented using a material with negative Kerr nonlinearity too. The nanocavities adjacent to the input waveguides are set to have resonance at wavelengths below the input wavelength of light. In this way, they are opaque when there is no input pulse. On the other hand, when a pulse is sent through the input waveguide, there is an increase in local refractive index due to optical Kerr effect, and as a result, the **L** cavities are set to resonance such that the light reaches the output waveguide through the **A** cavity. This can occur when a pulse is sent through both or either of the input waveguides. Thus, the arrangement of nanocavities in this manner helps mimic the operation of an OR gate in a PhC slab.

4.3. AND gate in a photonic crystal

For an AND gate, the output can go to logical one only, when all the inputs are at logical one. For the rest of the cases, the output of an AND gate turns out to be logical zero. The circuit symbol and truth table of an AND gate are shown in **Figure 8(a)**. For two inputs A and B, AND logical operation is expressed by $A \cdot B$. It is possible to achieve AND operation with a proper combination of NOT and OR gates. This is the usual practice in digital circuits wherein the universal gates such as NOR and NAND are used for creating other gates. This insight stems from De Morgan's theorem in Boolean algebra which is at the very basis of digital circuitry. Accordingly, the implementation of an AND gate using NOR gates is shown in **Figure 8(b)** [31]. An optical logic circuit can also be constructed along the same lines using nanocavities in a PhC slab.

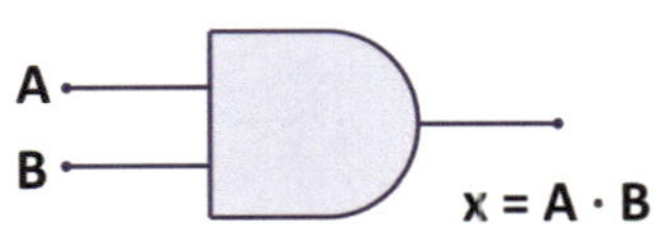

Circuit symbol of AND gate

A	B	x = A · B
0	0	0
0	1	0
1	0	0
1	1	1

Truth table of AND gate

(a)

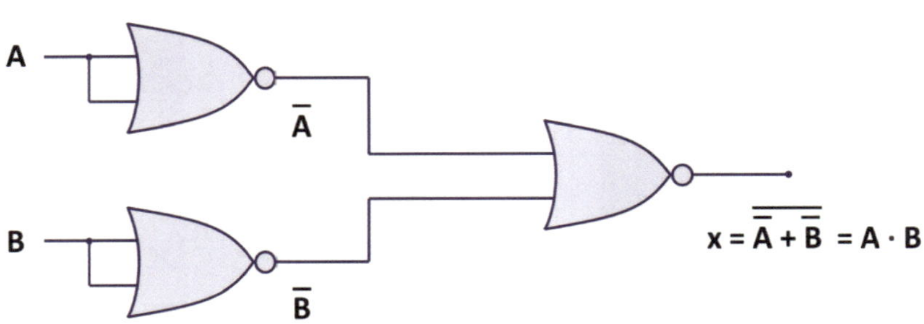

(b)

Figure 8. (a) Circuit symbol and truth table of AND gate and (b) implementation of AND operation using NOR gates.

http://dx.doi.org/10.5772/intechopen.71253

5. Optical memory using photonic crystals

Memory unit in a computer stores data and instructions required by a microprocessor for execution. For a modern digital computer, there are two types of memories, namely, a main memory which communicates directly with the microprocessor and an auxiliary memory which serves as a secondary storage [33]. Data and programs in auxiliary memory are brought into the main memory before being executed by the microprocessor. An efficient memory unit is required for increasing the performance of a computer. A microprocessor can locate memory addresses of a main memory using the address bus and transfer data through the data bus. Usually, the main memory which stores the data as electric charge in capacitors are volatile and cannot store data for more than few milliseconds due to the discharge of capacitor within this period. So, they need to be refreshed regularly to store data continuously. An optical analog of capacitors for storing data can be created in a PhC slab by using nanocavities [34, 35]. The scheme is similar to implementation of NOT gate discussed in the previous section. Here, the memory device is capable of achieving optical bistability as a result of optical Kerr nonlinearity-induced switching process [36]. The nanocavity can be set to resonance by the input pulse denoting a logical one, whereas the withdrawal of the pulse distorts the resonance which resets the cavity back to the logical zero state. Whenever the cavity is set to resonance, there is light at the output waveguide, denoting the logical one state of the cavity and vice versa. Light at the output waveguide can be used for identifying the instantaneous state of the cavity. In this way, nanocavities in a PhC slab can be used for storing digital data. The working of nanocavity-based memory device is illustrated in **Figure 9**.

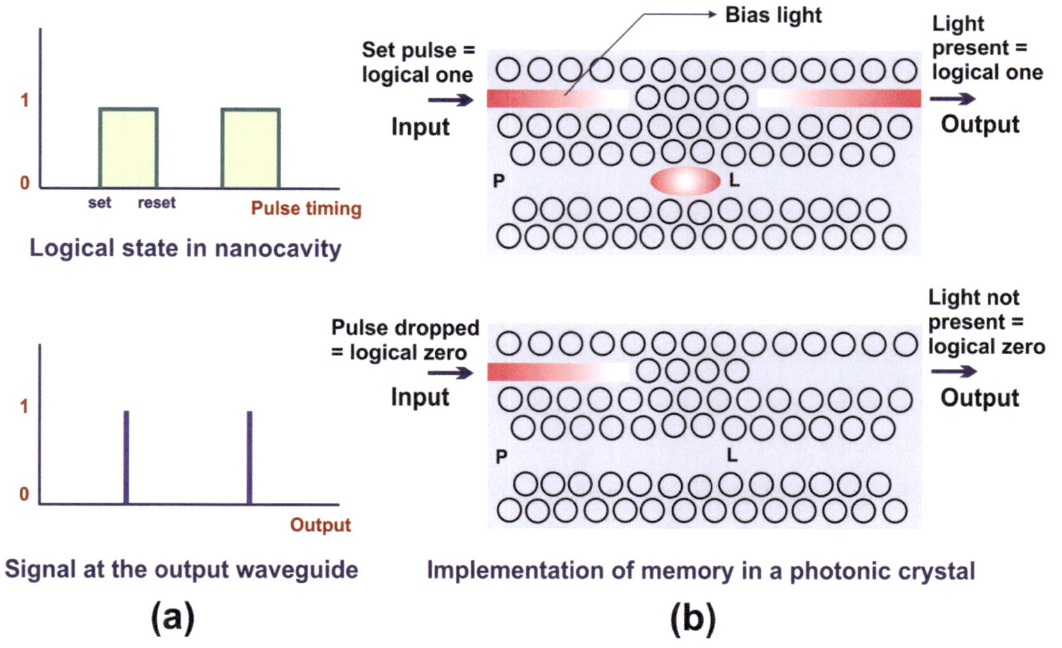

Figure 9. (a) Logical state of nanocavity and its indication in the output waveguide, and (b) implementation of memory in photonic crystal.

6. Conclusion

In this chapter, the implementation of an all-optical computational system has been delineated using photonic crystal slabs. Following the introduction to the computer architecture which comprises of microprocessor, memory, and input/output devices, the implementation of the same using photonic crystal slabs has also been discussed. Further, we have demonstrated optical logic operations in photonic crystals using optical nanocavities which can be created in these structures. The key feature which decides the working of photonic crystal slab-based logic gates is the change in refractive index which arises due to the optical Kerr effect upon the passage of laser pulses. An attempt has also been made for creating optical memory in photonic crystals, wherein resonance condition decides the logical state inside the cavity. Photonic crystal slabs with nanocavity-enabled logic gates and memory units can be used for constructing an all-optical information processor analogous to semiconductor microprocessors.

Acknowledgements

This work was inspired by the video "Photonic crystal optical bit memory" by NTT Basic Research Laboratories, Japan.

Author details

Renju Rajan, Padmanabhan Ramesh Babu and Krishnamoorthy Senthilnathan*

*Address all correspondence to: senthee@gmail.com

Department of Physics, School of Advanced Sciences, VIT University, Vellore, Tamil Nadu, India

References

[1] Null L, Lobur J. The Essentials of Computer Organization and Architecture. Sudbury: Jones and Bartlett Publishers; 2003

[2] Jiles D. Introduction to the Electronic Properties of Materials. Dordrecht: Springer Science + Business Media; 1994

[3] Brey BB. The Intel Microprocessors: 8086/8088, 80186/80188, 80286, 80386, 80486, Pentium, Pentium pro Processor, Pentium II, Pentium III, Pentium 4, and Core2 with 64-Bit Extensions: Architecture, Programming, and Interfacing. 8th ed. Columbus: Pearson Education; 2009

[4] Ambs P. Optical computing: A 60-year adventure. Advances in Optical Technologies. 2010;**2010**:1-15

[5] Stone HS. Introduction to Computer Architecture. Chicago: Science Research Associates; 1975

[6] Akama S. Elements of Quantum Computing. New York: Springer Science+Business Media; 2015

[7] Gilmore CM. Introduction to Microprocessors. New York: McGraw-Hill; 1981

[8] Gaonkar RS. Microprocessor Architecture, Programming, and Applications with the 8085. New York: Prentice-Hall Inc.; 1999

[9] Sukhoivanov IA, Guryev IV. Photonic Crystals: Physics and Practical Modelling. New York: Springer; 2009

[10] Prather DW, Shi S, Murakowski J, Schneider GJ, Sharkawy A, Chen C, Miao B. Photonic crystal structures and applications: Perspective, overview, and development. IEEE Journal of Selected Topics in Quantum Electronics. 2006;**12**:1416-1437

[11] Joannopoulos JD, Meade RD, Winn JN. Photonic Crystals: Molding the Flow of Light. Princeton: Princeton University Press; 1995

[12] Notomi M, Shinya A, Mitsugi S, Kuramochi E, Ryu HY. Waveguides, resonators and their coupled elements in photonic crystal slabs. Optics Express. 2004;**12**:1551-1561

[13] Notomi M. Optical phenomena in photonic crystal. In: Noda S, Baba T, editors. Roadmap on Photonic Crystals. Boston: Kluwer Academic Publishers; 2003. p. 13-43

[14] Joannopoulos JD, Villeneuve PR, Fan S. Photonic crystals: Putting a new twist on light. Nature. 1997;**386**:143-149

[15] Kittel C. Introduction to Solid State Physics. 7th ed. New York: John Wiley & Sons; 1996

[16] Loncar M, Nedeljkovic D, Doll T, Vuckovic J, Scherer A, Pearsall TP. Waveguiding in planar photonic crystals. Applied Physics Letters. 2000;**77**:1937-1939

[17] Viktorovitch P. Physics of slow Bloch modes and their applications. In: Sibilia C, Benson TM, Marciniak M, Szoplik T, editors. Photonic Crystals: Physics and Technology. Milan: Springer; 2008. p. 27-42

[18] Liscidini M, Andreani LC. Photonic crystals: An introductory survey. In: Comoretto D, editor. Organic and Hybrid Photonic Crystals. New York: Springer; 2015. p. 3-30

[19] Tanabe T, Kuramochi E, Shinya A, Notomi M. Ultrahigh-Q photonic crystal nanocavities and their applications. In: Matsko A B, editor. Practical Applications of Microresonators in Optics and Photonics. Boca Raton: CRC Press; 2009. p. 1-52

[20] Notomi M. Manipulating light with strongly modulated photonic crystals. Reports on Progress in Physics. 2010;**73**:1-57

[21] Bogaerts W, Wiaux V, Taillaert D, Beckx S, Luyssaert B, Bienstman P, Baets R. Fabrication of photonic crystals in silicon-on-insulator using 248-nm deep UV lithography. IEEE Journal of Selected Topics in Quantum Electronics. 2002;**8**:928-934

[22] Prather DW, Sharkawy A, Shi S, Murakowski J, Schneider G, Chen C. Design and fabrication of planar photonic crystals. In: Kemme S, editor. Microoptics and Nanooptics Fabrication. Boca Raton: CRC Press; 2010. p. 151-189

[23] Levinson HJ. Principles of Lithography. 3rd ed. Bellingham: SPIE Press; 2010

[24] Loncar M, Scherer A. Microfabricated optical cavities and photonic crystals. In: Vahala K, editor. Optical Microcavities. Singapore: World Scientific; 2004. p. 39-93

[25] Griffiths DJ. Introduction to Electrodynamics. 3rd ed. New Delhi: Prentice-Hall; 1999

[26] Quimby RS. Photonics and Lasers: An Introduction. Hoboken: John Wiley & Sons; 2006

[27] Bründermann E, Hübers HW, Kimmitt MF. Terahertz Techniques. Berlin: Springer; 2012

[28] Fox M. Optical Properties of Solids. 2nd ed. New York: Oxford University Press; 2010

[29] Gai X, Choi DY, Madden S, Luther-Davies B. Materials and structures for nonlinear photonics. In: Wabnitz S, Eggleton BJ, editors. All-Optical Signal Processing: Data Communication and Storage Applications. New York: Springer; 2015. p. 1-33

[30] Golubev VG. Ultrafast all-optical switching in photonic crystals. In: Limonov MF, De La Rue RM, editors. Optical Properties of Photonic Structures: Interplay of Order and Disorder. Boca Raton: CRC Press; 2012. p. 415-428

[31] Tocci RJ, Widmer NS, Moss GL. Digital Systems: Principles and Applications. 10th ed. Upper Saddle River: Pearson Education; 2007

[32] Butcher PN, Cotter D. The Elements of Nonlinear Optics. Cambridge: Cambridge University Press; 1991

[33] Mano MM. Computer System Architecture. 3rd ed. New Delhi: Pearson Education; 1993

[34] Notomi M. Manipulating light by photonic crystals. NTT Technical Review. 2009;**7**:1-10

[35] Shinya A, Matsuo S, Tanabe T, Kuramochi E, Sato T, Kakitsuka T, Notomi M. All-optical on-chip bit memory based on ultra high Q InGaAsP photonic crystal. Optics Express. 2008;**16**:19382-19387

[36] Nozaki K, Shinya A, Matsuo S, Suzaki Y, Segawa T, Sato T, Kawaguchi Y, Takahashi R, Notomi M. Ultralow-power all-optical RAM based on nanocavities. Nature Photonics. 2012;**6**:248-252

Chapter 7

One-Photon Absorption-Based Direct Laser Writing of Three-Dimensional Photonic Crystals

Dam Thuy Trang Nguyen, Mai Trang Do, Qinggle Li,
Quang Cong Tong, Thi Huong Au and
Ngoc Diep Lai

Additional information is available at the end of the chapter

http://dx.doi.org/10.5772/intechopen.71318

Abstract

A simple and low-cost technique called low one-photon absorption (LOPA) direct laser writing (DLW) is demonstrated as an efficient method for structuration of multidimensional submicrostructures. Starting from the diffraction theory of the electromagnetic field distribution of a tightly focused beam, the crucial conditions for LOPA-based DLW are theoretically investigated, and then experimentally demonstrated using a simple optical confocal microscope. Various 1D, 2D, and 3D submicrostructures were successfully fabricated in different materials, such as commercial SU8 photoresist and magnetic nanocomposite. The advantages and drawbacks of this LOPA-based DLW technique were also studied and compared with the conventional two-photon absorption based DLW. Several methods were proposed to overcome the existing problem of the DLW, such as the dose accumulation and shrinkage effect, resulting in uniform structures with a small lattice constant. The LOPA-based DLW technique should be useful for the fabrication of functionalized structures, such as magneto-photonic and plasmon photonic crystals and devices, which could be interesting for numerous applications.

Keywords: direct laser writing, one-photon absorption, photonic crystal, magnetic nanocomposite, magneto-photonic microstructures

1. Introduction

In recent years, various fabrication techniques have been proposed and implemented to realize structures at micro- and nanoscales, opening numerous applications such as micro-machining, optical data storage, nanophotonics, plasmonics, and bio-imaging, etc. [1–4]. Among those techniques, optical lithography, which includes mask lithography [5, 6], interference

INTECH
open science | open minds

© 2018 The Author(s). Licensee InTech. This chapter is distributed under the terms of the Creative Commons Attribution License (http://creativecommons.org/licenses/by/3.0), which permits unrestricted use, distribution, and reproduction in any medium, provided the original work is properly cited.

or holography photolithography [7–9], and direct laser writing [10–13], is the most popular because of its simplicity, flexibility and capability of producing different kinds of microstructures, addressing a variety of applications.

The fundamental working principle of the optical lithography involves the use of a photoresist, a light-sensitive material, which changes its chemical property when exposed to light. Based on the reactions of photoresists to light, they are classified into two types: positive photoresist and negative photoresist. With positive photoresists, the areas exposed to the light absorb one or more photons and become more soluble in the photoresist developer. These exposed areas are then washed away with the photoresist developer solvent, leaving the unexposed material. With negative resists, exposure to light causes the polymerization of the photoresist chemical structure, which is just the opposite of positive photoresists. The unexposed portion of the photoresist is then dissolved by the photoresist developer. Light sources of different wavelengths are used based on the purposes of the fabrication, which involve the absorption mechanisms of the used photoresist. There are two types of absorption mechanisms, namely one-photon absorption (OPA) and two-photon absorption (TPA). The OPA excitation method is an ideal way to fabricate one- and two-dimensional (1D and 2D) thin structures [14]. In this technique, a simple and low-cost continuous-wave (CW) laser operating at a wavelength located within the absorption band of the thin film material is used as the excitation source. Wavelengths in the UV range or shorter are commonly used to achieve high resolution [15]. This method is usually applied in mask lithography and interference techniques, where an entire pattern over a wide area is created in seconds. All structures are often realized at the same time; therefore, these techniques are called parallel processes. However, due to the strong absorption effect, light is dramatically attenuated from the input surface. Thus, it is impossible for OPA to address thick film materials or 3D optical structuring.

The TPA (or multi-photon absorption) technique presents a better axial resolution. In this case, two low energy photons are simultaneously absorbed inducing the optical transition from the ground state to the excited state of the material, equivalent to the case of linear absorption (OPA). Two-photon absorption is a nonlinear process, which is several orders of magnitude weaker than linear absorption, thus very high light intensities are required to increase the number of such rare events. In practice, the process can be achieved using a pulsed (picosecond or femtosecond) laser. The TPA method is commonly applied for the technique called direct laser writing (DLW), in which a pulsed laser beam is focused into a sub-micrometer spot, resulting in a dramatic increase of the laser intensity at the focusing spot. Hence, TPA-based 3D imaging or fabrication can be achievable [4, 11, 12] with high spatial resolution.

Indeed, DLW has been proved to be an ideal way to fabricate sub-micrometric arbitrary structures, offering flexibility, ease of use, and cost effectiveness. As opposed to mask lithography and interference techniques, DLW is a serial process in which a structure is realized by scanning the focusing spot following a desired pattern. Thus, any arbitrary 1D, 2D, and 3D periodic or non-periodic pattern can be fabricated on demand.

However, as mentioned above, the TPA-based DLW requires the use of a femtosecond or picosecond laser and a complicated optical system, making it a rather expensive fabrication technique. Recently, an original method called LOPA (low one-photon absorption) DLW has been demonstrated [16, 17], allowing one to combine the advantages of both OPA and TPA methods. Indeed, the LOPA method employs a simple, CW and low power laser, as in the case of conventional OPA, but it allows the optical addressing of 3D objects, as what could be realized by the TPA method, by using a combination of an ultralow absorption effect and a tightly focusing spot. The illustration of an absorption spectrum of photoresist shown in **Figure 1(a)** represents three absorption mechanisms, which are shown in **Figure 1(b)**: conventional OPA (solid ring), LOPA (dashed ring) and TPA (dash-dotted ring), respectively. If a laser beam, whose wavelength is positioned at the edge of the absorption band where the absorption is ultralow, is applied, the light intensity distribution remains almost the same as in the absence of material. In short, by tightly focusing an optical beam inside a thick material with a very low absorption at the operating wavelength, it is possible to address 3D imaging and 3D fabrication, as what realized by TPA method. As compared to the latter one, this LOPA-based DLW is very simple and inexpensive and it allows one to achieve very similar results.

This chapter presents theoretically and experimentally this original fabrication method, LOPA-based DLW technique, which allows the realization of multidimensional sub-micrometer photonic crystals. The advantages of this fabrication method will also be presented and compared with other fabrication techniques.

In Section 2, the theory of the LOPA-based DLW method and experimental conditions to realize 3D sub-microstructures are presented. In this case, the vectorial diffraction theory of a laser beam, tightly focused by a high numerical aperture objective lens, is extended taking into account the very low absorption of the propagating medium. Numerical calculations will also be shown.

In Section 3, it will be demonstrated experimentally that any sub-micrometer 1D, 2D, and 3D structures can be realized by the LOPA-based DLW technique, by choosing appropriate photoresist and excitation laser wavelength.

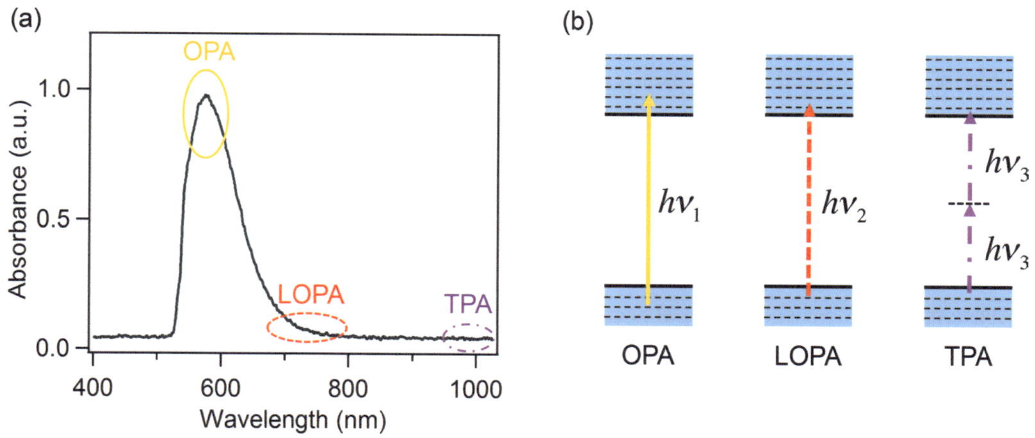

Figure 1. (a) Illustration of the absorption spectrum of a photoresist. The three rings illustrate three ranges of wavelength corresponding to (b) three absorption mechanisms: standard one-photon absorption (OPA), low one-photon absorption (LOPA), and two-photon absorption (TPA).

In Section 4, different additional methods to optimize the LOPA-based DLW are demonstrated in order to obtain sub-microstructures, which show high uniformity, less shrinkage, and with a lattice constant smaller than the diffraction limit.

To demonstrate the versatility of the LOPA-based DLW, in Section 5, the fabrication of magneto-photonic sub-microstructures for biomedical engineering realized by the LOPA-based DLW technique is presented. This opens many promising applications, such as tunable photonic structures based on magneto-optical effect and development of microrobotic tools for transport in biological systems.

In the last section, some conclusions of the newly developed LOPA-based DLW technique, advantages this technology brings to the photonic crystal field, as well as some prospects will be discussed.

2. Theory of LOPA-based DLW technique

In this section, the vectorial Debye approximation is presented, based on which a new mathematics representation is further established, where the absorption effect of the material is taken into account when a light beam propagates through an absorbing medium. Based on the new evaluation form of vectorial Debye theory, the influence of absorption coefficient of the studied material, the numerical aperture (NA) of the objective lens (OL), and the penetration depth of light beam on the formation of a tight focusing spot are investigated. From that, the crucial conditions for the realization of LOPA microscope and LOPA DLW are established.

2.1. Electromagnetic field distribution of a tight focused beam in an absorbing medium

The mathematical representation of the electromagnetic field distribution in the focal region of an OL was proposed by Wolf in the 1950s [18]. This theory based on the vectorial Debye approximation allows the calculation and prediction of the intensity and polarization distributions of a light beam focused inside a material by a high NA OL. Nonetheless, the influence of material absorption on the intensity distribution and the focused beam shape of a propagating optical wave have not been systematically investigated yet. In this section, the mathematical representation proposed by Wolf [18] will be employed, taking into account the absorption effect of the material when a focused light beam propagates through it, in order to investigate the intensity distribution, especially in the focal region.

The schematic representation of light focusing in an absorption medium is shown in **Figure 2**. D is the interface between the transparent material, such as a glass substrate or air, and the absorbing material. To simplify the problem, it is assumed that the refractive index mismatch problem arising at any interface is negligible. d represents the distance between the D interface and the focal plane. The electromagnetic field near the focal plane in Cartesian coordinates (x, y, z) [16, 19] is represented by,

$$\mathbf{E} = -\frac{ikC}{2\pi} \iint_{\Omega} \mathbf{T}(s) A(s)\, e^{[ik(s_x x + s_y y + s_z z)]}\, ds_x\, ds_y, \qquad (1)$$

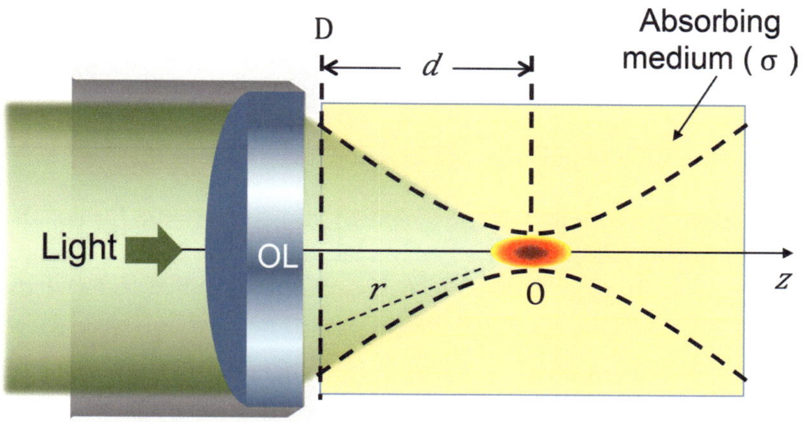

Figure 2. Schematic representation of a tightly focused light beam inside an absorbing medium. σ is the absorption coefficient of the medium, O is the focal point, D is the interface between the transparent and the absorbing media, r is the distance from an arbitrary point on the D plane to the focal point, and d is the distance between the D plane and the focal plane of the objective lens.

where C is a constant, k is the wave number ($k = 2\pi n/\lambda$), n is the refractive index of the absorbing medium, λ is the excitation wavelength, Ω is a solid angle corresponding to the objective aperture, $\mathbf{S} = (s_x, s_y, s_z)$ is the vector of an arbitrary optical ray, and $\mathbf{T}(s) = \mathbf{P}(s)\mathbf{B}(s)$is a transmission function where $\mathbf{P}(s)$ is the polarization distribution and $\mathbf{B}(s)$ is the amplitude distribution at the exit pupil. $A(s)$ represents the absorption effect of the material, which is expressed as $A(s) = \exp(-\sigma r)$, where σ is the absorption coefficient and r indicates the optical path of each diffracted light ray in the absorbing medium, which is defined as the distance from a random point located in the D plane to the focal point, as shown in **Figure 2**. For calculations, r is determined by

$$r = \sqrt{(x'-x)^2 + (y'-y)^2 + (d-z)^2}, \tag{2}$$

where (x', y', d) gives the position of an arbitrary diffracted light ray located on the D plane. Theoretically, Eq. (1) allows calculation of the light distribution resulting from the interference of all light rays diffracted by the exit pupil of the OL. However, in practice, the light intensity and the focus shape at the focus region depend strongly on the absorption term $A(s)$ since light is absorbed by the material in which it propagates and its amplitude decreases along the propagation direction.

The light intensity distribution, also called point spread function (PSF), in the focal region of the OL is defined as

$$I_{PSF} = \mathbf{E}\mathbf{E}^* \tag{3}$$

This theory is applicable for any cases, OPA, LOPA, or TPA. It can be seen that the EM field distribution in the focal region depends on various parameters, such as the polarization of incident light, the NA of OL, the absorption coefficient of the material, etc. It is impossible to have an analytical solution of the light field at the focusing spot of a high NA OL. However, it can be numerically calculated, which will be shown in the next part.

2.2. Numerical calculation of point spread function

In order to numerically calculate the light intensity distribution, the I_{PSF} equation was programmed by a personal code script based on Matlab software, with the influence of different input parameters including the absorption coefficient of the studied material, the NA of the OL, and the penetration depth of the light beam.

First, the influence of the absorption effect of the SU8 material, which will be used later for experimental demonstration in the next sections, was investigated. Based on the absorption spectrum of SU8 shown in **Figure 5(b)**, three typical wavelengths to calculate the intensity distribution in the focal region were chosen, representing three cases of interest: conventional OPA (308 nm), LOPA (532 nm) and TPA (800 nm), respectively. The corresponding absorption coefficients in each case are: $\sigma_1 = 240{,}720\ \mathrm{m}^{-1}$ ($\lambda_1 = 308$ nm), $\sigma_2 = 723\ \mathrm{m}^{-1}$ ($\lambda_2 = 532$ nm), and $\sigma_3 = 0\ \mathrm{m}^{-1}$ ($\lambda_3 = 800$ nm). The absorption interface (D) was arbitrarily assumed to be separated from the focal point O by a distance of 25 μm, the NA of the OL was chosen to be 0.6, the refractive index (n) of SU8 is 1.58. As seen in **Figure 3(a_1)**, the incoming light is totally attenuated at the interface D because of the strong absorption of SU8 at 308 nm, which explains why it is not possible to optically address 3D object with the conventional OPA method. However, when using an excitation light source emitting at 800 nm, the absorption coefficient is zero, thus light can penetrate deeply inside the material, resulting in a highly resolved 3D intensity distribution. The numerical calculation result derived from the quadratic dependence of the EM field is shown in **Figure 3(a_3)**. The size of the focusing spot (full width at half maximum, FWHM) is quite large due to the use of a long wavelength. However, in practice, two photons can only be simultaneously absorbed at an intensity above the polymerization threshold. Therefore, a small effective focusing spot below the diffraction limit can be achieved with the TPA method by controlling the excitation intensity.

The case where the linear absorption is very low (LOPA) was considered. At the wavelength $\lambda = 532$ nm, the absorption coefficient is only $723\ \mathrm{m}^{-1}$, which is much smaller than that at 308 nm. Simulation results show that light can penetrate deeply inside the absorbing material without significant attenuation thanks to this very low linear absorption. As shown in **Figure 3(a_2)**, the light beam can be tightly confined at the focusing spot, which can then be moved freely inside the thick material, exactly as in the case of TPA. Furthermore, LOPA requires a shorter wavelength as compared with TPA, the focusing spot size (FWHM) is therefore smaller. The diagram depicted in **Figure 3(a_4)** shows clearly the difference of the intensity distribution along the optical axis of three excitation mechanisms. It is important to note that there is no intensity threshold in the case of LOPA, because it is a linear absorption process. Therefore, LOPA requires a precise control of light dose in order to achieve high resolution optical addressing.

The NA of OL is also an important parameter to be taken into account for LOPA case. It was demonstrated that the use of a high NA OL is a crucial condition. The intensity distributions at the focusing spot obtained with OLs of different NA values (with the same low absorption

http://dx.doi.org/10.5772/intechopen.71318

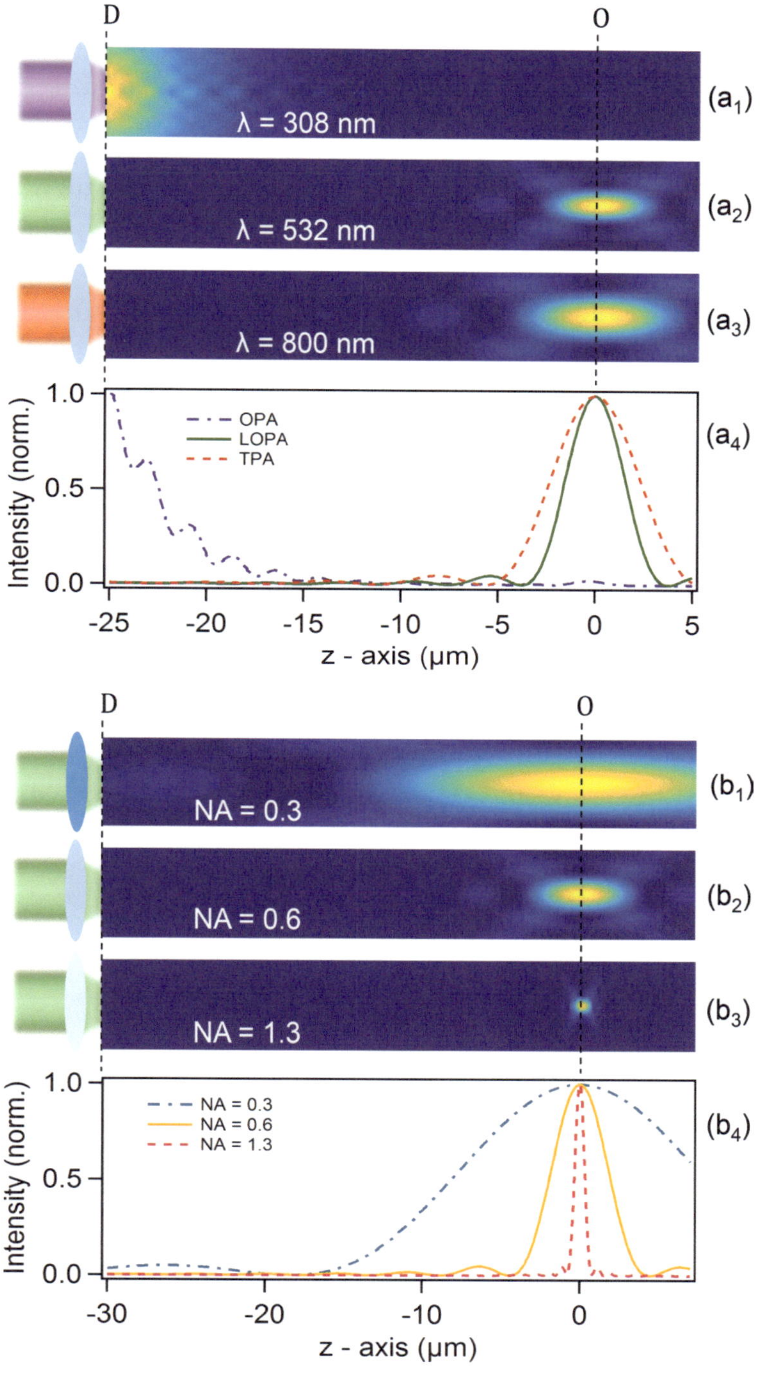

Figure 3. (a) Numerical calculations of light propagation inside SU8, by using different wavelengths, (a_1) 308 nm, (a_2) 532 nm, and (a_3) 800 nm, respectively. (a_4) Intensity distributions along z-axis of the light beams shown in (a_{1-3}). In this calculation, NA = 0.6, refractive index n = 1.58, and d = 25 μm. (b) Propagation of light (λ = 532 nm) inside SU8, with (b_1) NA = 0.3, (b_2) NA = 0.6, and (b_3) NA = 1.3, respectively, d = 30 μm. (b_4) Intensity distributions along z-axis of the light beams shown in (b_{1-3}).

coefficient $\sigma_2 = 723\ m^{-1}$) are shown in **Figure 3(b)**. It is clearly seen that with an OL of low NA (NA = 0.3), the light beam is not well focused, resulting in low contrast intensity distribution between the focal region and its surrounding. Therefore, the LOPA-based microscopy using a low NA OL cannot be applied for 3D optical addressing. However, in the case of tight focusing (for example, NA = 1.3), the light intensity at the focusing spot is a million time larger than that at out of focus, resulting in a highly resolved focusing spot (**Figure 3(b$_3$)**).

For all the above calculations, it can be concluded that the LOPA-based microscopy is promising for the realization of 3D imaging and 3D fabrication, similarly to what could be realized by TPA microscopy. By using the LOPA technique, 3D fluorescence imaging or fabrication of 3D structures can be realized by moving the focusing spot inside the material since fluorescence (for imaging) or photopolymerization (for fabrication) effects can be achieved efficiently within the focal spot volume only.

It is worth noting that in LOPA technique, the absorption exists, even if the probability is very small, the penetration depth is therefore limited to a certain level. This effect exists also in the case of TPA, but it is more important for LOPA. **Figure 4** represents the maximum intensity at the focusing spot as a function of the penetration length, *d*. For this calculation, the ultralow absorption coefficient of SU8 at $\lambda = 532$ nm, $\sigma = 723\ m^{-1}$ was considered. At the distance of 390 μm, the intensity was found to decrease by half with respect to that obtained at the input of absorbing material (D interface). This penetration depth of several hundred micrometers is fully compatible with the scanning range of piezoelectric stage (typically, 100 μm for a high resolution), or with the working distance of microscope OL (about 200 μm for a conventional high NA OL).

In summary, in order to realize the LOPA-based microscopy, two important conditions are required: (i) ultralow absorption of the studied material at the chosen excitation wavelength, and (ii) a high NA OL for tight focusing of the excitation light beam. To experimentally demonstrate the application of LOPA in DLW, in the next chapter, SU8 will be used as the material and a CW laser at $\lambda = 532$ nm as the excitation source.

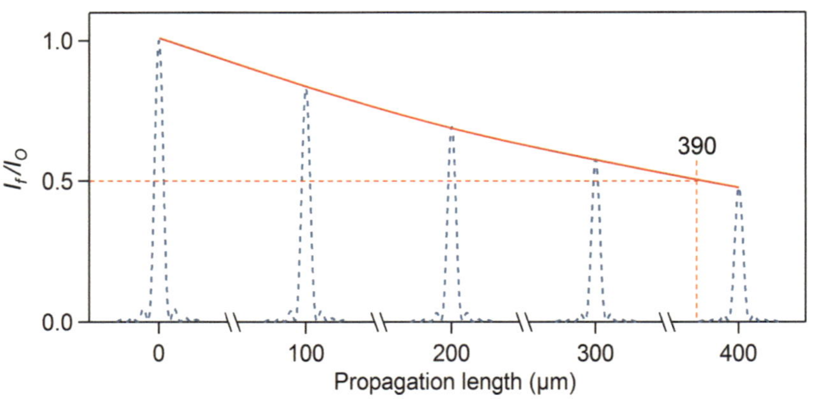

Figure 4. Red curve: normalization of intensity (I_f/I_0) at the focusing spot as a function of the propagation length. I_f and I_0 are the intensities obtained with and without absorption medium, respectively. Dot curves: zoom on intensity profiles of the focusing spot along the optical axis, calculated at different *d*. The results are simulated with: $\sigma = 723\ m^{-1}$; $\lambda = 532$ nm; NA = 1.3 ($n = 1.58$).

3. Experimental demonstration of LOPA-based 3D microfabrication

3.1. Experimental setup and fabrication procedure

The LOPA technique can obviously be used for all 3D applications, including 3D imaging and 3D fabrication. As mentioned in section 2, two conditions are required: a photoresist that presents an ultralow absorption at the wavelength of the excitation laser, and a high focusing confocal laser scanning (CLSM) system. For the first condition, SU8 photoresist is an excellent candidate, thanks to its ultralow absorption in the visible range, for example at 532 nm (**Figure 5(b)**), which is the wavelength of a very popular and low-cost laser. By using a high NA oil-immersion OL of NA = 1.3 to focus a laser beam into the photoresist, the second condition is then satisfied.

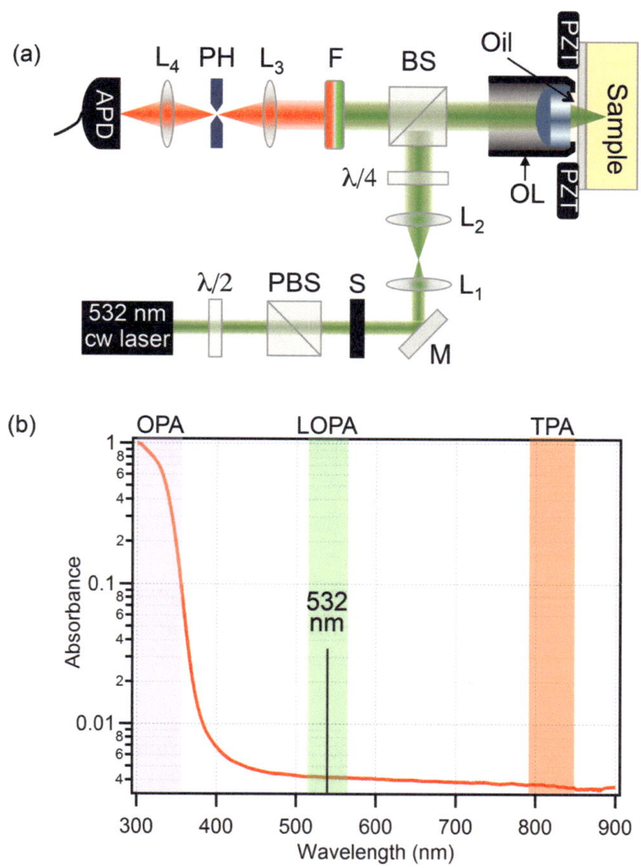

Figure 5. (a) A sketch of the experimental setup. PZT: piezoelectric translator, OL: oil immersion microscope objective, $\lambda/4$: quarter-wave plate, $\lambda/2$: half-wave plate, BS: beam splitter, PBS: polarizer beam splitter, M: mirror, S: electronic shutter, L_{1-4}: lenses, PH: pinhole, F: 580 nm long-pass filter, APD: avalanche photodiode. (b) Absorption spectrum of SU8 photoresist. The color bars indicate three cases: OPA (purple bar), LOPA (green bar), and TPA (red bar). To demonstrate LOPA DLW, a laser operating at 532 nm is used.

In order to demonstrate the LOPA DLW technique, a confocal optical system illustrated in **Figure 5(a)** was built. In this system, a CW laser operating at 532 nm is used. The laser power is monitored by a combination of a half-wave plate ($\lambda/2$) and a polarizer. The laser beam is directed and collimated by a set of lenses and mirrors. In order to realize mapping or fabrication, samples are mounted on a 3D piezoelectric actuator stage (PZT), which is controlled by a LabVIEW program. A quarter-wave plate ($\lambda/4$) placed in front of the OL is inserted and oriented to generate a circularly polarized beam for mapping and fabrication. The high NA oil-immersion objective (NA = 1.3) placed beneath the glass coverslip is used to focus the excitation laser beam. The fluorescence signal emitted by the samples is collected by the same objective, filtered by a 580 nm long-pass filter, and detected by an avalanche photodiode (APD).

For fabrication, SU8 photoresist is coated on a glass substrate. In order to remove all contamination on the surface, glass substrates must be treated with acetone and an ultra-sonication prior to the spin coating. After the cleaning process, SU8 of different viscosities (SU8 2000.5, SU8 2005, or SU8 2025) is spin coated on the glass substrates, depending on the types of the desired structure. For 1D and 2D structures, SU8 2000.5 and 2002, which give a layer thickness of 0.5 and 2 μm, are used. For 3D structures, other types of SU8 at higher viscosity, for example SU8 2005 or 2025, are required. The spin coating step is followed by a soft baking step at 65°C and 95°C, the soft baking time depends on the types of SU8 used.

Before writing the structure on the photoresist, the interface between the glass substrate and the photoresist layer must be determined. To determine the interface, the focusing spot is scanned along *xz* or *yz* plane at very low laser power to prevent polymerization, and the fluorescence signal can be collected by the APD. This step allows one to precisely write the structure at the desired position. Then the laser power is increased to several mW to fabricate any desired structures by scanning the focusing spot along a path programmed with Labview. Post exposure bake (PEB) is carried out after the fabrication, and followed by a development step.

3.2. Verification of LOPA-based fabrication

It has been demonstrated, by analyzing fluorescence emission, that the SU8 photoresist linearly absorbs the excitation laser at 532 nm-wavelength [17]. For the LOPA CLSM, the intensity at the focusing region (of the order of 10^7 W/cm^2) is much higher than that at the input of the optical system (10^{-2} W/cm^2), allowing the excitation and fluorescence detection of the focal spot volume only. The fluorescence measurements in which the emission is quite low are enabled by the use of an avalanche photodiode (APD).

By using the standard fabrication process described in the previous part, it was demonstrated that polymerization is achieved only at the focusing spot of the microscope objective, where the excitation intensity is sufficiently high to compensate the low linear absorption of the resist. In the case of TPA, there are two thresholds: the first one related to light intensity, above which two photons are simultaneously absorbed, and the second one related to dose, above which complete photopolymerization is achieved. However, in the case of LOPA, there exists only one threshold related to dose. Thanks to a high intensity at the focusing spot, the complete photopolymerization is only achieved in this region. **Figure 6** shows SEM images of experimental results. For each exposure, a solid structure, called "voxel," corresponding to

http://dx.doi.org/10.5772/intechopen.71318

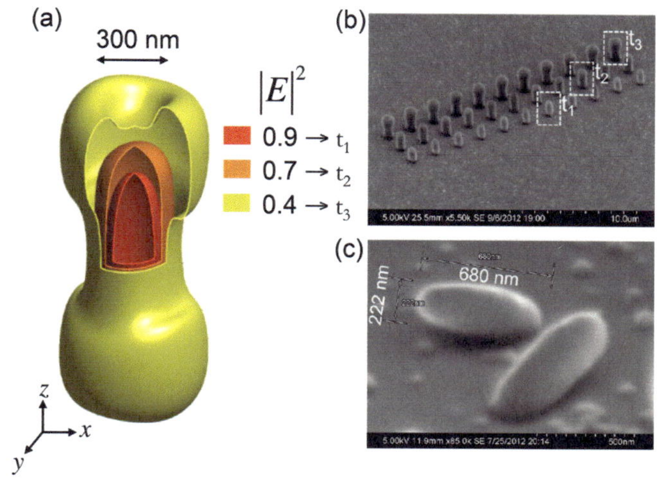

Figure 6. Fabrication of voxels by single-shot exposure. (a) Theoretical calculation of the contour plot of light intensity at the focusing region (NA = 1.3, n = 1.518, λ = 532 nm). (b) SEM image of a voxels array obtained by different exposure doses. Three ranges of voxels are fabricated correspondingly to the exposure time t_1, t_2 and t_3 whose corresponding doses are 0.9, 0.7 and 0.4 as indicated in (a). These experimental results explain the OPA nature where the formed voxel shape is determined by the exposure dose. (c) Complete voxel lying on the substrate indicated an ellipsoidal form (exposure time t_1 was applied). This form is similar to that obtained by the TPA method.

a focusing spot, was obtained. By changing either the excitation power or the exposure time, i.e. the dose, the voxel size and shape can be adjusted, as shown in **Figure 6(b)**. A CW green laser power of only 2.5 mW and an exposure time of about 1 second per voxel were required to create these structures. **Figure 6(b)** shows, for example, a voxel array realized with three different exposure times. The dose dependence can be explained theoretically by calculating the iso-intensity of the focusing spot at different levels. Three kinds of voxels obtained with t_1, t_2 and t_3 in **Figure 6(b)** correspond to three different iso-intensities illustrated in **Figure 6(a)**, namely 0.9, 0.7 and 0.4, respectively. The operation in the OPA regime is fully confirmed by the evolution of voxel size and shape observed experimentally. Indeed, in the case of TPA, the creation of bone-like voxel shape requires very high excitation intensity and could not be easily realized due to the TPA intensity threshold. In the case of LOPA, all these voxels shapes were obtained by simply adjusting the exposure time while the laser power is kept at a low value. Smaller voxels as shown in **Figure 6(c)** require shorter exposure times. Certainly, the exposure time required to create sub-micrometer structures varies as a function of the laser power.

Pillar arrays were also fabricated by scanning the focusing spot of the laser along the thickness of a 1 μm SU8 film with different writing speeds. **Figure 7** shows the pillar size as a function of writing velocity for three values of laser power, P = 7.5, 6, and 4.5 mW. The fabrication of smaller pillars down to 190 nm is possible [17]. The size of individual pillars is quite small when considering the wavelength (532 nm) used for the writing process. As for the linear dependence with intensity (OPA vs. TPA) [20, 21], the diameter-dose relationship agreement confirms this behavior, as the intensity I_0 is used for the fit, instead of I_0^2 in the case of TPA. This result confirms the fabrication of sub-microstructures by LOPA-based DLW method.

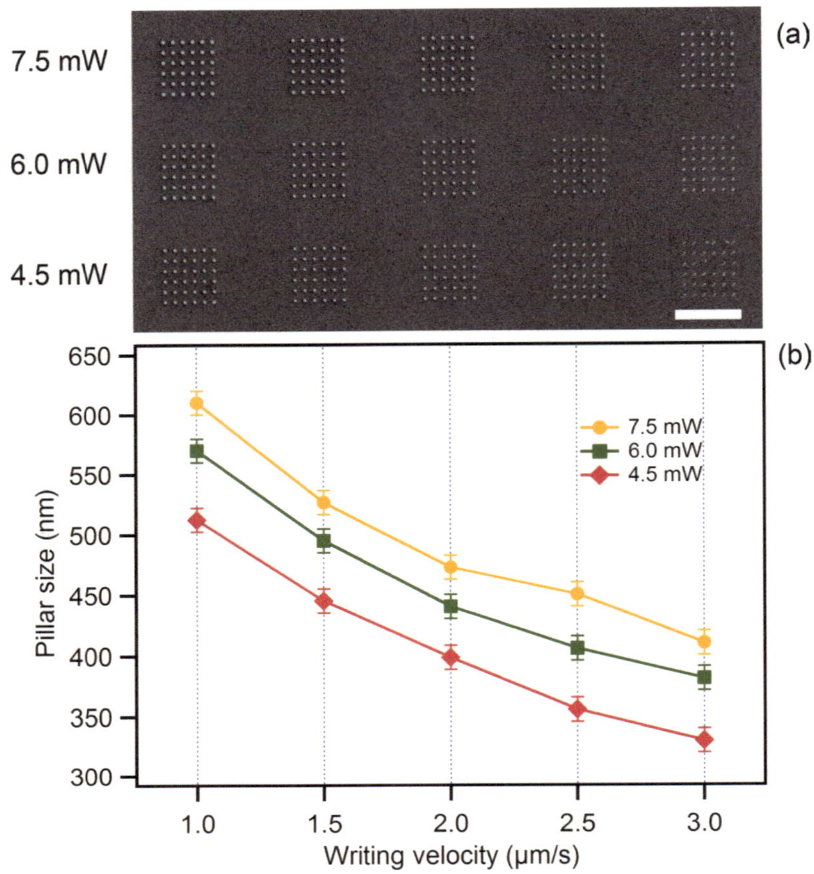

Figure 7. (a) SEM image of a pillar arrays fabricated at different writing velocity and laser powers. (b) Dose dependence of size of pillars shown in (a), with different laser power values, $P = 7.5, 6, 4.5$ mW, and different writing velocities, $v = 1.0$, 1.5, 2.0, 2.5, and 3.0 μm/s. Scale bar: 10 μm.

3.3. LOPA-based DLW for 3D microstructure fabrication

In order to demonstrate 3D fabrication, 3D arbitrary photonic crystals (PCs) have been realized. In the fabrication process, the dose was adjusted by changing the velocity of the PZT movement while the input power is fixed. The doses were varied from structure to structure depending on the size and separation (periodicity of PC). In the first experiment, a series of different size 3D woodpile PC structure on glass substrate was fabricated. It is noted that, SU-8 exhibits a strong shrinkage effect, which results in the distortion of the fabricated structure. After a number of experiments, acceptable parameters for woodpile, which are an input power of 2.5 mW and a velocity of 1.4 μm/s, were found. Applying these parameters, 3D diamond lattice-like-based PC structures such as woodpile, twisted chiral and circular spiral can be realized.

Figure 8(a–f) shows SEM images of 3D woodpile, spiral and chiral PCs fabricated with these optimum fabrication parameters. The woodpile structure (**Figure 8(a** and **b)**) consists

http://dx.doi.org/10.5772/intechopen.71318

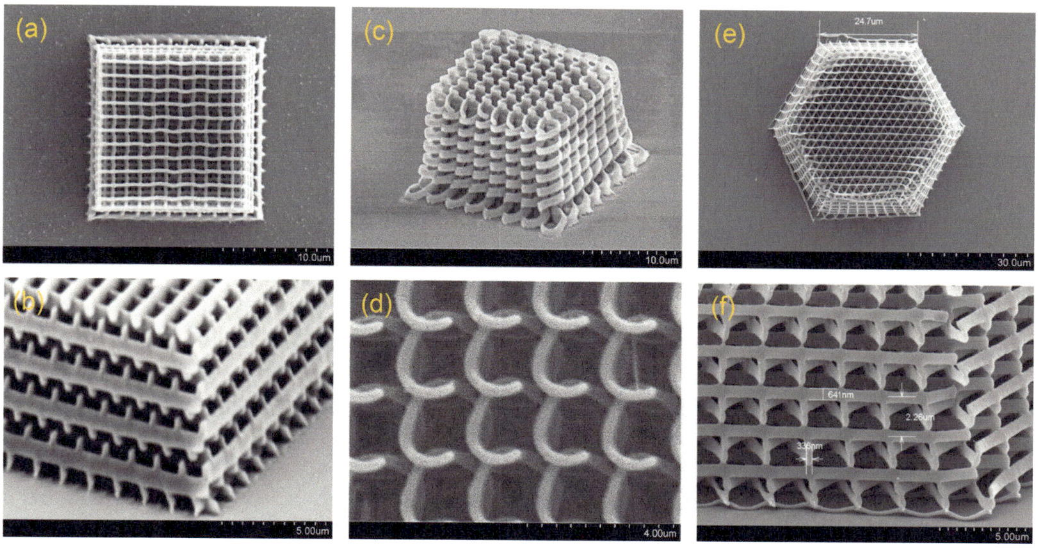

Figure 8. (a–b) SEM images of a woodpile structure fabricated with the following parameters: distance between rods = 1.5 μm; distance between layers = 0.7 μm; number of layers = 20; laser power P = 2.5 mW and scanning speed v = 1.4 μm/s. (c and d) SEM images of a chiral structure. Structure and fabrication parameters: distance between rods a = 2 μm; distance between layers c/3 = 0.75 μm; number of layers = 28; line width r = 300 nm; laser power P = 2.8 mW and scanning speed v = 1.34 μm/s. (e–f) SEM images of a spiral structure (shown in inset). Fabrication parameters: diameter of a spiral D = 2 μm; spiral pitch C = 2 μm; lattice constant a = 3 μm; spiral height equals to film thickness = 15 μm; laser power P = 2.6 mW and scanning speed v = 1.34 μm/s.

of stacked 20 layers. Each layers consists of parallel rods with period a = 1.5 μm. Rods in successive layers are rotated by an angle of 90° relative to each other. Second nearest-neighbor layers are displaced by a/2 relative to each other. Four layers form a lattice constant c = 2.1 μm. SEM measurement shows that the line width of rods on the top layer is about 320 nm, which is 1.5 times of the voxel standard size. This means that the size of the rod can be minimized further. Woodpile structure with measured separation between rods on the top layer of only 800 nm and the measured line width of 180 nm was also successfully fabricated. This result shows evidently that separation of rods and layers are comparable to the wavelength of visible light. **Figure 8(c–f)** shows, as examples, two other kinds of sub-micrometer 3D structures. Clearly, 3D chiral or spiral structures are well created, which are as good as those obtained by TPA DLW. The structures features are well separated, layer by layer, in horizontal and in vertical directions. The feature sizes are about 300 nm (horizontal) and 650 nm (vertical).

Experimental realization of LOPA-DLW in fabrication of 1D, 2D and 3D photonic crystal showed evidently the advantage of LOPA idea in combining with regular DLW. With a few milliwatts of a CW laser and in a moderate time, any kind of sub-micrometer structure with or without designed defect could be fabricated. However, some fabricated structures are not uniform or distorted. The physical causes of the distortion can be attributed to two main effects: dose accumulation effect [22] and shrinkage effects [23]. In the next section, some techniques to overcome those effects will be experimentally demonstrated.

4. Optimization of LOPA-based DLW technique

4.1. Dose accumulation effect

As a consequence of linear behavior of absorption mechanism, the dose accumulation effect is the inherent nature of linear absorption material [24, 25]. In contrast to a conventional TPA method, a photoresist operated in OPA regime does not have any threshold of polymerization [26], hence the voxel size can be controlled by adjusting the exposure dose. In principle, polymerization occurs at the focusing spot with a single-shot exposure resulting in a very small voxel (smaller than the diffraction limit) [27]. However, when two voxels are built side-by-side with a distance of about several hundred nanometers, two resulting voxels are no longer separated [17]. This issue evidently originates from the dose accumulation effect in OPA process.

Similar to OPA microscopy where the microscopy image cannot resolve two small objects which localize at about several hundred nanometers from each other, the fabricated voxels in DLW also cannot be separated. Abbe's criterion states that, the minimum resolving distance of two objects is defined as $0.61\lambda/NA$, where λ is the wavelength of incident light. This diffraction barrier thus imposes the minimum distance between different voxels, created by different exposures. Moreover, when multiple exposures are applied, although isolated voxel fabrication is ideally confined to the focal volumetric spot, the superposition of many out-of-focus regions of densely-spaced voxels leads to undesired and unconfined reaction. This results in the larger effective voxel size, even with a distance far from the diffraction limit. Indeed, in the case of OPA, photons could be absorbed anywhere they are, with an efficiency depending on the linear absorption cross-section of the irradiated material. The absorbed energy is gradually accumulated as a function of exposure time.

Figure 9 shows the theoretical calculations and experimental results of the dependence of the voxels size on distance. When two voxels are separated by a distance shorter that 1 μm, a clear accumulation effect is observed, resulting in a voxel of larger size. When the separation changes from 2 to 0.5 μm, the FWHM of each voxel is increased from 190 to 300 nm. Moreover, for a short separation, the voxels array is not uniform from the center to the edge part, as can be seen in case of a separation of 0.5 μm. In the parts below, two strategies to get rid of this accumulation effect are discussed.

4.1.1. Dose compensation strategy

According to the structures fabricated by LOPA DLW and the theoretical calculation, the dose accumulation effect should be compensated in order to get superior structure uniformity. The compensation technique idea is based on a balance of the exposure doses over the structure. In other words, a certain amount of the exposure dose should be reduced at the region (or a part or a division) where the dose accumulation effect strongly occurs and should be added to the region where the dose is lacking.

As demonstrated above, the accumulation effect depends on the separation distance, s. 2D micropillars array was fabricated by scanning the focusing spot along the z-axis to demonstrate

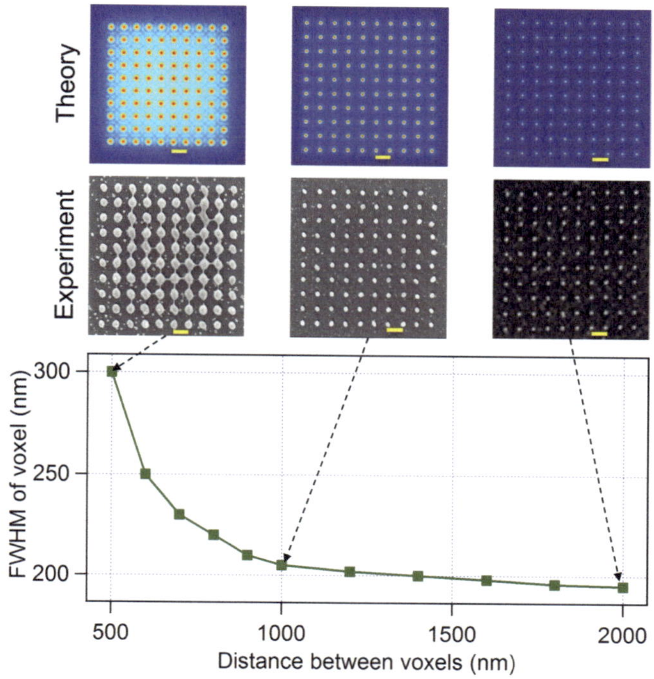

Figure 9. Dependence of voxel sizes on separation between voxels calculated at the iso-intensity = 0.5 ($I = I_0/2$), showing the influence of energy accumulation from one focusing spot to another. Insets show simulated images (blue background) and SEM images (gray background) of 2D voxel arrays fabricated with different distances between two voxels: 2, 1, and 0.5 μm. Scale bar (from left to right): 0.5, 1, and 2 μm.

the accumulation effect and the dose compensation technique. The variation of the pillars size as a function of s distance, from 0.3 to 3 μm, has been systematically investigated [22]. According to the pillars size variation, from the center to the edge of the structure, the size difference could be compensated by gradually increasing the exposure doses for outer pillars. The intensity distribution of a pillar array and the dose compensation strategy are shown in **Figure 10(a)**. Different dose compensation ratios have been experimentally applied, indicated by letters A, B, C, and D. Two sets of structures have been fabricated on the same sample in order to maintain the same experimental conditions: one without and the other with dose compensation. The accumulation effect in the case of s = 0.4 μm, obtained without compensation, is shown in **Figure 10(b)**. It can be seen that the structure is not uniform, and the outmost voxels collapse into central pillars. With dose compensation, i.e. the doses for outer pillars are increased, 2D micropillars arrays become uniform. Structures realized with different dose ratios, corresponding to A, B, C, and D schemes depicted in **Figure 10(a)**, are shown in **Figure 10(c–f)**. With high dose compensation amplitude, the size of outer pillars becomes even larger than those in the center. By using an appropriate dose compensation ratio, perfectly uniform 2D micropillars array is obtained, as shown in **Figure 10(c)**. The structure period is only 400 nm, the height of pillars is about 700 nm, and pillar's diameter is about 160 nm. Structures with periods of several hundred nanometers are very suitable for photonic applications in visible range.

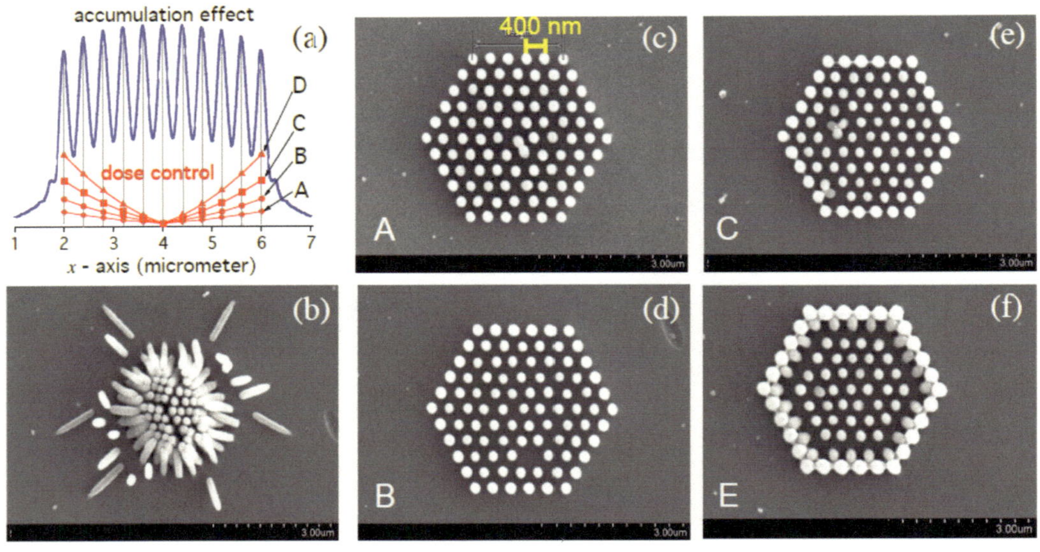

Figure 10. Compensation of dose accumulation for s = 0.4 μm. (a) Theoretical calculation of the intensity distribution as a function of x–position (blue color), and proposition of different doses, indicated by A, B, C, and D (red color) to compensate the dose accumulation effect. (b–f) SEM images of 2D hexagonal structures realized without compensation (b), and with compensation with different doses: A (c); B (d); C (e), and D (f). The excitation power was fixed at 2.5 mW and the dose was adjusted by changing the writing velocity.

4.1.2. Local PEB for small and uniform microstructures

The dose compensation technique are presented above and proved to be able to compensate the size difference in the structures. Nonetheless, this technique requires numerous calculation and tests in order to find out the appropriate dose compensation parameters. For example, for the fabrication of other 2D structures with a sub-micrometer period containing an arbitrary defect, such as a microcavity or a waveguide of arbitrary shape, the dose should be controlled for individual voxel (or pillar) as a function of its position in the structure and with respect to the defect. This requires many attempts to find out the optimized parameters since the dose compensation is different for different structures.

In this part, the thermal effect induced by a CW green laser in the LOPA-based DLW technique has been investigated, which plays a role as a heat assistant for completing the crosslinking process of the photopolymerization of SU8 [28]. The fabrication of sub-microstructures using LOPA DLW with a laser induced thermal effect, also called local PEB, was demonstrated, and it was also shown to alter the traditional PEB on a hot plate and help overcome the accumulation effect existing in standard LOPA DLW.

For the demonstration of local PEB, two sets of 2D structures were fabricated, each set contains pillar arrays written at different doses (by varying the laser power and the writing speed). One set was realized following the traditional process, i.e., after exposure, the sample was post-baked for 1 min at 65°C and then 3 min at 95°C using hot plate. For the other set, the PEB step was skipped, which means that the exposure process was followed directly by the development process. It has been observed that for the same dose, the pillars fabricated without PEB are smaller than those fabricated with PEB [28]. Due to the fact that the structures are not formed at low power (low light intensity), it is assumed that the applied laser power or intensity must be above a threshold in order to induce enough heat for the crosslinking process, and this threshold should be higher than that of standard LOPA (with PEB) for a complete photopolymerization.

Figure 11(a) shows the accumulation effect observed in the structures fabricated using traditional PEB process at different doses. The distance between two pillars is 500 nm, which is considered as a sufficiently short separation to induce noticeable accumulation effect. It can be observed that the pillar size decreases from the center to the edge, with a variation of 10–20%, resulting in non-uniform structures. **Figure 11(b)** shows the SEM images of structures obtained using local PEB. In this case, in order to induce sufficiently high temperature, the laser power (5 mW) was higher than that (3 mW) used with traditional PEB step. In contrast to the structures fabricated using traditional PEB, which shows accumulation effect (**Figure 11(a)** on the top), the structures obtained with local PEB show nearly perfect uniformity (**Figure 11(b)** on the top). Moreover, as compared to the dose compensation method, the LOPA-based DLW using local PEB does not require testing, since the dose applied is constant for all voxels and the range of applicable doses is large. In addition, the PEB step is skipped, which is a great advantage in terms of fabrication time. Furthermore, comparing to structures realized by TPA-based DLW, the period of these fabricated structures is much shorter (only 500 nm or even 400 nm) thanks to the use of short laser wavelength, which can be an additional advantage of this LOPA technique.

The heat equation [29] was also solved using finite element model realized by Matlab to demonstrate that the heat induced by high excitation intensity of a 532 nm CW laser confines the crosslinking reaction in the local region where the temperature is higher than the PEB temperature. This resulted in fine and uniform structures, since only the material within the effective temperature region was properly polymerized. Temperature-depth dependence calculation shows that this technique allows the fabrication of uniform 3D sub-micro structures with large thickness. This is then evident by an experimental demonstration of fabrication of a uniform 3D woodpile structure without PEB step, with a period as small as 400 nm [28]. Compared to the commonly used TPA method, LOPA-based DLW with local PEB shows numerous advantages such as simple, low-cost setup and simplified fabrication process, while producing same high-quality structures.

4.2. Shrinkage effect

Shrinkage is a fundamental issue for photopolymerization in the photopolymer. It is difficult to avoid the non-uniformity when the conventional polymerized microstructures are attached to substrates. The origin of this effect for TPA polymerization has been investigated [30]. It was suggested that the origin of the shrinkage is the collapse of this material during the development stage owing to the polymer not being fully cross-linked when they are made at the irradiation power close to the photopolymerization threshold. At fabrication intensities slightly above the photopolymerization threshold, the photopolymerization yield is not

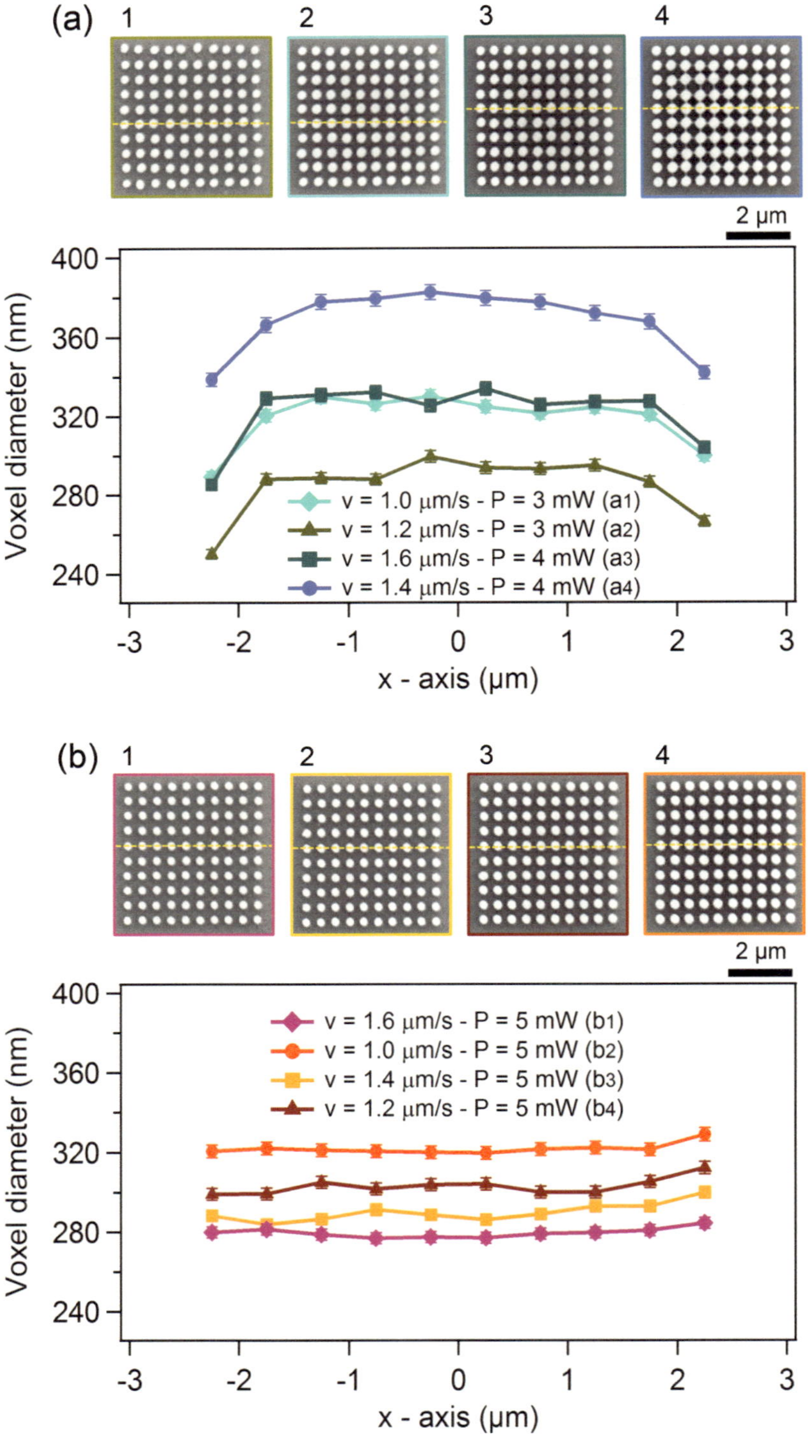

Figure 11. Comparison between structures fabricated with standard PEB (a) and local PEB (b) shows the advantage of "local PEB" in overcoming the accumulation effect. Left: SEM images of fabricated structures. Top: sizes of pillars indicated by the dashed line on SEM images plotted as a function of position. With standard PEB: the pillars are larger at center of pattern. With local PEB: pillars are very uniform. All structures are fabricated with a period of 500 nm.

http://dx.doi.org/10.5772/intechopen.71318

100%, resulting in a sponge-like material after the development process. Hence, the structural shrinkage observed at average laser powers is the result of the collapse of this material at a molecular level [30]. However, it has also been confirmed that, although the sponge-like materials are formed during the development because of the non-full polymerization at low laser power, the shrinkage indeed occurs due to the capillary forces and the dramatic change of surface tension during the drying process [31].

For LOPA 3D fabrication, this shrinkage effect is also observed, with different levels of distortion for different exposure doses, i.e., the lower the exposure dose is the higher the shrinkage is. (**Figure 12a–d**) shows the shrinkage effect observed in 3D woodpile structures fabricated at different doses (P = 9 mW, writing speed v = 4, 3, 2, and 1 μm/s, from left to right, respectively). It can be clearly seen that the degree of shrinkage at different exposure doses is different, i.e., from left to right, the writing speed decreases (the dose increases), the shrinkage decreases. This result is also in agreement with previous report on the shrinkage in the case of TPA polymerization [30], which suggests that in both cases, the shrinkage effect might have a similar origin.

Non-uniform shrinkage might destroy the structural periodicity of a photonic crystal, resulting in the degradation of its optical quality. While shrinkage is an intrinsic problem, which cannot be avoided, non-uniform shrinkage problem can be positively resolved. Several approaches have been reported on how to overcome this non-uniform shrinkage, including pre-compensation for deformation [32], single [33] and multi-anchor supporting method [30, 34], or freestanding microstructures trapped in cages [35]. All the above methods are investigated for the TPA case. In this work, the multi-anchor supporting method was employed to reduce the deformation caused by the non-uniform shrinkage of 3D microstructures by LOPA DLW. To demonstrate this idea, instead of fabricating woodpile structures directly on glass substrate, four "legs" at four corners of structures were created in order to avoid the attachment of the structures to the glass substrate, which is the direct cause of the non-uniform

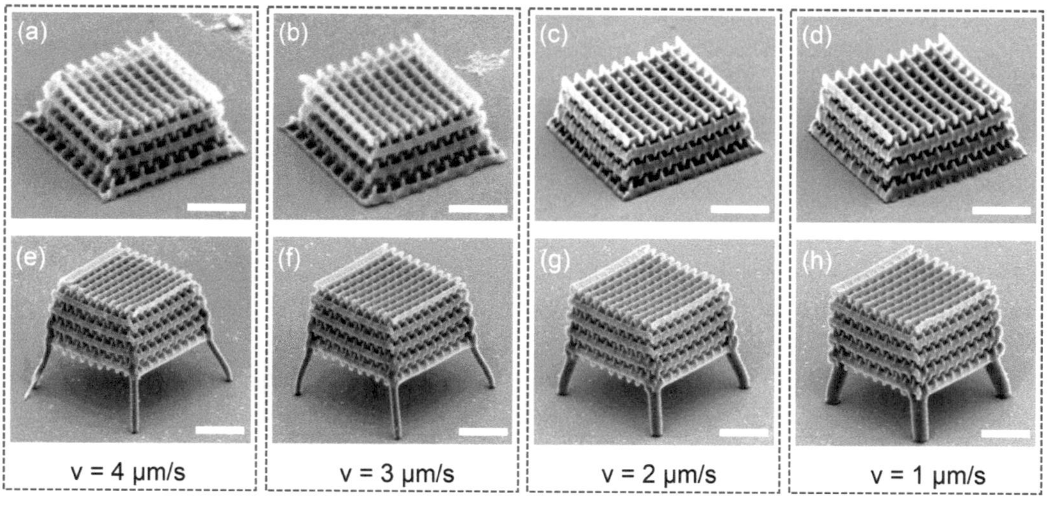

Figure 12. 3D woodpile structures (number of layers =16, rod spacing = 1.5 μm and layer spacing = 0.65 μm) fabricated without (a–d) and with "legs" (e–h), at laser power of 9 mW and different writing speeds: v = 4, 3, 2, and 1 μm/s, from left to right. Scale bars: 5 μm.

shrinkage. These anchors were designed and fabricated as rods of several hundred nanometers and did not affect the main structure. (**Figure 12e–h**) shows woodpile structures with anchors fabricated at different doses. It can be seen that at low dose, the non-uniform shrinkage still remains as shown in **Figure 12(e)**. However, when the dose is increased by a small amount, the distortion decreases and disappears. In this case, a laser power of 9 mW and a writing speed of 3 μm/s were sufficient for a considerable uniform structure. Compared to the non-anchored structures, the fabrication speed is reduced by three times to obtain a nearly deformation-free PC, which is remarkable especially for the fabrication of large structures. The shrinkage effect can also be exploited to produce PCs with small lattice constant.

To conclude, the multi-anchor supporting method was successfully applied to reduce the non-uniform deformation of 3D microstructure caused by attachment to the substrate. Since the support with four anchors allows uniform shrinkage of a polymeric microstructure by releasing it from the substrate, the fabricated microstructure shrinks isotropically. The combination of LOPA with local PEB and this supporting method is a promising way of producing small lattice constant photonic structures, which possess a photonic bandgap in the visible range.

5. Realization of magneto-photonic microstructures by LOPA-based DLW

LOPA-based DLW has been demonstrated to be not only applicable for SU8 but also other materials [36, 37], and be capable of fabricating structures incorporated with nanoparticles (NPs), such as plasmonic [38] or magnetic NPs [39]. In this section, as a proof of the versatility of LOPA DLW, the fabrication of 2D and 3D sub-microstructures from nanocomposite consisting of magnetite (Fe_3O_4) NPs and a commercial SU8 photoresist by employing LOPA DLW technique is presented [39]. The nanocomposite was synthesized by incorporating magnetic nanoparticles (MNPs) into SU8 matrix. Due to the magnetic force (mainly inter-particle interactions), MNPs tend to form large agglomerations, causing difficulties in achieving a homogeneous distribution of MNPs in polymer matrix. Hence, different concentrations of Fe_3O_4 MNPs and various types of SU8 with different viscosities have been investigated to obtain the best dispersion of MNPs in the polymer environment. Finally, the best compromise was achieved between SU8 2005, with moderate viscosity, and a MNP concentration of 2 wt%, which is low enough to achieve a homogeneous nanocomposite and high enough to give strong response to external magnetic field.

Arbitrary magnetic structures from magneto-polymer nanocomposite have been realized using LOPA DLW. **Figure 13(a)** shows an SEM image of an arbitrary 2D structure, the letter "LPQM," fabricated at the power of 15 mW with a writing speed of 3 μm/s. For this fabrication, a point matrix technique was used to shape the letter. The distance between pillars was set at 150 nm, which resulted in a continuous line due to the accumulation of exposure energy at the vicinity of each point. An SEM image of a 2D pillar array magneto-photonic structure, with a period of 1.5 μm is shown in **Figure 13(b)**. This structure was fabricated at

http://dx.doi.org/10.5772/intechopen.71318

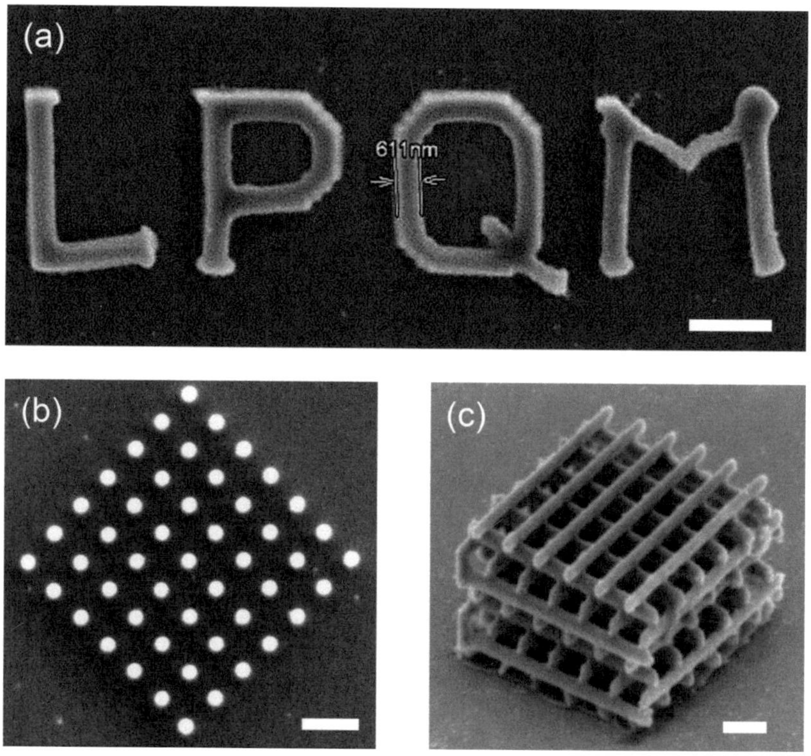

Figure 13. (a) SEM image of letter LPQM, fabricated at a laser power P = 15 mW, writing speed v = 3 μm/s. (b) SEM image of a 2D pillar array, fabricated at P = 12 mW, v = 2 μm/s. (c) SEM image of a 3D woodpile PC. Fabrication parameters: rod spacing = 2 μm; distance between layers = 1 μm; number of layers = 10; laser power P = 12 mW and scanning speed v = 2 μm/s. Scale bars: 2 μm.

a laser power of 12 mW and a writing speed of 2 μm/s. Similarly, various 3D structures have also been realized, with a laser power of about 12 mW. **Figure 13(c)** shows an SEM image of a woodpile structure, which consists of 10 alternating layers separated from each other by 1 μm. The rod spacing in x- and y-directions is 2 μm. These experimental results confirm that the LOPA-based DLW allows the fabrication of any magneto-photonic sub-micrometer structure or device on this magneto-polymer nanocomposite, which is very promising for a wide range of applications using magnetic microdevices and micro-robotic tools.

Indeed, magneto-photonic structures for remote actuation have attracted a great deal of attention recently [40]. With the aid of an external magnetic field, the displacement in three dimensions of magnetic structures can be controlled as desired. As shown previously, small magneto-photonic structures, which are adaptable to small targets, have been created. In order to prove the response of magnetic structures to a magnetic field, arrays of micro-pillars were fabricated as an example for demonstration [39]. To release the structures from the substrate, before coating the nanocomposite layer, an extra sacrificial layer of PMMA, which can be dissolved with acetone to release the structures into solution, was added. A magnetic field (only 8 mT) generated by a permanent magnet was then applied to examine the magnetic field response of the magnetic micro-swimmers. The whole process from structural development to the movement toward higher gradient of the external magnetic field was observed via optical microscope, in which all of the micro-pillars quickly moved toward the magnetic tip, confirming the presence of Fe_3O_4 MNPs inside the structures and their strong response to the applied magnetic field. The magnetic structures responded strongly to the external magnetic field, opening many promising applications, such as tunable photonic structures based on magneto-optical effect and development of microrobotic tools for transport in biological systems.

6. Conclusions and prospects

In this chapter, a new technique based on an already-well-known mechanism, one-photon absorption (OPA) direct laser writing (DLW), for fabrication of low-cost, high-quality 3D PCs was introduced. This technique was demonstrated through both theory and experiment on the ultra-low absorption regime (LOPA) of photosensitive material. It was pointed out that DLW based on LOPA microscopy enables 3D fabrication in any kind of photoresist material with flexible defect engineering.

Some additional methods, such as dose compensation and local PEB, have been also proposed and applied to optimize the structures fabricated by LOPA DLW. In particular, by using the optically induced thermal effect, the fabrication of accumulation-free sub-micrometer and uniform polymeric 2D and 3D structures was realized. Also, the multi-anchor supporting method was employed to reduce the deformation caused by the non-uniform shrinkage of 3D polymeric microstructures.

As a proof of the versatility of the LOPA-based DLW technique, the capability to fabricate magneto-photonic microstructures using a SU8/Fe_3O_4 nanocomposite has been demonstrated. The fabricated micro-swimmers showed strong response to an applied external magnetic field, which emphasizes the importance of free-floating structures as a robotic technology for magnetic devices such as sensors, actuators, magnetic labeling, and drug targeting. 3D magneto-photonic structures have also been successfully fabricated, potentially aiding the development of magnetic nanodevices and micro-robotic tools for a wide range of applications.

Author details

Dam Thuy Trang Nguyen, Mai Trang Do, Qinggle Li, Quang Cong Tong, Thi Huong Au and Ngoc Diep Lai*

*Address all correspondence to: nlai@lpqm.ens-cachan.fr

Laboratoire de Photonique et Moléculaire, Ecole Normale Supérieure de Cachan, CentraleSupélec, CNRS, Université Paris-Saclay, Cachan, France

References

[1] Lin SY, Fleming JG, Hetherington DL, Smith BK, Biswas R, Ho KM, et al. A three-dimensional photonic crystal operating at infrared wavelengths. Nature. 1998;**394**(6690):251-253

[2] Cumpston BH, Ananthavel SP, Barlow S, Dyer DL, Ehrlich JE, Erskine LL, et al. Two-photon polymerization initiators for three-dimensional optical data storage and microfabrication. Nature. 1999;**398**:51-54

[3] Strickler JH, Webb WW. Three-dimensional data storage in refractive media by two-photon point excitation. Optics Letters. 1991;**16**:1780-1782

[4] Denk W, Strickler JH, Webb WW. Two-photon laser scanning fluorescence microscopy. Science. 1990;**248**:73-76

[5] Alkaisi MM, Blaikie RJ, McNab SJ, Cheung R, Cumming DRS. Sub-diffraction-limited patterning using evanescent near-field optical lithography. Applied Physics Letters. 1999;**75**(22):3560-3562

[6] Mata A, Fleischman A, Roy S. Fabrication of multi-layer SU-8 microstructures. Journal of Micromechanics and Microengineering. 2006;**16**:276-284

[7] Campbell M, Sharp DN, Harrison MT, Denning RG, Turberfield AJ. Fabrication of photonic crystals for the visible spectrum by holographic. Nature. 2000;**404**(6773):53-56

[8] Lai ND, Liang WP, Lin JH, Hsu CC, Lin CH. Fabrication of two- and three-dimensional periodic structures by multi-exposure of two-beam interference technique. Optics Express. 2005;**13**(23):9605-9611

[9] Lai ND, Zheng TS, Do DB, Lin JH, Hsu CC. Fabrication of desired three-dimensional structures by holographic assembly technique. Applied Physics A. 2010;**100**(1):171-175

[10] Seet KK, Mizeikis V, Juodkazis S, Misawa H. Three-dimensional horizontal circular spiral photonic crystals with stop gaps below 1 μm. Applied Physics Letters. 2006;**88**(22): 221101-221103

[11] Sun HB, Matsuo S, Misawa H. Three-dimensional photonic crystal structures achieved with two-photon absorption photopolymerization of resin. Applied Physics Letters. 1999;**74**(6):786-788

[12] Gissibl T, Thiele S, Herkommer A, Giessen H. Two-photon direct laser writing of ultra-compact multi-lens objectives. Nature Photonics. 2016;**10**:554-560

[13] Straub M, Gu M. Near-infrared photonic crystals with higher-order bandgaps generated by two-photon photopolymerization. Optics Letters. 2002;**27**(20):1824-1826

[14] Rensch C, Hell S, Schickfus MV, Hunklinger S. Laser scanner for direct writing lithography. Applied Optics. 1989;**28**(17):3754-3758

[15] Bratton D, Yang D, Dai J, Ober CK. Recent progress in high resolution lithography. Polymers for Advanced Technologies. 2006;**17**:94-103

[16] Li Q, Do MT, Ledoux-Rak I, Lai ND. Concept for three-dimensional optical addressing by ultralow one-photon absorption method. Optics Letters. 2013;**38**(22):4640-4643

[17] Do MT, Nguyen TTN, Li Q, Benisty H, Ledoux-Rak I, Lai ND. Submicrometer 3D structures fabrication enabled by one-photon absorption direct laser writing. Optics Express. 2013;**21**(18):20964-20973

[18] Wolf E. Electromagnetic diffraction in optical systems I. An integral representation of the image field. Proceedings of the Royal Society A. 1959;**253**(1274):349-357

[19] Helseth LE. Focusing of atoms with strongly confined light potentials. Optics Communication. 2002;**212**(4-6):343-352

[20] Sun HB, Tanaka T, Kawata S. Three-dimensional focal spots related to two-photon excitation. Applied Physics Letters. 2002;**80**(20):3673-3675

[21] Lee CH, Chang TW, Lee KL, Lin JY, Wang J. Fabricating high-aspect-ratio sub-diffraction-limit structures on silicon with two-photon photopolymerization and reactive ion etching. Applied Physics A. 2004;**79**(8):2027-2031

[22] Do MT, Li Q, Ledoux-Rak I, and Lai ND. Optimization of LOPA-based direct laser writing technique for fabrication of submicrometric polymer two- and three-dimensional structures. Proc. SPIE 9127, Photonic Crystal Materials and Devices XI. 2014

[23] Matta JA, Outwater JO. The nature, origin and effects of internal stresses in reinforced plastic laminates. Polymer Engineering and Science. 1962;**2**(4):314-319

[24] Scott TF, Kloxin CJ, Forman DL, McLeod RR, Bowman CN. Principles of voxel refinement in optical direct write lithography. Journal of Materials Chemistry. 2011;**21**:14150-14155

[25] Chen KS, Lin IK, Ko FH. Fabrication of 3D polymer microstructures using electron beam lithography and nanoimprinting technologies. Journal of Micromechanics and Microengineering. 2005;**15**(10):1894-1903

[26] Correa DS, Boni LD, Otuka AJG, Tribuzi V, Mendonça CR. Two-photon polymerization fabrication of doped microstructures. In: De Souza Gomes A, editor. Polymerization. Rijeka: InTech; 2012

[27] Li L, Gattass RR, Gershgoren E, Hwang H, Fourkas JT. Achieving $\lambda/20$ resolution by one-color initiation and deactivation of polymerization. Science. 2009;**324**(5929):910-913

[28] Nguyen DTT, Tong QC, Ledoux-Rak I, Lai ND. One-step fabrication of submicrostructures by low one-photon absorption direct laser writing technique with local thermal effect. Journal of Applied Physics. 2016;**119**(1):013101-013106

[29] Carslaw HS, Jaeger JC. Conduction of Heat in Solids. 2nd ed. United Kingdom: Oxford University Press; 2000

[30] Ovsianikov A, Shizhou X, Farsari M, Vamvakaki M, Fotakis C, Chichkov BN. Shrinkage of microstructures produced by two-photon polymerization of Zr-based hybrid photosensitive materials. Optics Express. 2009;**17**(4):2143-2148

[31] Sun Q, Ueno K, Misawa H. In situ investigation of the shrinkage of photopolymerized micro/nanostructures: The effect of the drying process. Optics Letters. 2012;**37**(4):710-712

[32] Sun HB, Suwa T, Takada K, Zaccaria RP, Kim MS, Lee KS, et al. Shape precompensation in two-photon laser nanowriting of photonic lattices. Applied Physics Letters. 2004;**85**(17):3708-3710

[33] Maruo S, Hasegawa T, Yoshimura N. Single-anchor support and supercritical CO_2 drying enable high-precision microfabrication of three-dimensional structures. Optics Express. 2009;**17**(23):20945-20951

[34] Lim TW, Son Y, Yang SY, Pham TA, Kim DP, Yang BI, et al. Net shape manufacturing of three-dimensional SiCN ceramic microstructures using an isotropic shrinkage method by introducing shrinkage guiders. International Journal of Applied Ceramic Technology. 2008;**5**(3):258-264

[35] Sun Q, Juodkazis S, Murazawa N, Mizeikis V, Misawa H. Freestanding and movable photonic microstructures fabricated by photopolymerization with femtosecond laser pulses. Journal of Micromechanics and Microengineering. 2010;**20**(3):035004

[36] Mao F, Tong QC, Nguyen DTT, Huong AT, Odessey R, Saudrais F, et al. LOPA-based direct laser writing of multi-dimensional and multi-functional photonic submicrostructures. Proceedings of SPIE 10115, Advanced Fabrication Technologies for Micro/Nano Optics and Photonics X. 2017;1011509

[37] Tong QC, Luong MH, Remmel J, Do MT, Nguyen DTT, Lai ND. Rapid direct laser writing of desired plasmonic nanostructures. Optics Letters. 2017;**42**(12):2382-2385

[38] Do MT, Nguyen DTT, Ngo HM, Ledoux-Rak I, Lai ND. Controlled coupling of a single nanoparticle in polymeric microstructure by low one-photon absorption-based direct laser writing technique. Nanotechnology. 2015;**26**(10):105301

[39] TH A, Trinh DT, Tong QC, Do DB, Nguyen DP, Phan MH. Direct laser writing of magneto-photonic sub-microstructures for prospective applications in biomedical engineering. Nanomaterials. 2017;**7**(5):105

[40] Markides H, Rotherham M, El Haj AJ. Biocompatibility and toxicity of magnetic nanoparticles in regenerative medicine, biocompatibility and toxicity of magnetic nanoparticles in regenerative medicine. Journal of Nanomaterials. 2012;**2012**:614094

Chapter 8

1D Photonic Crystals: Principles and Applications in Silicon Photonics

Liangshun Han

Additional information is available at the end of the chapter

http://dx.doi.org/10.5772/intechopen.71753

Abstract

One-dimension (1D) photonic crystals have been widely used in silicon photonics due to its simple structure and multiple working regimes: diffraction, Bragg reflection, and sub-wavelength regimes. Thanks to recent development of photonic technologies and high-resolution lithography, many 1D photonic crystal-assisted silicon integrated devices have been proposed and demonstrated to further increase integration density and improve device performance. This chapter first presents some fundamentals of 1D photonic crystals. An overview of the applications of 1D photonic crystals in silicon photonics is then given including grating couplers, waveguide crossings, multimode interference couplers, polarization-independent directional couplers, hybrid lasers, polarizers, and high-order mode filters, among others. Particular attention is paid to providing insight into the design strategies for these devices.

Keywords: photonic crystals, diffractive gratings, sub-wavelength gratings, Bragg gratings, silicon photonics

1. Introduction

Extensive research has been performed to realize large-scale integration of silicon photonics for optical communication networks [1, 2], optical computing [3, 4], and biosensing [5, 6]. The impetus of this research lies in the fact that silicon-on-insulator (SOI) technology is fully compatible with complementary metal oxide semiconductor (CMOS) technology and possesses high refractive index contrast enabling compact devices. Recent efforts in silicon photonic devices, such as III–V/silicon hybrid lasers [7–9], modulators [10, 11], photodetectors [12], and switches [13, 14], have all paved a path toward realizing silicon-based high-density electronic and photonic integration circuits (EPICs) [15]. However, due to high-index contrast, photonic

INTECH
open science | open minds

© 2018 The Author(s). Licensee InTech. This chapter is distributed under the terms of the Creative Commons Attribution License (http://creativecommons.org/licenses/by/3.0), which permits unrestricted use, distribution, and reproduction in any medium, provided the original work is properly cited. (cc) BY

devices based on SOI waveguides always suffer from some issues including high polarization sensitivity, limited bandwidth, sever phase errors caused by fabrication, thermal sensitivity, and relatively large loss. Those issues greatly limit application range of silicon-based photonic devices. There has been considerable effort by the international community to improve device performances by designing new structures, introducing new materials, combining different effects, developing new fabrication processes, discovering new functionalities, etc. The application of 1D photonic crystals in silicon photonics is a representative example.

1D photonic crystals are the simplest structure in photonic crystal family [16]. Interestingly, 1D photonic crystals still possess many exciting properties such as adjustable dispersion and birefringence, acting as homogeneous materials. Compared with 2D or 3D photonic crystals, the simple structure of 1D photonic crystals makes them easy to be integrated with the existing photonic devices without changing fabrication procedures. Consequently, many high-performance silicon-based devices have been proposed and demonstrated by exploiting those exciting properties of 1D photonic crystals. This chapter aims to provide the readers with some fundamentals of 1D photonic crystals and present an extensive overview of their applications in silicon photonics. The chapter is organized as follows. Section 2 describes the fundamentals of 1D photonic crystals. For the sake of clarity, the overview of their applications is divided into three sections (Sections 3–5) on the basis of three operating regimes.

2. Fundamentals of 1D photonic crystals

In silicon photonics, the 1D photonic crystals can be formed by periodic strip structures as shown in **Figure 1**. Normally, those strips are high-index crystal silicon layer which is optical waveguide core layer as well. The high-index strips are surrounded by low-index material such as air, silicon oxide, and polymer to form a periodically modulated refractive index distribution. In general, the incident direction of light could be any direction relative to photonic crystals. In practical terms, two special cases are widely employed: (a) light propagates crosswise through the 1D photonic crystal (propagation along the y- or z-axis in **Figure 1**), and (b) light propagates lengthwise through the 1D photonic crystal (propagation along the x-axis in **Figure 1**). In this chapter, we will focus on lengthwise periodic structures since lengthwise structures can be easily integrated with sub-micrometer

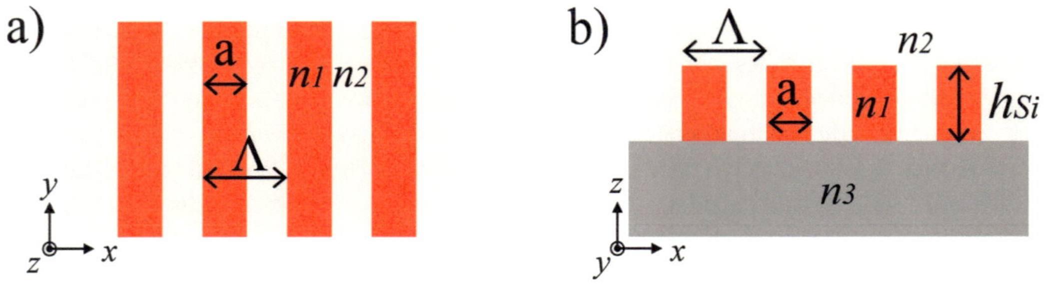

Figure 1. (a) Top view and (b) cross section of a 1D photonic crystal.

silicon waveguides and provide more flexibilities than that of crosswise periodic structures. Nowadays, extensive review papers and books are available for fully understanding the specific electromagnetic properties of the periodic structures considered here [16]. However, to provide intuitive guidelines for the design process, a much simpler theory will be used in this chapter.

A lengthwise 1D photonic crystal shown in **Figure 1** generally operates in the following three regimes, depending on the ratio between the structure's pitch (Λ) and the operating free-space wavelength (λ):

i. Diffraction regime. The incoming beam is scattered in different orders.

ii. Bragg reflection regime. The incoming beam is reflected backward.

iii. Sub-wavelength regime. The diffraction and reflection effects due to the periodicity of the structure are suppressed.

Figure 2 shows a schematic k–ω diagram of a 1D periodic structure with lengthwise propagation (along the x-axis) [16]. One can see that, for a given pitch periodic structure, the working regime is strongly related to free-space operating wavelength λ or operating frequency ω. When $\omega > \omega_2$ (above the first photonic bandgap), the waveguide becomes lossy for Bloch mode, and the light will be radiated out of the waveguide. This character has been utilized to design fiber-to-chip surface couplers (grating couplers). In the frequency range of $\omega_2 > \omega > \omega_1$ (the first photonic bandgap), light cannot propagate through the periodic structure and is reflected, and this is Bragg reflection regime. The propagation constant in this regime is

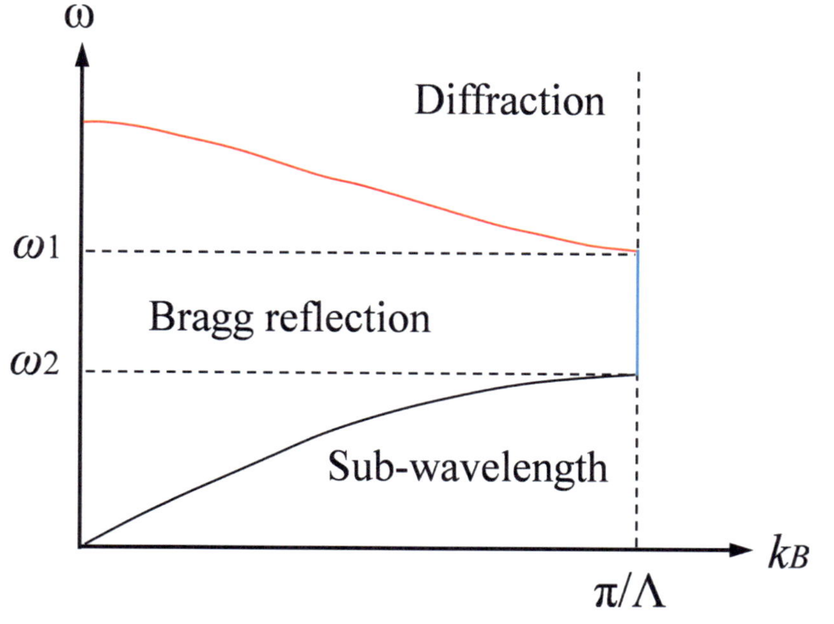

Figure 2. The schematic band diagrams a 1D photonic crystals with lengthwise propagation (along the x-axis).

constant: $k_B = \pi/\Lambda$. The bandgap has been extensively exploited to design distributed Bragg reflectors (DBRs) on different photonic platforms.

The last regime is located below the first photonic bandgap where the operating frequency is smaller than ω_1. From **Figure 2**, one can find that the propagation constant k_B grows monotonically as operating frequency increases, indicating that the periodic waveguide behaves as a conventional waveguide. Thus, in sub-wavelength regime, periodic structures work as homogeneous media. Consider a general case, light propagates (along the x-axis) through the structure (shown in **Figure 1**), with a linear polarization either parallel ($E_{\parallel}$) or perpendicular ($E_{\perp}$) to the interfaces between the two media. The equivalent effective index of the structure can be expressed with the zeroth-order approximation [17]:

$$n_{\parallel}^2 = \frac{w}{\Lambda}n_1^2 + \frac{\Lambda - w}{\Lambda}n_2^2 \tag{1}$$

$$\frac{1}{n_{\perp}^2} = \frac{w}{\Lambda}\frac{1}{n_1^2} + \frac{\Lambda - w}{\Lambda}\frac{1}{n_2^2} \tag{2}$$

where w/Λ is duty cycle of the structure. Thus, both effective index and effective birefringence can be engineered by adjusting the geometry of the structure. This allows for completely new design approaches and can be exploited in a variety of devices. Eqs. (1) and (2) also apply to crosswise structures.

3. Applications of diffraction regime

For the sake of clarity, the discussion on the applications of 1D photonic crystals in silicon photonics is divided into three parts. Hereinafter, we will use the general expression "diffractive gratings" and "sub-wavelength gratings (SWG)" to indicate the 1D photonic crystals operating in diffraction regime and in sub-wavelength regime, respectively. This section is devoted to applications of the diffraction regime. Applications of sub-wavelength regime and Bragg reflection regime are discussed in Sections 4 and 5, respectively. This section focuses on the key components in silicon photonics: grating couplers. To further improve the performance, combination of diffractive gratings and crosswise sub-wavelength structures is also presented.

A grating coupler is a 1D periodic structure that can diffract light from propagation in the waveguide (in plane) to free space (out of plane). By placing an optical fiber above the chip, part of the radiated light is collected. It is normally used as an I/O device to couple light between fiber and sub-micrometer silicon waveguides. And, grating couplers are defined lithographically and can be placed anywhere on the chip surface to enable inputs and outputs, which are particularly useful for massive production. **Figure 3** shows the cross section view of a shallow-etched grating coupler on a SOI wafer. The thickness of the core silicon layer and the thickness of the buried oxide (BOX) layer are determined by the wafer type. Normally, the thickness of core silicon layer (h_{Si}) is 150–300 nm. The cladding material is usually air ($n_2 = 1$), silicon dioxide, or an index-matching liquid ($n_2 \sim 1.45$).

http://dx.doi.org/10.5772/intechopen.71753

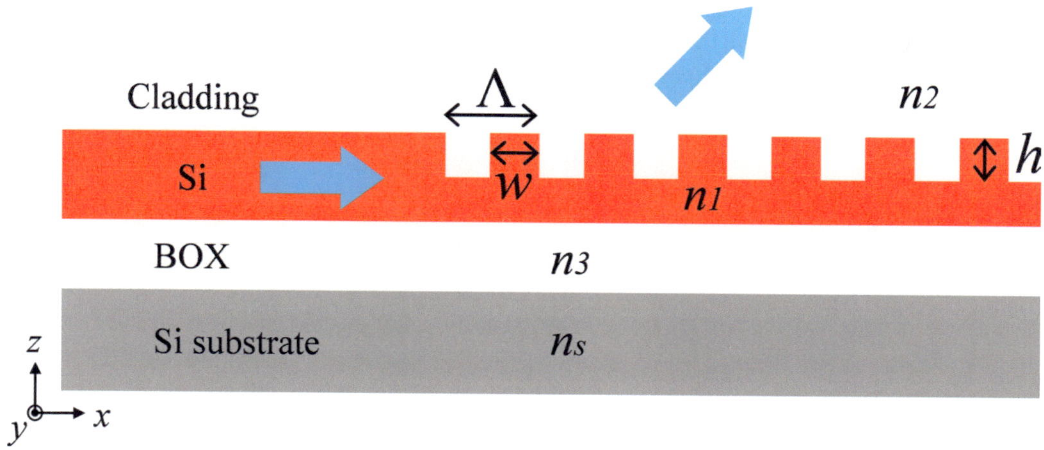

Figure 3. The cross section view of a shallow-etched grating coupler.

Figure 4 illustrates the concept of a grating coupler working as an output coupler. The operation of a grating coupler can be understood by the constructive and destructive interference arising from the wave fronts generated by the diffraction of light from the grating strips [18, 19]. First, we can check **Figure 4a**. In this case, the operating wavelength matches the pitch of the gratings. The first-order diffraction will vertically propagate, and the second-order diffraction will be back to the waveguide. However, the back-traveling light will be considered as reflection for the silicon waveguide, which is undesired since the light path between two grating couplers can form a Fabry-Perot oscillation cavity. This oscillation will modulate the transmission of other components. One efficient way to address this issue is detuning the grating and tilting the fiber (see **Figure 4b**) with a small angle to the grating surface. When the wavelength inside waveguide is smaller than the pitch of the gratings, the diffracted light waves will propagate at an angle, and reflection caused by second-order diffraction will be strongly suppressed.

The grating coupler could be well described by Bragg Law. If the grating parameters are fixed as constants (such as pitch and refractive indices of layers), then the radiation angle is given by the equation [18]:

$$n_2 \sin\theta_k = n_{eff} + \frac{k\lambda}{\Lambda} \tag{3}$$

where n_2 is the refractive index of the cladding material (see **Figure 3**) and k is an integer which represents the diffraction order. In some literatures, the effective index n_{eff} is substituted by Bloch-Floquet mode index (n_B). The main reason is that the relatively short light extraction length between a fiber and a grating coupler generally causes a strong perturbation in the waveguide. But for preliminary design process, using effective index is enough to qualitatively understand the property. In Ref. [18], only when $\sin\theta_k$ is real, the gratings diffract light from the waveguide. Consequently, we can play with the pitch of gratings so that only the order $k = -1$ occurs or dominates.

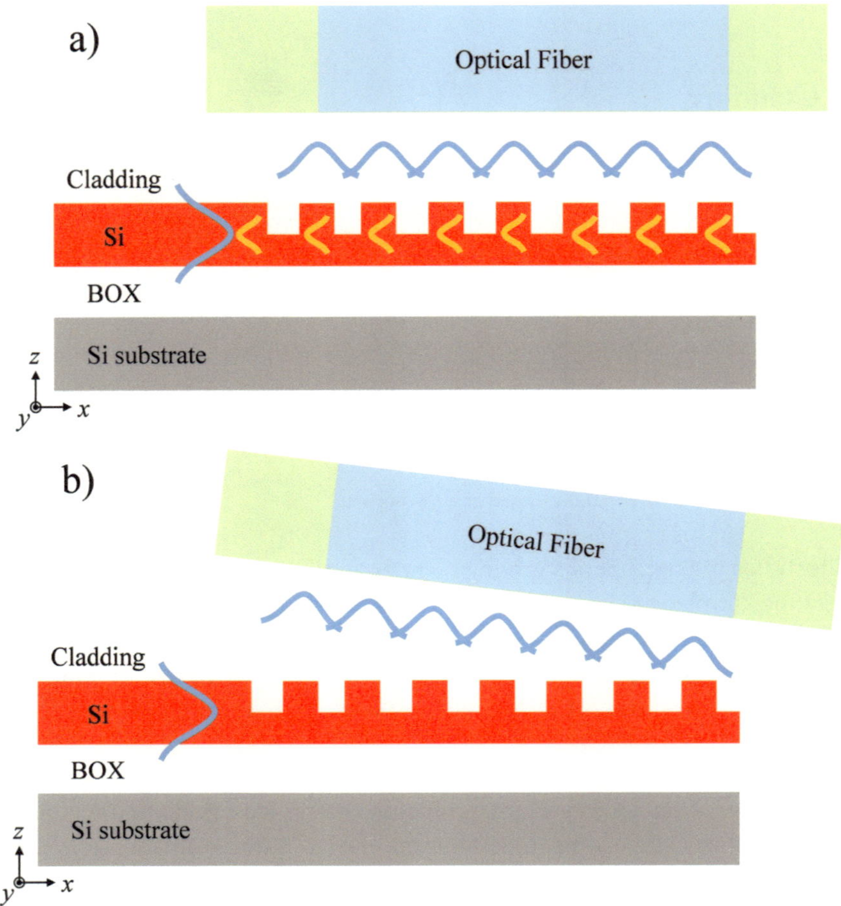

Figure 4. Diagrams describing the concept of a grating coupler working as an output coupler. (a) Vertical first-order diffraction and backward second-order diffraction and (b) tilted first-order diffraction without second-order diffraction.

A straightforward way to realize a grating coupler is using straight gratings connected to a taper to convert the fiber mode to a narrow waveguide mode. To obtain high conversion efficiency, the taper length needs to be more than 100 μm [19–21], which is undesirable for compact and high-density integration schemes. Alternatively, confocal gratings enable more compact designs. The whole grating is shaped to an ellipse with a common focal point, which coincides with the optical focusing point where the single-mode silicon waveguide is connected [22, 23].

One vital property of grating couplers is coupling efficiency. There are three main factors that contribute to a reduced efficiency of a grating coupler:

i. Penetration loss, that is, the fraction of waveguide power that escapes into the substrate. For a shallow-etch structure, there is about 30% of power lost in the substrate; and the penetration loss can be more than 50% in a full-etch structure [19]. This can be improved by using reflectors imbedded in the substrate [24–26]. As a result, the upward radiated

http://dx.doi.org/10.5772/intechopen.71753

optical power is improved, and radiation downward the substrate is reduced. Those reflectors can be a metal layer or multiple-layer-distributed Bragg reflector.

ii. Back reflection, that is, the optical power reflected from the coupler into the waveguide. For a well-designed shallow-etched grating coupler, the back reflection is less than -20 dB, so this does not contribute a significant loss to the coupler. However, this loss can be much higher in a full-etched grating coupler. Coupling the light under a small angle also helps to eliminate the first-order Bragg reflection, which is another reason why grating couplers are often designed with a coupling angle.

iii. Mode mismatch, that is, the overlap integral between the diffracted light beam and the near-Gaussian optical fiber mode [27]. In theory, the overlap integral is determined by the fraction of waveguide power diffracted from each pitch of the gratings. Approximately, this fraction is in direct proportion to refractive index difference between the strips of the gratings and the grooves. Therefore, in comparison with full-etched structures, a shallow-etched structure possesses a lower index difference which produces a relative wide radiated light beam. Consequently, conventional grating couplers with shallow-etched gratings help enhancing the field overlap.

It is should be noted that the power loss caused by mode mismatch can be further improved by apodizing or chirping the gratings [20, 28, 29] in which the index difference is varied along the structure to obtain a radiated beam that resembles the optical fiber mode. Duty cycle varied crosswise SWG structures can also enable the design of apodized grating couplers fabricated by a single, full-etch step. Consider a crosswise SWG structure enabled grating coupler shown in **Figure 5**. x direction is the in plane propagation direction of a conventional grating coupler, and the original grooves are occupied by crosswise SWG structures acting as artificial homogeneous media. The equivalent index of SWG can also be approximately estimated by Eqs. (1) and (2). Thereby, a grating can be apodized by varying the duty cycle (w_y/Λ_y) of the SWG structure located in each groove. However, the introduction of crosswise SWG structures also causes a variation of the radiation angle θ along the grating. To compensate

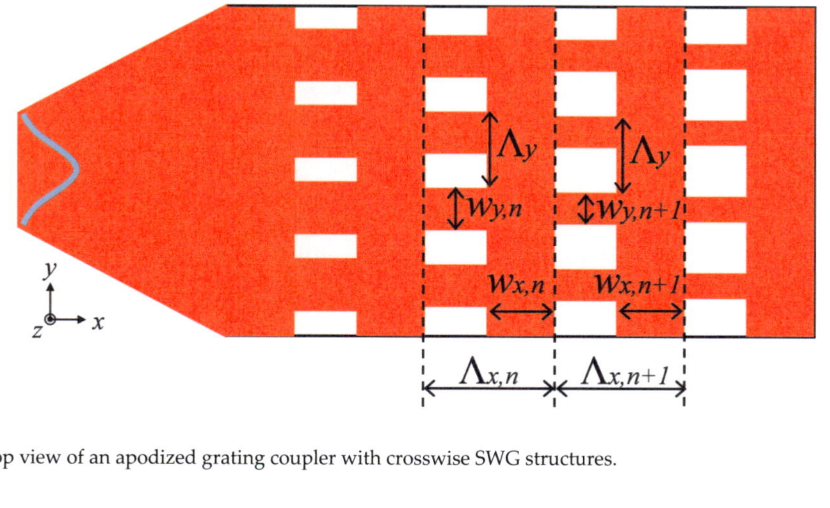

Figure 5. Top view of an apodized grating coupler with crosswise SWG structures.

this variation, chirping the grating along x direction is necessary. An apodized grating coupler with focusing design exhibiting a coupling efficiency of −1.8 dB was reported in Ref. [30].

The combination of diffractive gratings and crosswise SWG structures is also exploited to obtain polarization-independent grating couplers [31, 32]. Conventional grating couplers are highly polarization sensitive, which can be used as polarization filters or polarizers. The effective index in the grating region is highly birefringent, causing different radiation angles for TE and TM polarization states. When we consider a grating coupler with SWG structures shown in **Figure 5**, the effective index of the TE (z-polarized) mode will be much larger than that of the TM (y-polarized) mode in the silicon slab regions. However, in the sub-wavelength regions, from Eqs. (1) and (2), it was found that the effective index of the TE mode $n_{\perp}$ is smaller than that the TM mode $n_{\parallel}$. It is possible to design grating region so that these two effects cancel each other resulting in an identical Bloch mode effective index for both polarization states. Consequently, the grating coupler becomes polarization independent. Ref. [32] demonstrated a similar design possessing a coupling efficiency of −6.5 dB for both polarization states.

Broadband grating couplers with SWG structures were also proposed and demonstrated. A grating coupler with a bandwidth of ~100 nm and a coupling efficiency of −5.6 dB was presented in Ref. [33]. A simulated bandwidth of ~300 nm and a coupling efficiency of −2.8 dB were obtained by employing fully etched triangular holes in Ref. [34].

4. Applications of sub-wavelength regime

This section further presents applications of lengthwise SWG structures in which light propagates perpendicularly to the interfaces of the structure (along the x direction in **Figure 1**). As mentioned in Section 2, it is possible to engineer both effective index and effective birefringence which can be exploited in many silicon photonic devices for optical bandwidth and polarization-sensitivity improvement. In this section, we first discuss the mode convertor between SWG-based waveguides and conventional waveguides, which is the basic component for integration of these two kinds of technologies. Then, we move to some other key components such as waveguide crossings, multimode interference (MMI) couplers, and wideband and polarization-independent directional couplers.

4.1. Mode converters

To fully utilize the properties afforded by SWG-based waveguides and other components, integration and interconnect with conventional waveguides are needed. There are two key points to design such mode converters: low loss and low reflection. To attain both targets, several works have been done in theories and experiments [35, 36]. **Figure 6** shows the schematic diagram of a mode converter which contains three parts: conventional waveguide section, transition section between conventional waveguide and SWG-based waveguide, and SWG-based waveguide section. The transition section determines the performance of this mode converter; a clear approach is chirping the pitch and duty cycle and incorporating bridging

http://dx.doi.org/10.5772/intechopen.71753

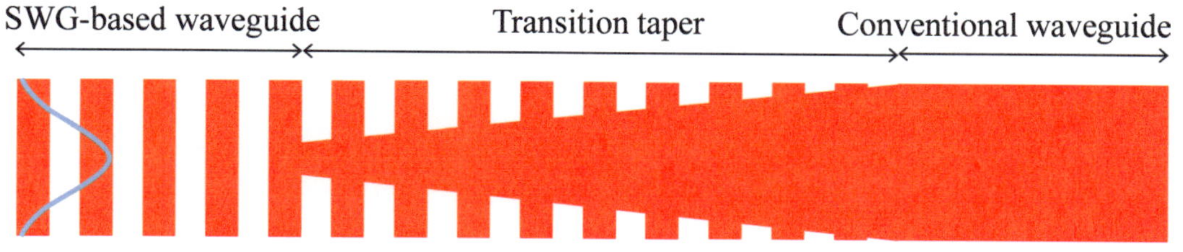

Figure 6. Top view of a SWG-based mode converter with section indications.

elements between the two waveguide sections [36]. Similar to conventional inverse taper, the mode effective index in transition section also gradually changes along propagation; thus, particular care should be taken for avoiding Bragg reflection regime within target operating optical bandwidth.

Mode converters can also be used for fiber-to-chip coupling [35, 37] and coupling between continuous fields and discretized fields [38]. For fiber-to-chip coupling applications, mode converter is carefully designed to match the on-chip waveguide mode size, shape, and effective index to that of a single-mode fiber, thereby increasing coupling efficiency. The insertion loss of such mode converter was extracted from experiments: −0.23 dB for TE mode and −0.47 dB for TM mode [38]. The application for coupling between continuous fields and discretized fields can be found in arrayed waveguide grating (AWG). In an AWG, the main loss and backward reflection are from transition between the slab waveguide (where continuous fields are from) and the array waveguides (where discretized fields appear). The loss is from the gap between arrayed waveguides, and reflection is caused by high-index difference between slab region and free propagation region. By placing well-designed SWG structures between arrayed waveguides, the index difference and fields coupling can be further optimized. Ref. [39] shows some experimental results about AWGs with SWG-based mode converters.

4.2. Waveguide crossings

An efficient waveguide crossing is highly desired to materialize the full potential of silicon photonics for on-chip optical interconnects. Due to the high-index contrast of the silicon platform such as SOI, the insertion loss of a conventional waveguide crossing is around 0.15 dB. To reduce optical loss, waveguide crossing designed by using particle swarm optimization has been proposed and demonstrated, with a loss of −0.028 ± 0.009 dB for 1550 nm operating wavelength [40]. In this part, we introduce some waveguide crossing designs assisted by SWG structures. The experimentally confirmed insertion loss is comparable with or even lower than that of particle swarm-optimized designs. A further advantage of using SWG structures is improving optical bandwidth and polarization sensitivity.

First, we go through some fundamentals of conventional waveguide crossings. **Figure 7a** gives the top view of a conventional waveguide crossing, and cross section of the multimode waveguide region is shown in **Figure 7b**. A conventional design is composed of MMI regions, single-mode access waveguides, and a crossing section. It was found that the intrinsic loss of such design is attributed to three aspects:

i. Phase error in the MMI region. When wide multimode waveguides are utilized, it is possible to find more than two modes excited in MMI regions. Some phase errors will appear, and the perfect self-imaging position cannot be found by adjusting MMI length. To avoid exciting high-order modes, narrow multimode waveguides supporting three modes (fundamental, first order, and second order) are applicable. Among these modes, only fundamental and second-order modes are excited in an symmetric interference MMI structure. Then, the self-imaging length is only related to two modes, and high-order mode phase errors could be thoroughly eliminated.

ii. Loss caused by sharp transitions between single-mode waveguides and MMI regions. As suggested by [41], a linear or nonlinear taper can be applied to reduce the transition loss. It is worth to underline that the taper also affects the power portion in the second-order mode in MMI regions [42]. Through simulation, we can find the optimal power ratio between fundamental mode and second-order mode. If the power ratio is too small, the MMIs even lose the capability of focusing light beam at the center of the crossing section, causing mode mismatch between the diverged beam after passing through the crossing section and the guided modes of the multimode waveguide. When the power ratio is larger than the optimal value, a large amount of the second-order mode possessing a wide angular spectrum scatters through the crossing section, and some of it radiates into the orthogonal MMI regions, resulting in substantial loss and crosstalk. Consequently, when choosing the length of the taper, a trade-off between modal transition loss and two-mode power ratio needs to be found.

iii. Mode mismatching between multimode waveguides and the crossing section. The crossing section is much wider than the multimode waveguide width, which can be considered as a slab waveguide that supports pure TE modes. However, the mode in the single-mode access waveguide and the second-order mode in the multimode waveguide are both quasi-TE modes with a considerable amount of TM polarization. Then, the TM mode power will be dissipated in the crossing section. It is found that the power portion of TM polarization can be effectively suppressed by increasing the lateral cladding index (n_{lc}) in MMI regions. One straightforward way is depositing a high-index material such

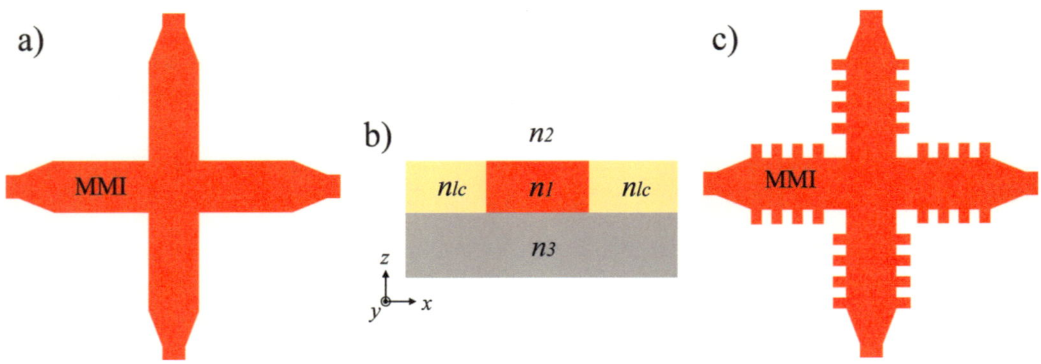

Figure 7. (a) Top view of a conventional waveguide crossing, (b) cross section view of MMI sections, and (c) top view of a SWG-assisted waveguide crossing design.

as Si_xN_{1-x}. Also, using a tightly controlled shallow-etched structure is feasible. Both approaches complicate the fabrication process.

In Ref. [42], researchers developed a new waveguide crossing assisted by SWG structures to engineer the lateral cladding index, simplifying fabrication procedures. **Figure 7c** gives the schematic of such design. A 101 × 101 MMI crossing matrix was demonstrated on the SOI platform. An insertion loss of 0.019 dB and a crosstalk lower than −40 dB at 1550 nm operating wavelength were obtained for each crossing. It is worth to note that the optical bandwidth (90 nm) is wider than that of conventional designs.

Figure 8 shows another waveguide crossing design that lies on SWG-based waveguides [43]. In comparison with the designs shown in **Figure 7a** and **7c** this waveguide crossing has no multimode waveguides. At the center of the crossing, a squared segment is used to enable a symmetrical structure. Loss per crossing was measured as 0.023 dB with polarization dependent loss of <0.02 dB and crosstalk below −40 dB [43]. An improvement from this design is polarization insensitivity. More discussions can be found in Ref. [43].

4.3. MMI couplers

MMI couplers are widely used as basic building blocks in many advanced photonic devices including MZ modulators [10], polarization-handling devices [44], mode handling devices [45, 46], etc. Before reviewing some high-performance MMI coupler assisted by SWG structures, we will briefly recall the fundamental limitations of conventional MMI couplers.

A MMI coupler consists of single-mode in−/output waveguides and a multimode waveguide section, as shown in **Figure 9**. The input light beam excites multiple modes in the central section with different excitation coefficients. The coefficients are determined by input mode profile and the structure of the central section (MMI section). Those modes propagate through the central section with different propagation constants (β_m) and interfere with each other to form the N-fold self-imaging of the input mode [47]. Placing output waveguides at the imaging position, beam splitting and combining function can be achieved. The MMI or central section length (L_{MMI}) has a certain relationship with beat length (L_π) of the two lowest excited modes in the central section. The beat length at operating wavelength is given by

$$L_\pi = \frac{\pi}{\beta_1(\lambda) - \beta_2(\lambda)} \tag{4}$$

Figure 8. Top view of a SWG-based waveguide crossing.

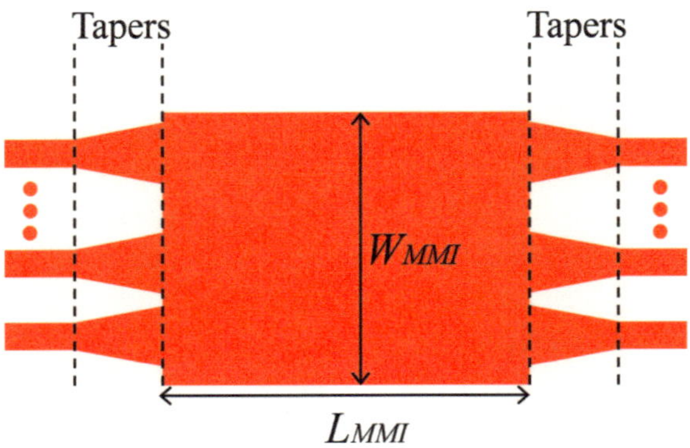

Figure 9. Schematic of a MMI coupler with indications.

Furthermore, to obtain clear images for low loss and power imbalance of MMI couplers, the propagation constant of each excited mode needs to meet the following relation:

$$\beta_m = \beta_1 - \frac{(m^2 - 1)}{3 L_\pi} \pi \tag{5}$$

where m is the mode order and β_1 is the propagation constant of the lowest-order mode. Unfortunately, this condition is only satisfied for the lower order modes, which causes MMI's phase error and excess loss. An obvious approach is to suppress excitations of higher order modes, which can be done by increasing modal width of input mode and reducing the number of modes that MMI supports. Inserting tapered input waveguides between single-mode waveguides and the central section (as shown in **Figure 9**) in return increases modal width ratio between input mode and the excited fundamental mode, resulting in a limited number of (the lower order) modes excited. About reducing supported modes in the central section, reducing the width of MMI section (same effect as that of the use of tapered input) and increasing lateral cladding index (decrease core-cladding refractive index contrast) are feasible methods.

In Eq. (4), the beat length is described only for a certain wavelength. We rewrite the equation as

$$L_\pi = \frac{\lambda}{2\left(n_1(\lambda) - n_2(\lambda)\right)} \tag{6}$$

It can be seen that the beat length decreases with increasing wavelength. This is the major limitation of the bandwidth of a MMI coupler. To achieve broadband MMI couplers, we need to find a way to suppress high-order modes excited in MMI section and keep $\left(n_1(\lambda) - n_2(\lambda)\right)$ increasing as wavelength increases.

Now, we review two high-performance MMI couplers assisted by SWG structures. The first case is 4 × 4 MMI coupler serving as a 90° hybrid for coherent detections. One vital parameter of this MMI coupler is phase error. As discussed before, the use of input taper is a common practice to obtain small phase errors. But for SOI platform, only using input taper is not enough to meet the target since the large core-cladding index contrast in SOI waveguides leads higher-order modes to not properly fulfill Eq. (4). The large lateral index contrast can be tailored by an (some) extra shallow-etch step(s) to define shallow-etched multimode region. This was experimentally confirmed as an efficient way [48]. However, extra etch steps and tightly controlled etching depth are required, complicating the fabrication process and not suitable for massive production. Instead of introducing shallow-etch process, laterally integrating SWG structures also allows for lateral cladding index engineering with single full-etch step. The design process is straightforward: (1) calculate the equivalent effective index of SWG structures under different pitch and duty cycles; (2) sweep the cladding index of a 4 × 4 MMI coupler to find an optimum value; (3) use the data calculated in Step 1 as a lookup table and, then, geometry parameters of the sub-wavelength structure that yields the desired value of lateral cladding index can be found; and (4) use FDTD to further verify the structure. The schematic diagram is depicted in **Figure 10a**. Simulated common mode rejection ratio and phase error of less than 24 dBe and 2°, respectively, are presented in Ref. [49]. This technology can be also exploited in any $N \times N$ MMI coupler.

The second case is broadband 2 × 2 MMI coupler. In practices, broadband MMI couplers are demanded to simplify system and reduce cost. Similar to the above case, optimization of tapered input broadens the bandwidth. In Ref. [50], researchers proposed a broadband 2 × 2 MMI coupler in which the conventional multimode section is replaced with a length-wise SWG-based waveguide of 6.0 μm width, 198.0 nm pitch, and 50% duty cycle. The schematic diagram is shown in **Figure 10b**. To fully understand how the SWG structures can be exploited to significantly increase the bandwidth of MMI couplers, it is useful to replot the sub-wavelength regime in **Figure 2** in an equivalent way as shown in **Figure 11**. One can see that the sub-wavelength regime can be divided into two regions: high dispersion region (λ is close to Bragg wavelength) and low dispersion region ($\lambda >> \Lambda$). The exciting part is that the dispersion value is adjustable through changing pitch, which can be utilized to design broadband MMI couplers. In results, the bandwidth of the MMI coupler in **Figure 10b** is determined by the pitch. To connect with conventional waveguide structures, mode converters similar as the one described in Section 4.1 were placed at in−/output ports as shown in **Figure 10b**. The mode converters also help to expand mode field to excite only lower-order modes. Due to the relatively low equivalent refractive index of the SWG-based multimode region, the MMI length is also shortened in return. Additionally, this technology can be applied to design broadband directional couplers [51].

a) b)

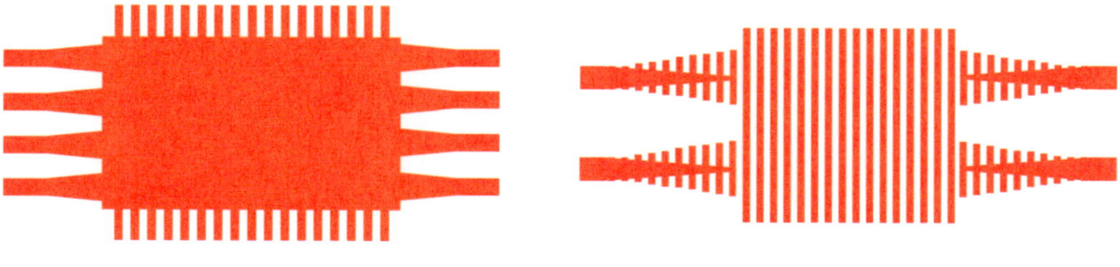

Figure 10. Schematics of MMI couplers: (a) a 4 × 4 MMI coupler with lateral SWG structures and (b) a 2 × 2 MMI couplers with SWG-based multimode section.

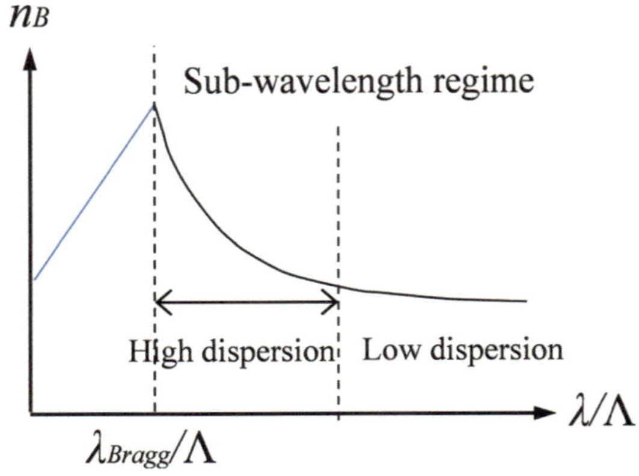

Figure 11. Dispersion in sub-wavelength regime.

Figure 12. Schematic of a polarization-independent directional coupler based on slot waveguides with SWG structures.

4.4. Polarization-independent directional couplers

With its superior performances, directional coupler (DC) has attracted considerable attentions and been widely used in many applications, including on-chip sensors, filters, switches, and polarization beam splitters. In comparison with other couplers, such as MMI couplers and Y-branch couplers, DCs process valuable properties in terms of ultra-low loss, low reflection, and arbitrary power coupling ratio obtained by adjusting the length and gap of the coupling region. However, by using sub-micrometer SOI waveguides, the beat length of TE mode greatly differs from that of TM mode due to the high birefringence of silicon waveguides, resulting in power ratio sensitivity to polarization states. Polarization-independent DC is useful for some applications where the amount of collected or split optical power is concerned. Some techniques have been proposed and experimentally confirmed to address this issue. One solution is to use polarization diversity schemes where polarization splitting and rotating devices are needed, which increases system size and complexity. Another approach is to make the directional coupler inherently polarization insensitive [52–54].

Recently, a SWG-assisted slot waveguide polarization-independent DC with relatively large fabrication tolerance has been demonstrated on SOI chip [55]. **Figure 12** describes the schematic diagram of this design. The SWG structures are extended into the gap between two parallel slot waveguides. The SWG structures are considered as homogenous media. By calculating the relation between beat length of both polarization states and refractive index of SWG structures (n_k), it was found that the two curves are of different slopes. By adjusting the duty cycle, if the effective indices of the TE mode $n_\perp$ and TM mode $n_\parallel$ in SWG structures meet the n_{kTE} and n_{kTM} for same beat length for TE and TM modes, then the optimum duty cycle is attained. FDTD simulations for further confirmation are necessary. The measured coupling efficiency is 97.4 and 96.7% for TE and TM modes, respectively, at a wavelength of 1550 nm. It is worth noting that an additional advantage of this design is its wide bandwidth over 120 nm (1475–1595 nm) theoretically and exceeding the entire C-band (1525–1570 nm) experimentally. To interconnect such design with a conventional waveguide, a slot-to-strip waveguide converter is required.

5. Applications of Bragg reflection regime

This section is devoted to applications of Bragg reflection regime of 1D photonic crystals. For the sake of clarity (unless otherwise stated), hereinafter we will use the general expression "Bragg gratings" to indicate the 1D photonic crystals operating in Bragg reflection regime as shown in **Figure 2**. One prominent feature of Bragg gratings is the linear relation $n_{Bragg} \sim \lambda/\Lambda$ within the photonic bandgap ($\omega_1 > \omega > \omega_2$). The bandgap has been extensively exploited to design DBRs on different photonic platforms as key elements of DBR lasers [56, 57], distributed feedback (DFB) lasers [58, 59], fiber Bragg gratings (FBGs) [60], VCSELs [61], etc. On silicon platform, silicon waveguide-based Bragg gratings have been integrated with III–V gain elements to provide optical feedback for lasers on a silicon chip.

This section starts with a brief discussion of Bragg gratings serving as mirrors in hybrid lasers on a silicon chip (Section 5.1). Section 5.2 is dedicated to Bragg gratings designed as a polarizer and higher-order mode pass filter.

5.1. Hybrid lasers with Bragg gratings

The realization of an efficient, reliable, and electrically pumped laser on a silicon wafer is still a scientific challenge. In Ref. [15], researchers first demonstrated a high-performance electrically pumped quantum-dot laser through epitaxial growth of III–V materials on a silicon substrate. Another common-practice approach to realize light source on a silicon wafer is wafer bonding. Several bonding process schemes have been proposed [9, 62–66]. To fabricate a single-wavelength laser and multiwavelength laser array, prefabricated single-wavelength laser diodes can be directly bonded on a silicon wafer [62]. Alternatively, embed III–V gain sections to silicon photonic circuits and the wavelength-selections are done by some silicon-based wavelength-sensitive devices such as ring resonator, slotted feedback structure, and Bragg reflector. In Ref. [66], a four-channel multiwavelength DFB evanescent laser array was designed and fabricated. Beneath each III–V gain section, Bragg gratings were fabricated on the surface of the silicon waveguide. The ASE from a gain section propagates and evanesces into a silicon waveguide and is modulated by gratings, forming a DFB laser. For realizing different channel wavelengths, the width of silicon waveguide is varied to obtain different Bragg propagation constants.

5.2. TM-pass polarizers

Photonic devices based on SOI waveguides always suffer from severe polarization-sensitivity problems due to high birefringence, which greatly limits their application range. Various polarization-handling devices have been attracting attentions, including polarization beam splitters, polarization rotators [67], and polarizers. Among them, a polarizer is used to achieve linearly polarized light with a high extinction ratio. In practices, a low loss, high extinction ration, and compact footprint polarizer is highly desired. Many designs have been proposed and demonstrated on silicon wafers [68–70]. The basic rationale of designing a polarizer is to lose or filter unexpected polarization state out from the optical propagation paths with negligible effect on desired polarization state and maintain the device size as small as possible.

In Ref. [71], a waveguide-based TM-pass polarizer is fabricated with 1D photonic crystals. The schematic of this design is depicted in **Figure 13**. The device is composed of three parts: in-/output waveguides, 1D photonic crystal with teeth and bridges, and transition tapers. The pitch was well designed to make the waveguide supporting Bloch mode for TM polarization state, so that the incident TM-polarized light goes through the waveguide with very low excess loss. On the other hand, for TE polarization state, the waveguide works as a Bragg reflector, and consequently the incident TE-polarized light is reflected with very high efficiency. Therefore, the following conditions should be satisfied approximately [71]:

$$n_b^{TE}(\Lambda - w) + n_{ac}^{TE} w = \lambda/2 \tag{7}$$

$$n_b^{TM}(\Lambda - w) + n_{ac}^{TM} w < \frac{\lambda}{2} \tag{8}$$

where n_b and n_{ac} are the effective indices of the TE or TM modes in the narrow and wide sections, respectively. Intrinsically, for a strip waveguide well-supported TE mode, the effective index of TM mode is smaller than that of TE mode, and thereby once Eq. (1) is satisfied, Eq. (2) is also satisfied. Thus, the design process flow can be as follows: (1) use Bragg reflection regime condition to roughly estimate parameters of the photonic crystal; (2) simulate the k–ω diagrams for both TE and TM modes by the use of 3D FDTD; and (3) calculate the transmission responses for both TE and TM modes to further check the design. In Ref. [71], a measured extinction ratio of 40 dB was obtained with a period number of 40.

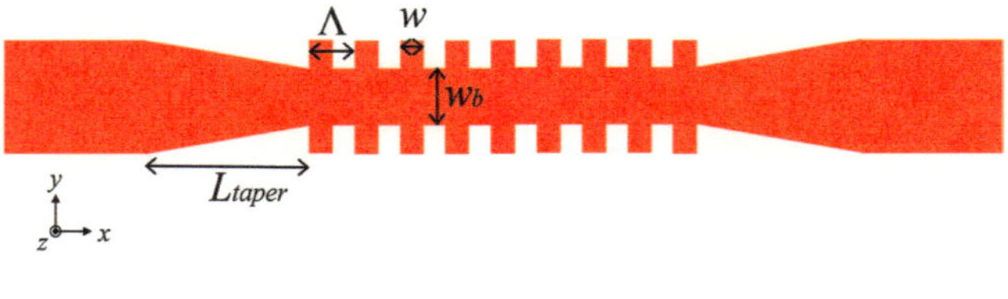

Figure 13. Schematic of a TM-pass polarizer assisted by 1D photonic crystals.

Adapting the same mechanism discussed above, higher-order mode pass filters have been demonstrated on a silicon wafer in Ref. [72], which are building blocks for mode-division multiplexing (MDM) systems. A 15.0-μm-long TE1 mode pass filter exhibits an extinction ratio of ~48 dB and an insertion loss of ~1.8 dB at 1550 nm.

6. Conclusion

1D photonic crystals have been well developed in silicon photonics and other material platforms as well. Three operating regimes allow 1D photonic crystals to be exploited in many different functional photonic devices with compact footprints. In diffraction regime, light beam can be diffracted into different directions by controlling configuration of photonic crystals in terms of pitch, duty cycle and effective index. In sub-wavelength regime, material refractive index, birefringence, and dispersion can be simply controlled by lithographic patterning to enable new design methods that lead to high-performance photonic devices. Propagation constant is proportional to operating wavelength and effective index in Bragg reflection regime. Thus, many wavelength-, mode-, and polarization-sensitive devices have been demonstrated with the assistance of Bragg gratings. In all these three regimes, when the operating wavelength range is fixed, the sub-wavelength gratings require tightest lithography process control that is high patterning resolution and is the most versatile one. Normally, sub-wavelength gratings are fabricated by electron-beam lithography which is not only expensive but also not suitable for mass production. And, the patterning resolution of DUV stepper lithography widely used in CMOS technology is not enough for accurately controlling sub-wavelength gratings. Therefore, continuing improvement in advanced lithography techniques is one important way to realize large-volume integration of sub-wavelength gratings with other electronic and photonic devices.

Author details

Liangshun Han

Address all correspondence to: liangshunhan@gmail.com

University of California, San Diego, USA

References

[1] Pantouvaki M et al. 50Gb/s Silicon Photonics Platform for Short-Reach Optical Interconnects, in Optical Fiber Communication Conference, OSA Technical Digest (online) (Optical Society of America, 2016), paper Th4H.4

[2] Heck MJR, Bowers JE. Energy efficient and energy proportional optical interconnects for multi-Core processors: Driving the need for on-Chip sources. IEEE Journal of Selected Topics in Quantum Electronics. Jul 2014;**20**(4):332-343

[3] Yu R et al. A scalable silicon photonic chip-scale optical switch for high performance computing systems. Optics Express. Dec 2013;**21**(26):32655

[4] Yang T et al. Experimental observation of optical differentiation and optical Hilbert transformation using a single SOI microdisk chip. Scientific Reports. Feb 2014;**4**:3960

[5] Juan-Colás J, Parkin A, Dunn KE, Scullion MG, Krauss TF, Johnson SD. The electrophotonic silicon biosensor. Nature Communications. Sep 2016;**7**:12769

[6] De Vos K, Bartolozzi I, Schacht E, Bienstman P, Baets R. Silicon-on-insulator microring resonator for sensitive and label-free biosensing. Optics Express. 2007;**15**(12):7610

[7] Chen S et al. Electrically pumped continuous-wave III–V quantum dot lasers on silicon. Nature Photonics. Mar 2016;**10**(5):307-311

[8] Duan G-H et al. Hybrid III–V on silicon lasers for photonic integrated circuits on silicon. IEEE Journal of Selected Topics in Quantum Electronics. Jul 2014;**20**(4):158-170

[9] Huolei W, Kim D, Harfouche M, Satyan N, Rakuljic G, and Yariv A. Narrow-Linewidth Oxide-Confined Heterogeneously Integrated Si/III-V Semiconductor Laser, in Conference on Lasers and Electro-Optics, OSA Technical Digest (online) (Optical Society of America, 2017), paper AM2B.2

[10] Thomson DJ et al. 50-Gb/s silicon optical modulator. IEEE Photonics Technology Letters. Feb 2012;**24**(4):234-236

[11] Xiao X et al. High-speed, low-loss silicon Mach–Zehnder modulators with doping optimization. Optics Express. Feb 2013;**21**(4):4116

[12] Novack A et al. Germanium photodetector with 60 GHz bandwidth using inductive gain peaking. Optics Express. Nov 2013;**21**(23):28387

[13] Lu L et al. 16 × 16 non-blocking silicon optical switch based on electro-optic Mach-Zehnder interferometers. Optics Express. May 2016;**24**(9):9295

[14] Tanizawa K et al. Ultra-compact 32 × 32 strictly-non-blocking Si-wire optical switch with fan-out LGA interposer. Optics Express. Jun 2015;**23**(13):17599

[15] Sun C et al. Single-chip microprocessor that communicates directly using light. Nature. Dec 2015;**528**(7583):534-538

[16] Joannopoulos JD, Johnson SG, Winn JN, Meade RD. Photonic Crystals: Molding the Flow of Light. Princeton, NJ: Princeton University Press; 2008

[17] Rytov SM. Electromagnetic properties of a finely stratified medium. Journal of Experimental and Theoretical Physics. 1956;**2**(3):466

[18] Tamir T, Peng ST. Analysis and design of grating couplers. Applied Physics. 1977;**14**(3): 235-254

[19] Chrostowski L, Hochberg M. Silicon Photonics Design. United Kingdom: Cambridge University Press; 2015

[20] Vermeulen D et al. High-efficiency fiber-to-chip grating couplers realized using an advanced CMOS-compatible silicon-on-insulator platform. Optics Express. Aug 2010; **18**(17):18278-18283

[21] Taillaert D et al. Grating couplers for coupling between optical fibers and Nanophotonic waveguides. Japanese Journal of Applied Physics. 2006;**45**(8R):6071

[22] Van Laere F et al. Compact focusing grating couplers for silicon-on-insulator integrated circuits. IEEE Photonics Technology Letters. 2007;**19**(23):1919-1921

[23] Waldhäusl R, Schnabel B, Dannberg P, Kley E-B, Bräuer A, Karthe W. Efficient coupling into polymer waveguides by gratings. Applied Optics. Dec 1997;**36**(36):9383-9390

[24] Van Laere F et al. Compact and highly efficient grating couplers between optical fiber and nanophotonic waveguides. Journal of Lightwave Technology. 2007;**25**(1):151-156

[25] Alonso-Ramos C, Zavargo-Peche L, Ortega-Moñux A, Halir R, Molina-Fernández I, Cheben P. Polarization-independent grating coupler for micrometric silicon rib waveguides. Optics Letters. Sep 2012;**37**(17):3663-3665

[26] Zaoui WS et al. Bridging the gap between optical fibers and silicon photonic integrated circuits. Optics Express. Jan 2014;**22**(2):1277-1286

[27] Taillaert D, Bienstman P, Baets R. Compact efficient broadband grating coupler for silicon-on-insulator waveguides. Optics Letters. Dec 2004;**29**(23):2749-2751

[28] Mekis A et al. A grating-coupler-enabled CMOS photonics platform. IEEE Journal of Selected Topics in Quantum Electronics. 2011;**17**(3):597-608

[29] Chen X, Li C, Fung CKY, Lo SMG, Tsang HK. Apodized waveguide grating couplers for efficient coupling to optical fibers. IEEE Photonics Technology Letters. 2010;**22**(15):1156-1158

[30] Ding Y, Ou H, Peucheret C. Ultrahigh-efficiency apodized grating coupler using fully etched photonic crystals. Optics Letters. 2013;**38**(15):2732-2734

[31] Chen X, Tsang HK. Polarization-independent grating couplers for silicon-on-insulator nanophotonic waveguides. Optics Letters. 2011;**36**(6):796-798

[32] Cheng Z, Tsang HK. Experimental demonstration of polarization-insensitive air-cladding grating couplers for silicon-on-insulator waveguides. Optics Letters. 2014; **39**(7):2206-2209

[33] Chen X, Xu K, Cheng Z, Fung CKY, Tsang HK. Wideband subwavelength gratings for coupling between silicon-on-insulator waveguides and optical fibers. Optics Letters. 2012;**37**(17):3483-3485

[34] Qin K et al. High efficiency and broadband two-dimensional blazed grating coupler with fully etched triangular holes. Journal of Lightwave Technology. 2012;**30**(14):2363-2366

[35] Cheben P et al. Refractive index engineering with subwavelength gratings for efficient microphotonic couplers and planar waveguide multiplexers. Optics Letters. 2010;**35**(15): 2526-2528

[36] Cheben P, Xu D-X, Janz S, Densmore A. Subwavelength waveguide grating for mode conversion and light coupling in integrated optics. Optics Express. 2006;**14**(11):4695-4702

[37] Almeida VR, Panepucci RR, Lipson M. Nanotaper for compact mode conversion. Optics Letters. 2003;**28**(15):1302-1304

[38] Bock PJ et al. Subwavelength grating periodic structures in silicon-on-insulator: A new type of microphotonic waveguide. Optics Express. 2010;**18**(19):20251-20262

[39] Bock PJ et al. Sub-wavelength grating mode transformers in silicon slab waveguides. Optics Express. 2009;**17**(21):19120-19133

[40] Ma Y et al. Ultralow loss single layer submicron silicon waveguide crossing for SOI optical interconnect. Optics Express. 2013;**21**(24):29374-29382

[41] Chen CH, Chiu CH. Taper-integrated multimode-interference based waveguide crossing design. IEEE Journal of Quantum Electronics. 2010;**46**(11):1656-1661

[42] Zhang Y, Hosseini A, Xu X, Kwong D, Chen RT. Ultralow-loss silicon waveguide crossing using Bloch modes in index-engineered cascaded multimode-interference couplers. Optics Letters. 2013;**38**(18):3608-3611

[43] Bock PJ et al. Subwavelength grating crossings for silicon wire waveguides. Optics Express. 2010;**18**(15):16146-16155

[44] Han L, Liang S, Zhu H, Zhang C, Wang W. A high extinction ratio polarization beam splitter with MMI couplers on InP substrate. IEEE Photonics Technology Letters. Apr 2015;**27**(7):782-785

[45] Han L, Liang S, Zhu H, Qiao L, Xu J, Wang W. Two-mode de/multiplexer based on multimode interference couplers with a tilted joint as phase shifter. Optics Letters. Feb 2015;**40**(4):518

[46] Han L, Liang S, Xu J, Qiao L, Zhu H, Wang W. Simultaneous wavelength- and mode-division (de)multiplexing for high-capacity on-Chip data transmission link. IEEE Photonics Journal. Apr 2016;**8**(2):1-10

[47] Soldano LB, Pennings ECM. Optical multi-mode interference devices based on self-imaging: Principles and applications. Journal of Lightwave Technology. 1995;**13**(4):615-627

[48] Halir R, Roelkens G, Ortega-Moñux A, Wangüemert-Pérez JG. High-performance 90° hybrid based on a silicon-on-insulator multimode interference coupler. Optics Letters. 2011;**36**(2):178-180

[49] Ortega-Monux A et al. High-performance multimode interference coupler in silicon waveguides with subwavelength structures. IEEE Photonics Technology Letters. 2011;**23**(19):1406-1408

[50] Maese-Novo A et al. Wavelength independent multimode interference coupler. Optics Express. 2013;**21**(6):7033-7040

[51] Halir R et al. Colorless directional coupler with dispersion engineered sub-wavelength structure. Optics Express. 2012;**20**(12):13470-13477

[52] Cheng N-C, Ma Y-F, Fu P-H, Chin C-C, Huang D-W. Horizontal slot waveguides for polarization branching control. Applied Optics. 2015;**54**(3):436-443

[53] Alam MZ, Aitchison JS, Mojahedi M. Polarization-independent hybrid plasmonic coupler for a silicon on insulator platform. Optics Letters. 2012;**37**(16):3417-3419

[54] Xiao J, Liu X, Sun X. Design of polarization-independent optical couplers composed of three parallel slot waveguides. Applied Optics. 2008;**47**(14):2687-2695

[55] Liu L, Deng Q, Zhou Z. Subwavelength-grating-assisted broadband polarization-independent directional coupler. Optics Letters. 2016;**41**(7):1648-1651

[56] Han L et al. DBR laser with over 20 nm wavelength tuning range. IEEE Photonics Technology Letters. 2016:1-1

[57] Han L et al. Electroabsorption-modulated widely tunable DBR laser transmitter for WDM-PONs. Optics Express. Dec 2014;**22**(24):30368

[58] Zhang C, Liang S, Zhu H, Han L, Wang W. Multichannel DFB laser arrays fabricated by upper SCH layer SAG technique. IEEE Journal of Quantum Electronics. Feb 2014;**50**(2): 92-97

[59] Hou L, Xu J, Eddie I, Han L, Zhu H, and Marsh J. DWDM Source Based on Monolithic Side-Wall Sample Grating DFB Laser Array, in Conference on Lasers and Electro-Optics, OSA Technical Digest (2016) (Optical Society of America, 2016), paper SW4M.1

[60] Kashyap R. Fiber Bragg Grating. San Diego: Academic Press; 1999

[61] Jewell JL, Harbison JP, Scherer A, Lee YH, Florez LT. Vertical-cavity surface-emitting lasers: Design, growth, fabrication, characterization. IEEE Journal of Quantum Electronics. 1991;**27**(6):1332-1346

[62] Roth JE, Palermo S, Helman NC, Bour DP, Miller DA, and Horowitz M. 1550nm Optical Interconnect Transceiver with Low Voltage Electroabsorption Modulators Flip-Chip Bonded to 90nm CMOS, in Optical Fiber Communication Conference and Exposition and The National Fiber Optic Engineers Conference, OSA Technical Digest Series (CD) (Optical Society of America, 2007), paper JThA38

[63] Kreissl J, Bornholdt C, Gaertner T, Moerl L, Przyrembel G, Rehbein W. 1550 nm flip-chip compatible electroabsorption-modulated laser with 40 Gb/s modulation capability. In: IPRM 2011 – 23rd International Conference on Indium Phosphide and Related Materials (IEEE, 2011); pp. 1-4

[64] Roelkens G, Brouckaert J, Van Thourhout D, Baets R, Nötzel R, Smit M. Adhesive bonding of InP/InGaAsP dies to processed silicon-on-insulator wafers using DVS-bis-Benzocyclobutene. Journal of the Electrochemical Society. 2006;**153**(12):G1015

[65] Liang D, Roelkens G, Baets R, Bowers JE. Hybrid integrated platforms for silicon photonics. Materials (Basel). 2010;**3**(3):1782-1802

[66] Tao L et al. 4-λ InGaAsP-Si distributed feedback evanescent lasers with varying silicon waveguide width. Optics Express. Mar 2014;**22**(5):5448

[67] Dai D, Bowers JE. Novel concept for ultracompact polarization splitter-rotator based on silicon nanowires. Optics Express. May 2011;**19**(11):10940

[68] Alam MZ, Aitchison JS, Mojahedi M. Compact and silicon-on-insulator-compatible hybrid plasmonic TE-pass polarizer. Optics Letters. 2012;**37**(1):55-57

[69] Avrutsky I. Integrated optical polarizer for silicon-on-insulator waveguides using evanescent wave coupling to gap plasmon-polaritons. IEEE Journal of Selected Topics in Quantum Electronics. 2008;**14**(6):1509-1514

[70] Dai D, Wang Z, Julian N, Bowers JE. Compact broadband polarizer based on shallowly-etched silicon-on-insulator ridge optical waveguides. Optics Express. 2010;**18**(26): 27404-27415

[71] Guan X, Chen P, Chen S, Xu P, Shi Y, Dai D. Low-loss ultracompact transverse-magnetic-pass polarizer with a silicon subwavelength grating waveguide. Optics Letters. 2014;**39**(15):4514-4517

[72] Guan X, Ding Y, Frandsen LH. Ultra-compact broadband higher order-mode pass filter fabricated in a silicon waveguide for multimode photonics. Optics Letters. 2015;**40**(16): 3893-3896

Chapter 9

Optical Diode Based on Two-Dimensional Photonic Crystal

Han Ye, Yumin Liu and Zhongyuan Yu

Additional information is available at the end of the chapter

http://dx.doi.org/10.5772/intechopen.71053

Abstract

The integrated optical diodes have been a thriving research theme due to their potential on-chip applications in photonic circuits for all-optical computing and information processing. Analogous to electronic counterparts, the unidirectional light propagation is characterized by the high contrast between forward and backward transmissions. In this chapter, we demonstrate the proposed schemes and designs for reciprocal and non-reciprocal optical diodes based on two-dimensional (2D) photonic crystal (PhC). The reciprocal devices are built by linear and passive PhC, and the spatial asymmetric mode conversion is utilized to achieve the unidirectionality. The presented nonreciprocal optical diodes rely on the optical nonlinearity of cavity. New 2D PhC optical diodes with high contrast ratio, low insertion loss, large operational bandwidth, small device footprint, and ease of fabrication are highly desirable and still pursued.

Keywords: two-dimensional photonic crystal, optical diode, unidirectional transmission, time-reversal symmetry, mode conversion, nonlinear effect

1. Introduction

The optical diodes, as the counterpart of electron diodes in optics, have attracted huge interests due to its capability of asymmetric light propagation. The unidirectionality invokes potential applications in integrated optical circuits for all-optical computing and information processing [1–5]. Highly desirable key performances of an optical diode are high contrast ratio, low insertion loss, large operational bandwidth, small device footprint, and ease of fabrication. The two types of diodes, reciprocal and nonreciprocal optical diodes, have been proposed based on different mechanisms. The conventional nonreciprocal designs utilize magneto-optical effect [6–8] or nonlinearity [9–12]. The reciprocity of the Lorentz theorem is broken in such designs. More importantly, the optical isolator which prohibits the propagation of any possible mode in one direction can be achieved [13]. Meanwhile, the reciprocal optical diodes,

INTECH
open science | open minds

© 2018 The Author(s). Licensee InTech. This chapter is distributed under the terms of the Creative Commons Attribution License (http://creativecommons.org/licenses/by/3.0), which permits unrestricted use, distribution, and reproduction in any medium, provided the original work is properly cited. (cc) BY

in which the time-reversal symmetry holds, have drawn increasing attention in recent years. The underlying mechanism is the mode conversion in linear structures with spatial asymmetry [14]. In such kind of the optical diode, the unidirectional transmission naturally exists for only certain modes. Due to the flexibility of designing coupling structures and tuning the photonic properties, the two-dimensional (2D) photonic crystal (PhC) has been shown as a promising building block to realize the optical diode. In this chapter, we will provide a summary of proposed schemes of both reciprocal and nonreciprocal optical diodes based on the 2D PhC slabs. This chapter is organized as follows: Section 2 will focus on the reciprocal optical diode based on the 2D PhC composed of dielectric rods or air-holed dielectric slab; Section 3 will introduce nonreciprocal optical diode based on nonlinear effects.

2. Reciprocal optical diode

Any reciprocal device based on passive and linear structure holds the time-reversal symmetry because the materials are described by scalar permittivity and permeability. The function of such reciprocal device has been proven and can be described by converting one set of orthogonal input modes to a matching set of orthogonal output modes [14]. In waveguide structure, the conversion between low-order even and odd modes is usually utilized. For an ideal reciprocal optical diode, complete conversion of a specific mode in one direction and meanwhile complete reflection of the same mode in opposite direction are required. An example of optical diode is illustrated in **Figure 1**. Modes A and B are used to denote the even mode and odd mode, respectively. The even mode A incident from the left port is converted into odd mode B on the right, while the odd mode B from the left is completely reflected. Meanwhile, the transmission of even mode A is not allowed from right to left, while the odd mode B from the right is converted to even mode A. It can be seen that the diode effect for even mode is achieved in this device. The symmetric even-to-odd mode and odd-to-even mode conversions reflect the time-reversal symmetry. It should be noted that this linear and passive device cannot block all possible modes in one direction, which means it is not an optical isolator. In this section, we will focus on the proposed schemes and designs in the last decades, which are based on the 2D PhC composed of dielectric rods or air-holed dielectric slab.

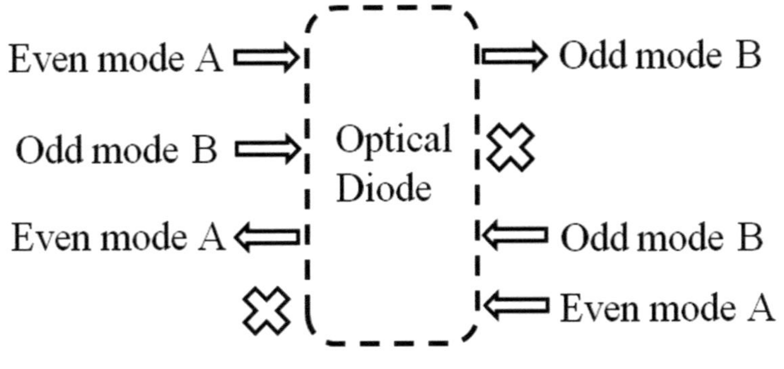

Figure 1. The functionality of a simple reciprocal optical diode of even mode.

2.1. Rod-type PhC optical diode

The rod-type 2D PhC represents the crystal consisting of periodic dielectric rods immersed in air. The rod material is commonly set as silicon in the below designs. The polarization of light is usually the chosen TM mode whose electric field is parallel to the rod. This is determined by the bandgap of such kind of PhC. The optical properties can be simulated by finite element method (FEM) or finite-difference time-domain (FDTD). Almost all simulations of rod-type designs discussed here are within two-dimensional. The length of rod is assumed infinite. Moreover, when estimating the performance of diode effect, two terms "unidirectionality" and "contrast ratio" may be used. The unit of unidirectionality is dB, and the expression is $U = 10\log_{10}(T_{for}/T_{back})$. The contrast ratio is defined as $CR = (T_{for} - T_{back})/(T_{for} + T_{back})$. The contrast ratio of an ideal optical diode is 1.

Feng and Wang [15] proposed a device with the abilities of wavelength filtering and unidirectional light propagation based on the 2D square-lattice PhC with a rectangular defect. In the L-type device shown in **Figure 2**, the left W1-typed waveguide formed by removing one row of rods only sustained an even-mode guiding band spanning the frequencies from 0.3020 c/a to 0.4160 c/a. The under waveguide constructed by removing one row of rods and shifting the adjacent two rows of rods outward for the distances of 0.7a and 0.35a could hold the odd-mode guiding bands from 0.3308 c/a to 0.4160 c/a and even-mode guiding bands within the frequencies from 0.2746 c/a to 0.4160 c/a. The rectangular defect was located at the corner to connect two waveguides. The two side lengths of the rectangle were 1.367a and

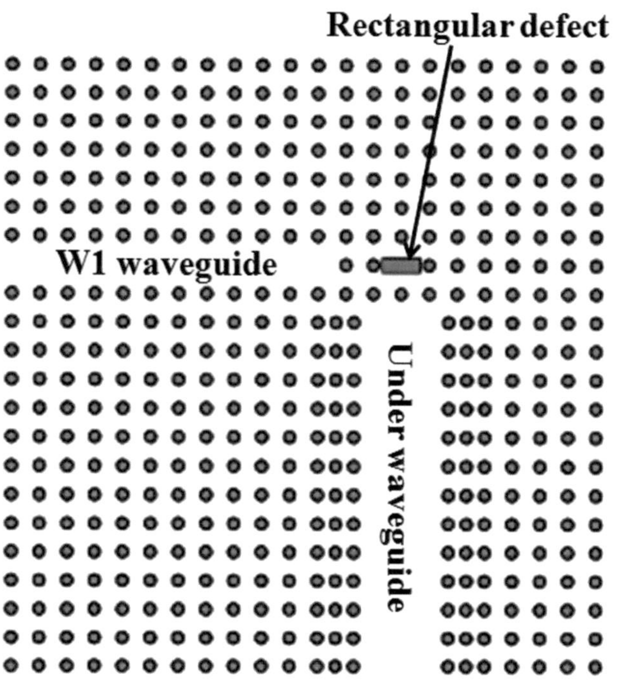

Figure 2. Coupling structure consists of PhC waveguides with the rectangular defects [15].

0.5a, respectively, and the long side was placed along the x-direction of the PhC. There were four defective modes localized by this defect. The adopted mode was located at 0.3977 c/a, and the symmetry was odd in the horizontal direction and even along the vertical direction. The unidirectional propagation of the fundamental mode (even mode) for TM polarization was realized at 0.3977 c/a. The insertion loss was about −6 dB in forward direction, and about 25 dB unidirectionality was achieved. Moreover, the resonant frequency can be linearly tuned by adjusting the coupling region between the defect and the input waveguide. Due to the coupling between defect/cavity and waveguide, the operational bandwidth is narrow.

Ye et al. [16] realized the optical diode effect for even mode by a 2D PhC structure comprising two line-defect waveguides (W1, W2) and a simple defect cavity, as shown in **Figure 3**. Here, the numbers of rows for removed rods were 1 and 2 in W1 waveguide and W2 waveguide, respectively, and the cavity was made by removing two central rods. From left to right, the structure was divided into five parts: W2 waveguide, four-column length matching stage, W1 waveguide, cavity, and W2 waveguide. The mode conversion was impelled by the similarity between the odd mode in the W2 waveguide and the second cavity mode with even symmetry in the horizontal direction and odd symmetry in the vertical direction at the resonant frequency 0.3930 c/a. When the even mode was inputted from the left port at this frequency, it would couple to the second mode in the cavity and then coupled to odd mode in the W2 waveguide. The unidirectional propagation of even mode was achieved thanks to the W1-cavity-W2 design breaking spatial symmetry. The full width at half maximum of the forward transmission spectra of the proposed design with incidence of even mode was about 0.0002 c/a. The transmission efficiency reached a peak 92.5% at the resonant frequency in forward direction, while the value stayed below 0.1% within the frequency range from 0.3900 c/a to 0.3960 c/a for the backward transmission, and the unidirectionality reached approximately 35 dB at resonant frequency. The resonant frequency could be tuned by changing the radii of the nearest rods surrounding the defect and shape of the cavity according to the equation

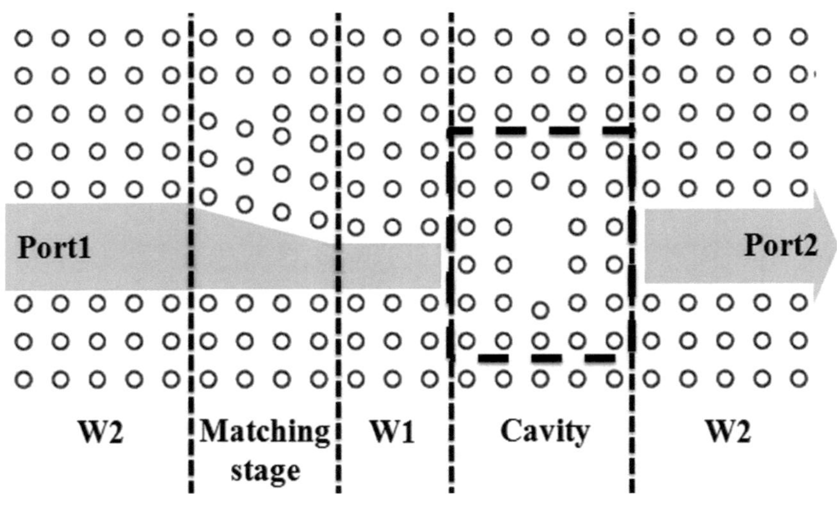

Figure 3. Coupling structure consists of PhC waveguides with the cavity [16].

$$f = 0.3889 + 0.464(r/a) - 2.2167(r/a)^2(0.18a < r < 0.22a)$$
$$f = 0.4112 - 0.0999(\Lambda/a) + 0.0459(\Lambda/a)^2(0.1a < \Lambda < 0.3a), \quad (1)$$

where r denoted the radius of the nearest rods surrounding the defect and Λ was the movement of two neighboring rods. The maximum forward transmission efficiencies of all of the tuned structures exceeded 90%. Due to the narrow bandwidth, the proposed design could be used for mode converter, optical diode, and waveguide filter.

Khavasi et al. [17] reported a broadband optical diode based on pure PhC waveguide with spatial asymmetry as illustrated in **Figure 4**. The components of the design mainly included a mode converter, a mode filter, a four-column length matching stage, and line-defect waveguides formed by removing two rows (W2) and one row (W1) of rods. Along the forward direction, the PhC structure could be divided into seven parts: W2, matching stage, W1, mode converter, W2, mode filter, and W2. The two different routes, U-shaped and straight paths, introduced the phase shifting in the mode converter with 5 × 5 unit cells. The geometrical lengths of the left, upper, and right arms of the U-shaped path were 3a, 2a, and 2a, respectively. And, two $\pi a/2$ components were added to account for the two quarter-circle paths to be traveled at each bend. The length of the straight path was around 4a. The required phase difference for mode conversion was induced by the different length according to the equation $\Delta\phi = \beta_1 \cdot \Delta L \approx 3\pi$, where $\beta_1 = 1.535/a$ was the propagation constant of the waveguide at the normalized frequency $\omega_n = a/\lambda = 0.367$. The mode filter was designed by placing a row of rods with $r = 0.18a$ in the middle of W2. The electric field profile of the odd mode was zero at the middle of the waveguide, while the

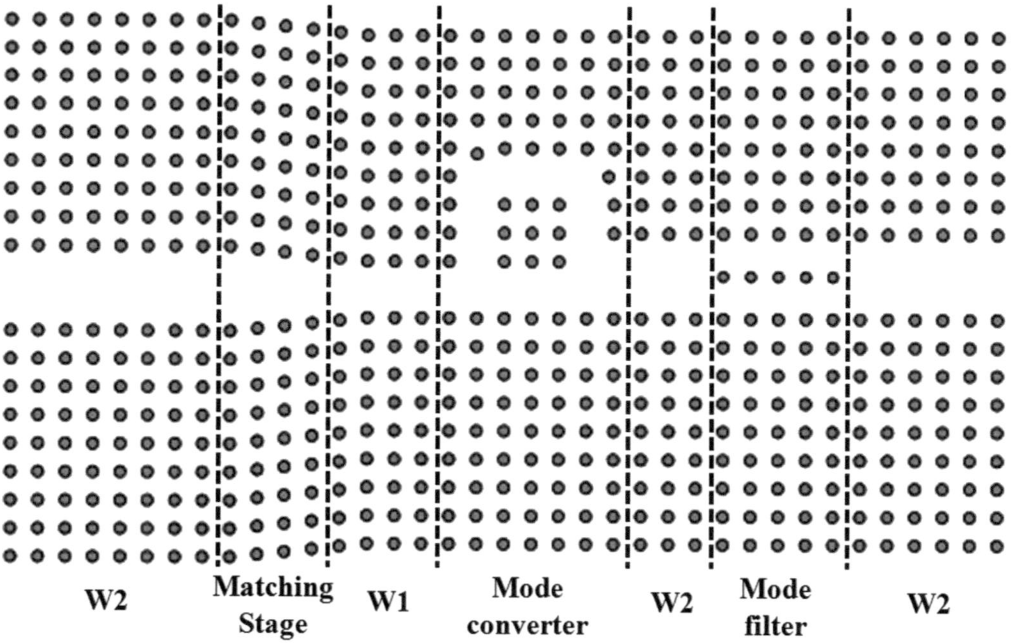

Figure 4. The broadband optical diode with U-shaped and straight split paths [17].

electric field profile of the even mode reached its maximum and was reflected back. Thus, the even mode excited at the left port converted into odd mode by the mode converter and pass through the mode filter, while the even mode was reflected back by the mode filter in the opposite direction, which achieved the optical diode effect. The transmission efficiency kept higher than 94% in the range of frequency $0.366 < \omega_n < 0.381$, and the achieved level of unidirectionality was approximately 78 dB.

Lu et al. [18] realized the optical diode effect based on two-dimensional square-lattice PhC, in which a directional coupler and 90° bend were utilized to realize the mode converter and mode filter functions. The scheme is depicted in **Figure 5**. The components in leftward direction were, respectively, W2 waveguide formed by removing two rows of rods, mode filter and mode converter, and W1 waveguide created by removing one row of rods, a four-column length adiabatic-matched region and W2. A perpendicular W1 waveguide and two 90° bends were combined together to achieve a mode converter. The conversion principle was as follows: when the light transmitted in the leftward direction, the odd mode was even symmetric along the horizontal direction in the mode filter, while the fundamental even mode in the perpendicular W1 waveguide was also symmetric about the parallel direction. Therefore, the odd mode could be thoroughly converted into the even mode after the 90° waveguide bend. On the other hand, the even mode would be converted into odd mode based on the same theory in the rightward direction. The mode filter was designed by placing a row of rods in the middle of W2 as several other works. For the device, the odd mode was completely transmitted in forward direction (left to right) and converted to the fundamental even mode but was blocked in the opposite direction. Approximate 95% forward transmission efficiency, 0.024 c/a operational bandwidth, and maximum 80 dB unidirectional extinction ratio were achieved in 2D simulations.

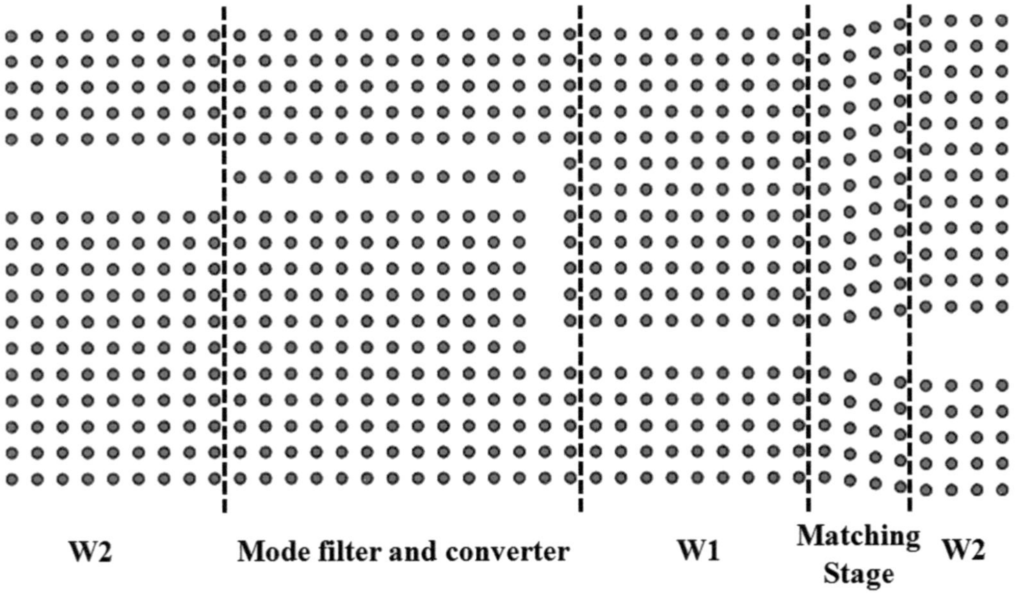

Figure 5. The broadband optical diode with a directional coupler and double 90° bends [18].

Kurt et al. [19] reported a photonic structure formed by using a chirped PhC waveguide to realize asymmetric light propagation as shown in **Figure 6**. A distance increment between each neighboring column of the dielectric rods, which caused that the waveguide mode moved toward the outside of the bandgap, was introduced to break the spatial symmetry. Here, the distance between each unit cell was still a constant along the y-direction (Δy = a), while the interval of rods linearly increased along the x-direction (0.5a < Δx < 2.5a), and the total length of the chirped PhC waveguide was approximately 16a. Thus, the structure became sparse along the x-direction, and the spatial distribution was uniform along the y-direction. When the light traveled in forward direction, the light mainly propagated at the center and leaked toward the two sides of the waveguide after propagating a section of distance; then, it traveled again along the x-direction and reached the end of the structure. The amount of light reaching the waveguide centerline was very small. However, the central lobe strongly appeared at the waveguide exit in the backward light transmission. The contrast ratio was around 0.75 at the operating frequency $a/\lambda = 0.3288$.

Giden et al. [20] from the same research team as the above work theoretically and experimentally realized the asymmetric light transmission based on PhC with a gradient index distribution along the x-direction, in which the waveguide was formed by introducing point and line defects. The total length and width of the proposed structure were L = 16a and W = 20a, respectively. Continuously graded refractive index distribution could be approximated by relocating the PhC dielectric rods. In order to achieve a desired index profile within certain wavelength, they utilized an equation $n(x) = n_0 \operatorname{sech}(\alpha x)$, to formulate the hyperbolic secant index profile and then to modulate the spatial distributions of PhC cells, where the constant $\alpha = 0.0373a^{-1}$ was the gradient factor and n_0 was the effective refractive index at $x = 0$. The effective refractive indices at $x = 0$ and 16a were set as $n(x = 0) = n_0 = 2.20$ and $n(x = L = 16a) = 1.41$. The distance between each rod along x-direction was modified to find the targeted longitudinal index distribution, and the lateral interval was fixed at a. After the spatial distributions of the silicon rods were determined, a line defect was introduced by removing one row of rods to

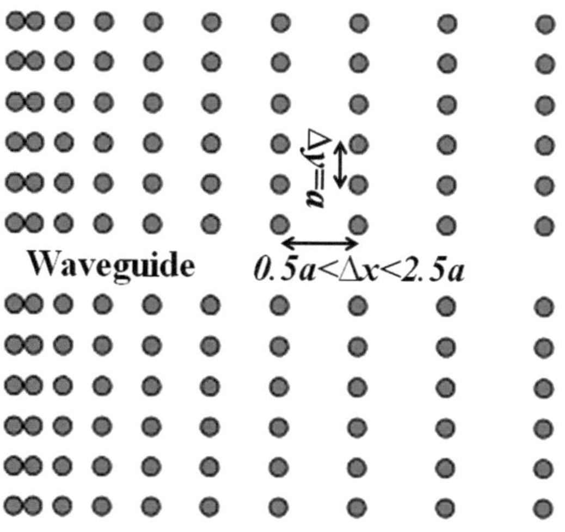

Figure 6. The optical diode in PhC waveguide with chirped PhC structure [19].

form a waveguide. The asymmetric propagation was achieved due to the broken spatial symmetry and the forward (left-to-right) and backward (right-to-left) transmission efficiencies reached 7.26% and 24.8% at the peak, respectively. The contrast ratio was in the range of 0.412–0.773 within the telecom wavelength range from 1523.5 nm to 1576.1 nm in simulation. Moreover, the point defects were inserted by shifting the positions of adjacent rods along the y-direction to increase the difference between forward and backward transmissions. In experiment validation (microwave regime), the proposed design had 4.11% and 49.8% forward and backward transmission efficiencies, respectively, exceeding 0.80 contrast ratio within the operating frequency band from 12.8 to 13.3 GHz.

Soltani et al. [21] proposed a 2D PhC made of elliptic silicon rods to realize the unidirectional propagation as depicted in **Figure 7**. The major and minor axes of the ellipse were 0.31a and 0.22a, respectively. The waveguide was created by removing one row of the dielectric rods along the x-direction. The distance between neighboring columns was modulated according to the equation, $d = 0.6a + (n - 1) \times 0.2a$, where n represented the nth interval in the x-direction and over all the structure $0.6a \leq d \leq 2.2a$. Meanwhile, a reflector between two successive columns was introduced to control losses of the scattered light. Thus, the asymmetric spatial distribution of the unit cells ensured the one-way light transmission. When the light was inputted from the left port, it propagated along the centerline of the structure, and then the flow of light leaked into the larger space between rods after several periods. Thus, the light was blocked in this direction. In the opposite direction, the light source was positioned on the right side of the structure; the light emitted was widely spread in the beginning until it reached the center of the structure. The transmission efficiency was lower than 10% in forward direction (left to right), while the value reached 80% in the opposite propagation within the frequency region from 0.28a/λ to 0.32a/λ. Finally, they studied the effects of the orientation of ellipses on the transmission of light and found that the orientation angle may introduce shift and oscillation to the transmission spectra.

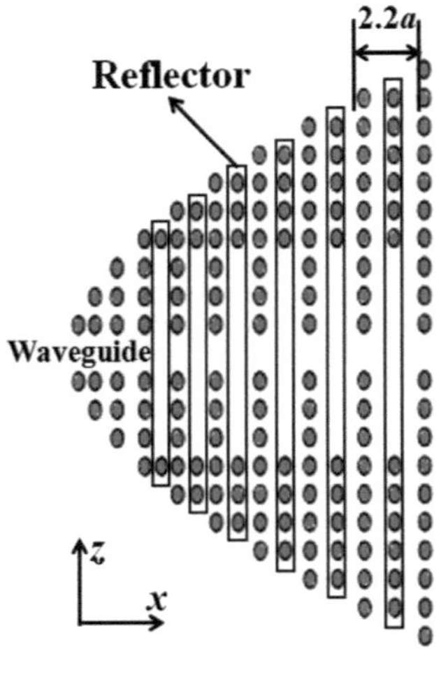

Figure 7. The optical diode in PhC waveguide with elliptic rods [21].

Singh et al. [22] realized the optical diode effect based on a hexagonal and square-lattice PhC structures with silicon rods immersed in air. The structure is demonstrated in **Figure 8**. The waveguide was formed by removing one row of rods and chirping the neighboring rods, which was realized by changing the radii of the rods. Two kinds of devices were proposed according to the different lattice. For the regular hexagonal lattice PhC, the lattice constant a and radius of the rods r were 0.68 μm and 0.23 μm, respectively, and the radii of the chirped rods were r_1 = 0.220 μm, r_2 = 0.209 μm, and r_3 = 0.197 μm. When light propagated from left to right (forward direction), it mainly propagated along the central region of PhC waveguide, and a high power was obtained from output port. In the opposite direction, the light leaked in the designed structure so that the field amplitude along the propagation direction was weak. Approximate 82% forward transmission efficiency and lowest value 35% in the backward direction were obtained at 1.55 μm. The relative extinction ratio was nearly 5.5 dB. For the square arrangement-based optical diode, the lattice constant a and radius of the rods r were 0.71 μm and 0.22 μm, respectively, and the radii of the chirped rods were r_1 = 0.21 μm, r_2 = 0.20 μm, r_3 = 0.19 μm, and r_4 = 0.17 μm. Here, around 52% forward transmission efficiency and 4.1 dB extinction ratio were achieved at 1.55 μm. Then, they showed that the engineering of chirping parameters not only in terms of rods' radii, but also other parameters such as lattice spacing and refractive index parameters may give rise to the superior performance of the optical diode-like photonic structures.

Liu et al. [23] devised an ultracompact mode converter based on a two-dimensional square-lattice PhC made of dielectric rods in air by a two-phase numerical optimization method and applied it to the optical diode. In numerical design and optimization, the design was evaluated by the error metric, $J = \sum_{\omega} (1 - P_{\omega})^2$, where P_{ω} was the conversion efficiency at frequency ω. A region with rods located on the 20 possible lattice sites was considered to limit the design space in the optimization process. In the first phase, a combinatorial search is performed using simulation method to analyze every possible combination of rods (presence or absence) on the 20 lattice sites in the region. Any structures with desirable conversion behavior ($J \leq 0.5$) were considered potential candidates for further optimization. In the second phase, a further optimization adjusted the radii of the existing rods in the initial combinatorial to minimize the error metric J. They obtained a compact mode converter with above 99% conversion efficiency, which occupied an area of 4 × 10 unit cells' large functional region (spatial inverse symmetry was hold). For optical diode design, they adopted the same optimization strategy, and the

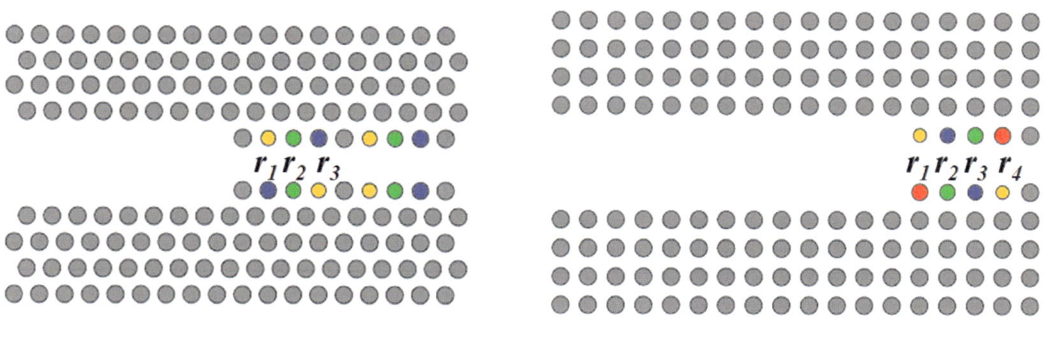

Figure 8. The optical diode in PhC waveguide with chirped rods with different symmetries [22].

more than ten structures with ideal performance were found, which occupied 4×5 unit cells' large functional region as demonstrated in **Figure 9**. The forward mode conversion efficiency exceeded 97%, and backward reflection was about 3% at frequency 0.4025 c/a. The operational bandwidth was larger than 0.015 c/a.

Ye et al. [24] demonstrated a compact broadband optical diode based on linear 2D rod-type PhC as shown in **Figure 10**. The proposed structure consisted of a simple waveguide made by removing two rows of rods and a 4×1 unit cells' large functional region, the spatial dielectric distribution in which was numerically optimized by finite element method (FEM) combining

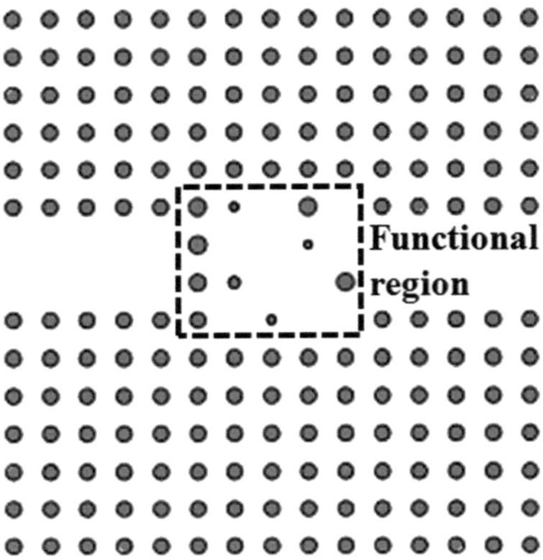

Figure 9. The broadband optical diode in PhC waveguide with 4×5 unit cells' compact-optimized functional region [23].

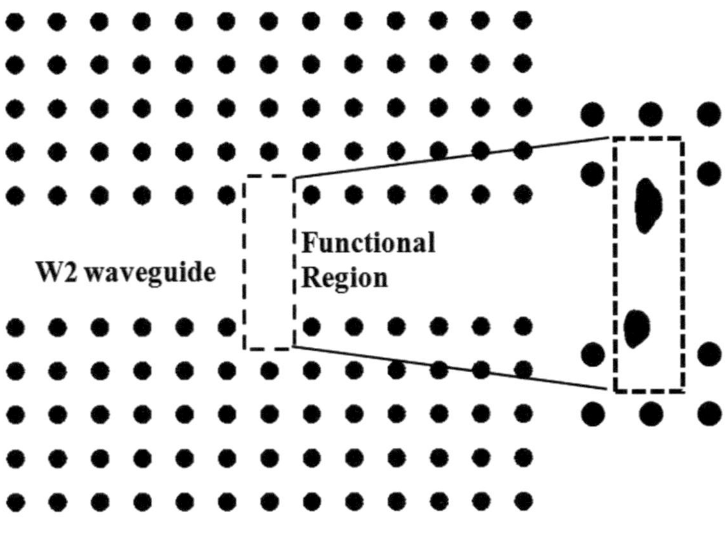

Figure 10. The broadband optical diode in PhC waveguide with 4×1 unit cells' compact-optimized functional region [26].

the geometry projection method (GPM) [25] and the method of moving asymptotes (MMA) [26]. Two dielectric rods inside the functional region were still kept during the optimization process. The GPM employed 6 × 20 control points whose heights can be fitted as a three-dimensional surface and the intersection line of level plane and three-dimensional surface could be used to define the rod shape. When level plane was set at zero, the dielectric constant at coordinate x could be formulated as $\varepsilon(x) = \varepsilon_{air} + \frac{(\varepsilon_{si} - \varepsilon_{air})}{2}\left(\tanh\left[\frac{sign[S(x)]d(x)}{\xi}\right] + 1\right)$, where S(x) was the surface function, d(x) was the minimum distance from the refined surface to the nearest intersection curve, and ξ was the control parameter of vanishing of intermediate dielectric at interface. The GPM controlled the rod shape, size, and position by adjusting the heights of the control points. The FEM and MMA were utilized to determine the heights of control points to find the solution. The optimization process proceeded in two phases to obtain large operating bandwidth. First, they searched a candidate at the center frequency 0.40 c/a; in the second phase, they optimized further by the new objective function, which was the weighted average of objective function values at five frequencies within the targeted frequency domain. The integrals of power flux for desired optical behavior were set as objective functions during the optimization process. The maximum forward transmission efficiency was 83%, and the above 60% transmission efficiency within bandwidth 0.01 c/a was achieved. The above 19 dB unidirectionality (maximum 30 dB) for even mode was achieved within bandwidth 0.01 c/a as well.

Wang et al. [27] demonstrated a graded photonic crystal (GPhC) design with 11a × 25a large scale, which was obtained from a conventional square-lattice PhC with $R = 0.2a$ by linearly changing the radius of dielectric rods, to achieve unidirectional light propagation. The structure is demonstrated in **Figure 11**. The radii of the dielectric cylinders in the x-y plane were given by$R_{(x+1)a} - R_{xa} = 0.01a$ (in x-direction, $R_{1a} = 0.2a$) and $R_{(y+1)a} - R_{ya} = 0.01a$ (in +y-direction),

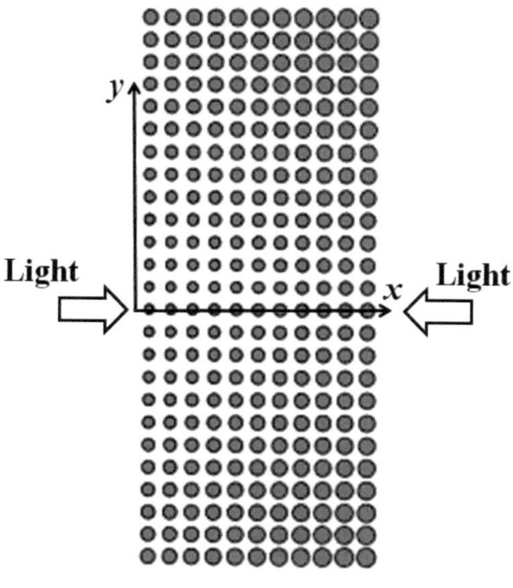

Figure 11. Asymmetric light propagation based on graded PhC [27].

where R_{xa} was the radius of dielectric rods of xth period. The structure in +y-direction was symmetrical along the x-axis with the structure in the −y-direction. Here, the x-axis was located at the middle of the structure along the y-direction. The transverse field profiles demonstrated a significant difference. When the light propagated from left to right (forward direction), most of the light at the x-axis propagated across the structure, and the light on the two sides of the x-axis could not propagate after several periods because of the directional bandgap. The light almost converged into the middle position at the end face of the structure. On the other hand, when the light propagated in backward direction (right to left), only the light along the optical axis transmitted into the GPhC structure, and the light propagating into GPhC spread to both ends of the structure after a few periods, the light could be scarcely detected at the end face. The intensity at the convergence spot of the light propagated in forward direction was almost 14 times more than the maximum intensity of the backward transmission light. Maximum 0.7 contrast ratio was achieved at the frequency of 0.395 (a/λ). If the numbers of period were extended to 13 in the horizontal direction, higher contrast ratio (near 1) was attained at the frequency of 0.394 (a/λ).

2.2. Air-holed PhC optical diode

Most of the designs based on rod-type PhC discussed above were confined in simulation because of the difficulties in fabricating. The only experimental validation was reported in Giden's work [20], but operating in the microwave regime (10 GHz). Moreover, the absence of light confinement along the rod direction is another limitation. Compared with rod-type PhC, the fabrication technique of air-holed PhC slab is more mature and compatible with conventional CMOS processing. Air-holed PhC is formed by etching periodic air holes in dielectric slab. The optical diode based on air-holed PhC can be directly connected to standard silicon waveguide, showing its ability of integration with other silicon-based photonic devices. However, unfortunately, lack schemes for air-holed 2D PhC optical diode can be found. One scheme is based on the PhC heterostructure with directional bandgap [28–31], and the other scheme is based on topology optimization [32]. The TE mode whose magnetic field is perpendicular to the slab is considered in all designs discussed in this section and most simulations are performed in 3D FDTD.

Cicek et al. [31] realized the unidirectional light transmission in a 2D PhC heterostructure with air holes in an AlGaAs host. The proposed structure was constructed of two kinds of photonic crystals (PhC1, PhC2) with the same orientation and lattice constant but different radii of air holes, as illustrated in **Figure 12**. The interface was located along its body diagonals. The radius of air holes in PhC1 with 16 × 16 unit cells was set as 0.460a, while the value in PhC2 with 15 × 15 unit cells was 0.352a. Moreover, PhC2 was offset by $d=0.2a$ to decrease the forward reflection losses introduced by effective index variation. In order to achieve the asymmetric light propagation, stop band for blocking light transmission in backward direction (right to left) was studied. When the light propagated in forward direction (left to right), it transmitted toward the PhC2 and eventually the output port. The waves were self-collimated in both PhC1 and PhC2 at the frequency 0.37ωa/2πc. While, in the opposite direction, the incident light was reflected at the air-PhC2 interface, the deflected wave components at the air-PhC2 interface could reach PhC1 and eventually the output port. However, several lights would leak out in the reverse direction. The transmission efficiency in forward direction was around 50% within the frequency region from 0.32ωa/2πc to 0.42ωa/2πc, while the value in backward direction kept lower than 20%. The maximum of contrast ratio was achieved approximately 0.90. Lu et al. [28] experimentally fabricated similar structure using poly[2-methoxy-5-(2-ethylhexyloxy)-1,4-phenylenevinylene] MEH-PPV PhC. In the experiment, the lattice constant was 440 nm, and the radii of air holes for PhC1 and PhC2 were 110 nm and 170 nm, respectively. The angle between the hetero-interface and the horizontal direction was 56°. The transmissions of 50% and 0.02% were achieved for the rightward and leftward incidence within the wavelength region from 640 nm to 660 nm.

Wang et al. [29, 30] proposed an optical diode design with functional region $6 \times 6\ \mu m^2$ based on silicon PhC slab. The radii of air hole for PhC1 and PhC2 were 0.24*a* and 0.36*a*, respectively. The hetero-interface was placed between PhC1 and PhC2 along the Γ-M direction. Two 4*a*-wide waveguides were located at the two sides of the structure and were used to enter or exit the light beam. There exists an isolation band ranging from 0.2649 to 0.2958 (a/λ), where the transmission efficiency in forward direction (left to right) reached a peak about 6%, while the backward transmission kept between 0.5% and 1% within the frequency region from 0.2649*a*/λ to 0.2958*a*/λ. The maximum contrast ratio 0.846 was obtained at the peak. Then, the original diode was optimized in order to decrease the backward transmission to be zero and increase the peak of forward transmission as much as possible. The revised structure is demonstrated in **Figure 13**. The backward transmission generally decreased a half compared with that of the original structure, and the forward peak reached approximately 13%. The maximum contrast ratio of the revised structure increased to 0.92 at the peak. In the experiment, the patterns were

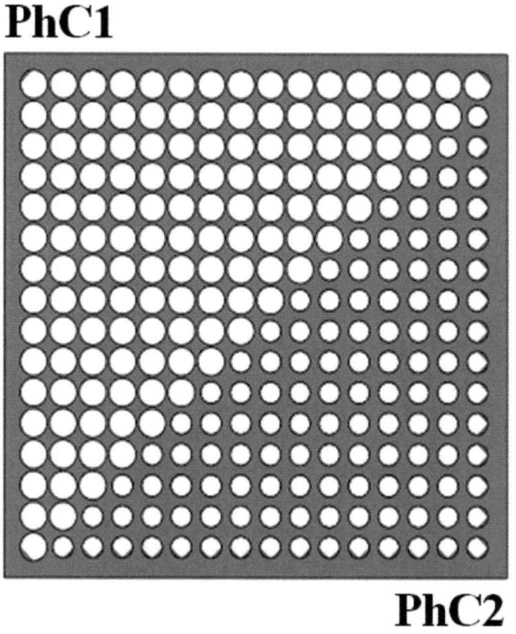

Figure 12. Refraction-based optical diode with PhC heterostructure [31].

defined in resist using the electron beam lithography (EBL) on the silicon-on-insulator (SOI) chip and then transferred to silicon layer using the inductively coupled plasma (ICP) reactive ion etching. To form an air-bridged structure, the SiO_2 underneath the patterned PhC was finally removed. The maximum forward transmission efficiency in experiment with asymmetric input/output waveguides approached 21.3%, and the best contrast ratio of the diode structure reached 0.885 at the peak.

Ye et al. [32] realized the optical diode effect in 2D silicon PhC slab waveguide with 1.2a × 2.8a large functional region, in which there was only one deformed air hole. The structure is illustrated in **Figure 14**. The waveguide was constructed by removing one row of air holes along x-direction and moving the first and second nearest row with 0.15a and 0.1a in y-direction, respectively. The spatial dielectric distribution in the functional region was numerically optimized by finite element method (FEM) combining the geometry projection method (GPM) and the method of moving asymptotes (MMA). The GPM employed 13 × 29 control points to fit the three-dimensional surface. The optimization algorithm MMA was utilized to determine the heights of all control points in GPM. In order to find a broadband optical diode, three optimization steps were taken: first, they searched a bidirectional mode converter with mirror symmetry in functional region at the center frequency 0.36 c/a. Next, the symmetry was broken to find the solution for optical diode working at center frequency. The last step was to extend the operational frequency range from single frequency. Here, the integrals of power flux for desired optical behavior were set as objective functions. The above 89.9% forward transmission efficiency, 0.01 c/a operational bandwidth, and maximum 24 dB unidirectionality were achieved in 2D FEM simulations. In the air-bridged 3D FDTD model, the maximum unidirectionality was about 19.6 dB, while the value kept higher than 12.7 dB for slab thickness 0.54a. It should be noted that identical performance of 3D device and 2D optimized structure remains a challenge.

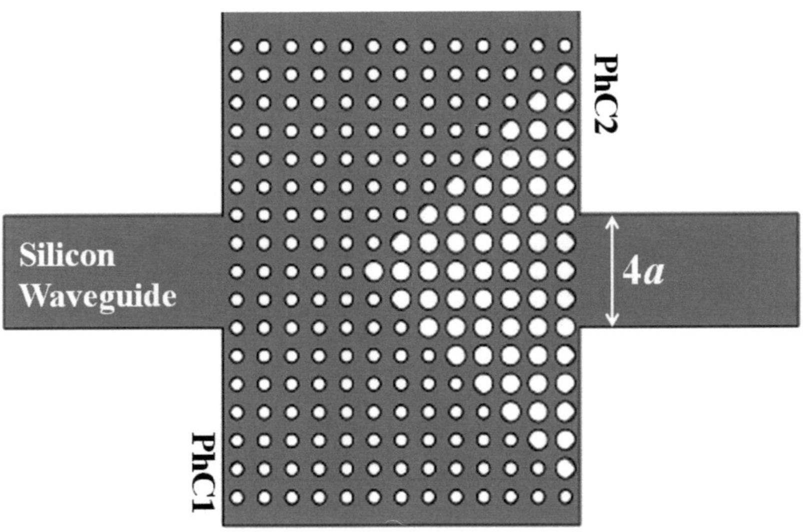

Figure 13. Optical diode with PhC heterostructure connecting standard silicon waveguide [29].

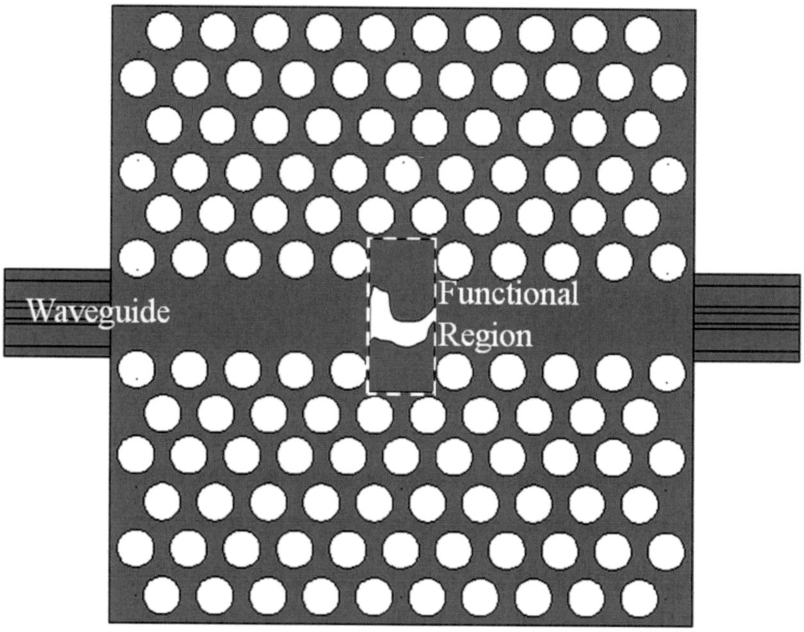

Figure 14. Optical diode with topological optimized functional region connecting silicon waveguide [32].

3. Nonreciprocal optical diode

The optical diode effect achieved in linear and passive PhC (discussed in Section 2) cannot break the reciprocity of the Lorentz theorem and consequently cannot play as an important optical device: the optical isolator. The conventional nonreciprocal designs utilize magneto-optical effect or nonlinearity. Unfortunately, common waveguide materials, especially Si, SiO_2, and Si_3N_4, present only relatively weak or even no magneto-optical effect. Meanwhile, the prerequisite external magnetic field would limit the integration for on-chip applications. Therefore, we will focus on the nonlinearity in this section. The major effects may include the second-order optical nonlinearity, the Kerr effect, two-photon absorption, free-carrier effect, and thermo-optic effect [33]. These effects have been pervasively employed in on-chip all-optical processing such as wavelength conversion, optical switch, modulation, and optical isolation [34–36].

Zhang et al. [37] proposed a silicon optical diode based on optical nonlinearity in cascaded PhC L3 cavities, which was formed by removing a line of three holes as demonstrated in **Figure 15**. The lattice constant was 420 nm, and the air-holed radius was 126 nm. The three air holes adjacent to the right cavity were shifted by 0.175*a*, 0.025*a*, and 0.175*a*, respectively. And, the footprint of the whole device was about 20 μm × 10 μm. The resonant wavelengths of two cascaded L3 cavities did not match exactly due to the different distribution of the holes in periphery. When the input power was low, the two L3 cavities were both working in the linear regime, and asymmetric transmission was not realized. As input power increased, the resonance dips redshifted due to the onset of nonlinearity induced by the thermo-optic effect in the silicon cavities. A nonreciprocal transmission ratio (NTR) of 30.8 dB and insertion loss of 8.3 dB was realized in the device. The device had a relatively broad 17 dB operation bandwidth of 0.08 nm, and at least 16 dB of NTR was achieved when input power varied between −6.25 and −2.95 dBm.

Bulgakov and Sadreev [38] proposed an all-optical diode in asymmetric L-shaped PhC waveguide with a single nonlinear Kerr microcavity at the corner. The structure is depicted in **Figure 16**. The linear L-shaped PhC waveguide was formed by removing one row of dielectric rods, and the microcavity was constructed by three linear rods and one nonlinear rod. The radius and dielectric constant of the GaAs rods were 0.18a and 11.56 in the PhC, where the lattice unit was set as 0.5 μm. The nonlinear defect rod had the radius 0.4a, and the dielectric constant was 6.5. The nonlinear refractive index was 2×10^{-12} cm^2/W. In order to achieve the asymmetric design of the microcavity, the dielectric constant of the third additional rod inside the right leg of the waveguide was chosen as 5. The two dipole eigenfrequencies of the microcavity were $0.3658 \times 2\pi c/a$ and $0.3650 \times 2\pi c/a$, respectively. When both dipole modes were excited simultaneously, the design could be opened for light transmission at some magnitudes of frequency and injected intensity because of the nonlinear coupling between them. The opening region mostly depended on the ratio between the nonlinearity constant and the coupling strengths of the dipole modes with the propagating waveguide modes. Therefore, the asymmetric transmission in the design was achieved due to the different threshold for light transmissions from left to right and backward. This came from different coupling strengths of dipole modes with the propagating mode of the left and right waveguide legs. In the design, the direction of optical diode transmission could be controlled by power or frequency of injected light.

Liu et al. [39] realized the optical diode effect based on asymmetrical coupling by a microcavity with nonlinear Kerr medium and a FP cavity at two sides of PhC waveguide. In the system, a Fano resonant microcavity with ultrahigh-quality factor was designed in rod-type PhC. An elliptical point defect was placed in the first layer below a line-defect PhC waveguide, and a FP

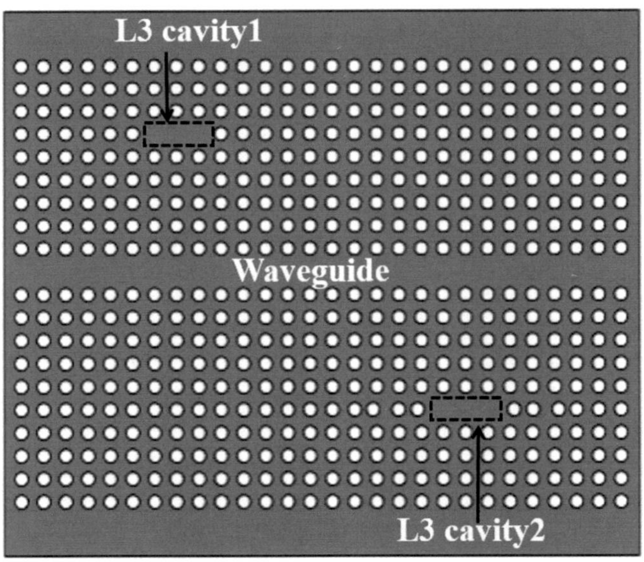

Figure 15. Silicon optical diode based on cascaded photonic crystal cavities [37].

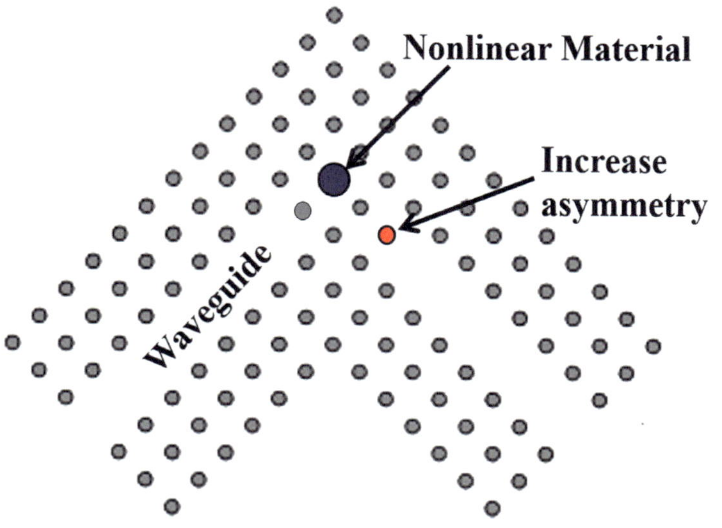

Figure 16. Optical diode based on dipole modes of Kerr microcavity in asymmetric L-shaped PhC waveguide [38].

cavity was located in the second layer above PhC waveguide, as shown in **Figure 17**. Moreover, a reflector layer was set below the left end of FP cavity to form asymmetric structure. The radius of dielectric cylinder was 0.3a; the major and minor axes of elliptical point defect were set as 0.534a and 0.3a, respectively. Because of the asymmetric design, backward (right to left) propagation required stronger incidence light to excite Kerr effect of microcavity than forward (left to right) transmission, so the unidirectional transmission could be realized. The high transmission in forward direction was approximately about 80%, and maximum contrast ratio was about 0.8.

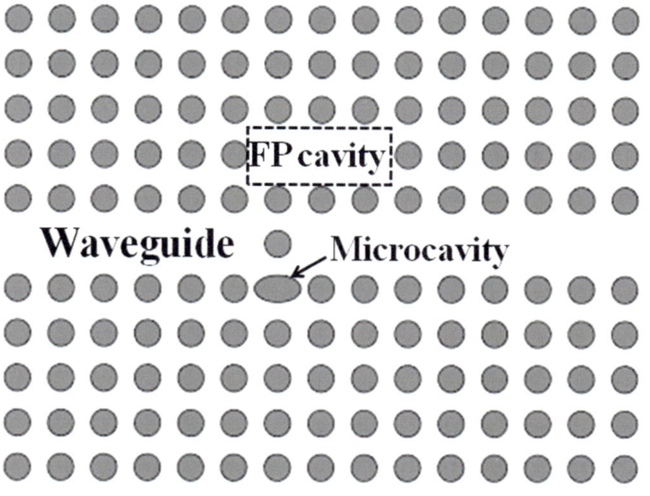

Figure 17. Optical diode based on asymmetrical coupling by a microcavity and FP cavity at two sides of PhC waveguide [39].

4. Conclusion

In this chapter, the proposed schemes for optical diodes based on 2D PhC have been briefly summarized. For reciprocal optical diode, we introduce 11 designs based on rod-type PhC and 4 designs based on air-holed PhC slab. Since the linear and passive structures hold the time-reversal symmetry, the unidirectional transmission of certain optical mode is achieved by spatial asymmetric mode conversion. For nonreciprocal optical diode, we mainly focus on the PhC cavity with optical nonlinear effects. The three designs can be used as optical isolator since it can block all possible modes in one direction. Considering the performances, new schemes for 2D PhC optical diodes with high contrast ratio, low insertion loss, large operational bandwidth, small device footprint, and ease of fabrication are still highly desirable.

Author details

Han Ye*, Yumin Liu and Zhongyuan Yu

*Address all correspondence to: han_ye@bupt.edu.cn

State Key Laboratory of Information Photonics and Optical Communications, Beijing University of Posts and Telecommunications, Beijing, PR China

References

[1] Fedotov VA, Mladyonov PL, Prosvirnin SL, Rogacheva AV, Chen Y, Zheludev NI. Asymmetric propagation of electromagnetic waves through a planar chiral structure. Physical Review Letters. 2006;**97**(16):167401

[2] Hwang J, Song MH, Park B, Nishimura S, Toyooka T, Wu JW, et al. Electro-tunable optical diode based on photonic bandgap liquid-crystal heterojunctions. Nature Materials. 2005;**4**(5):383-387

[3] Bi L, Hu JJ, Jiang P, Kim DH, Dionne GF, Kimerling LC, et al. On-chip optical isolation in monolithically integrated non-reciprocal optical resonators. Nature Photonics. 2011;**5**(12):758-762

[4] Xu J, Zhuang X, Guo P, Huang W, Hu W, Zhang Q, et al. Asymmetric light propagation in composition-graded semiconductor nanowires. Scientific Reports. 2012;**2**:820

[5] Fan L, Wang J, Varghese LT, Shen H, Niu B, Xuan Y, et al. An all-silicon passive optical diode. Science. 2012;**335**:447-450

[6] Haldane FDM, Raghu S. Possible realization of directional optical waveguides in photonic crystals with broken time-reversal symmetry. Physical Review Letters. 2008;**100**(1):013904

[7] Yu Z, Veronis G, Wang Z, Fan S. One-way electromagnetic waveguide formed at the interface between a plasmonic metal under a static magnetic field and a photonic crystal. Physical Review Letters. 2008;**100**(2):023902

[8] Wang Z, Chong Y, Joannopoulos JD, Soljacić M. Observation of unidirectional backscattering-immune topological electromagnetic states. Nature. 2009;**461**:772-775

[9] Lin XS, Wu WQ, Zhou H, Zhou KF, Lan S. Enhancement of unidirectional transmission through the coupling of nonlinear photonic crystal defects. Optics Express. 2006;**14**(6):2429-2439

[10] Wang J, Fan L, Varghese LT, Shen H, Niu B, Qi MH. A theoretical model for an optical diode built with nonlinear silicon microrings. Journal of Lightwave Technology. 2013;**31**:313-321

[11] Soljacic M, Luo C, Joannopoulos JD, Fan S. Nonlinear photonic crystal microdevices for optical integration. Optics Letters. 2003;**28**(8):637-639

[12] Sahoo PK, Joseph J. Optical diode using nonlinear polystyrene ring resonators in two-dimensional photonic crystal structure. Applied Optics. 2013;**52**:8252-8257

[13] Jalas D, Petrov A, Eich M, Freude W, Fan SH, Yu ZF, et al. What is—and what is not—an optical isolator. Nature Photonics. 2013;**7**(8):579-582

[14] Miller DAB. All linear optical devices are mode converters. Optics Express. 2012;**20**:23985-23993

[15] Feng S, Wang Y. Unidirectional reciprocal wavelength filters based on the square-lattice photonic crystal structures with the rectangular defects. Optics Express. 2013;**21**:220-228

[16] Ye H, Zhang JQN, Yu ZY, Wang DL, Chen ZH. Realizing mode conversion and optical diode effect by coupling photonic crystal waveguides with cavity. Chinese Physics B. 2015;**24**:094214

[17] Khavasi A, Rezaei M, Fard AP, Mehrany K. A heuristic approach to the realization of the wide-band optical diode effect in photonic crystal waveguides. Journal of Optics. 2013;**15**:075501

[18] Lu J, Ren HL, Guo SQ, Gu DY, Wen H, Qin YL. Ultra-wideband optical diode based on photonic crystal 90° bend and directional coupler. Chinese Optics Letters. 2014;**12**:102301

[19] Kurt H, Yilmaz D, Akosman AE, Ozbay E. Asymmetric light propagation in chirped photonic crystal waveguides. Optics Express. 2012;**20**(18):20635-20646

[20] Giden IH, Yilmaz D, Turduev M, Kurt H, Çolak E, Ozbay E. Theoretical and experimental investigations of asymmetric light transport in graded index photonic crystal waveguides. Applied Physics Letters. 2014;**104**:031116

[21] Soltania A, Ouerghia F, AbdelMaleka F, Haxhab S, Ademgilc H, Akowuahd EK. Effect of the elliptic rods orientations on the asymmetric light transmission in photonic crystals. Optics Communications. 2017;**392**:147-152

[22] Singh BR, Rawal S, Sinha RK. Chirped photonic crystal with different symmetries for asymmetric light propagation. Applied Physics A: Materials Science & Processing. 2016;**122**:605

[23] Liu V, Miller DAB, Fan S. Ultra-compact photonic crystal waveguide spatial mode converter and its connection to the optical diode effect. Optics Express. 2012;**20**:28388-28397

[24] Ye H, Wang D, Yu Z, Zhang J, Chen Z. Ultra-compact broadband mode converter and optical diode based on linear rod-type photonic crystal waveguide. Optics Express. 2015;**23**:9673-9680

[25] Frei WR, Johnson HT, Choquette KD. Optimization of a single defect photonic crystal laser cavity. Journal of Applied Physics. 2008;**103**:033102

[26] Svanberg K. The method of moving asymptotes-a new method for structural optimization. International Journal for Numerical Methods in Engineering. 1987;**24**(2):359-373

[27] Wang LH, Yang XL, Meng XF, Wang YR, Chen SX, Huang Z, et al. Asymmetric light propagation based on graded photonic crystals. Japanese Journal of Applied Physics. 2013;**52**:122601

[28] Lu CC, Hu XY, Zhang YB, Li ZQ, Xu XA, Yang H. Ultralow power all-optical diode in photonic crystal heterostructures with broken spatial inversion symmetry. Applied Physics Letters. 2011;**99**(5):051107

[29] Wang C, Zhou CZ, Li ZY. On-chip optical diode based on silicon photonic crystal heterojunctions. Optics Express. 2011;**19**(27):26948-26955

[30] Wang C, Zhou CZ, Li ZY. Linear and passive silicon optical isolator. Scientific Reports. 2012;**2**:674

[31] Cicek A, Yucel MB, Kaya OA, Ulug B. Refraction-based photonic crystal diode. Optics Letters. 2012;**37**(14):2937-2939

[32] Ye H, Yu Z, Liu Y, Chen Z. Realization of compact broadband optical diode in linear airhole photonic crystal waveguide. Optics Express. 2016;**24**(21):24592-24599

[33] Leuthold J, Koos C, Freude W. Nonlinear silicon photonics. Nature Photon. 2010;**4**(8):35-544

[34] Driscoll JB, Astar W, Liu X, Dadap JI, Green WM, Vlasov YA, et al. All-optical wavelength conversion of 10 Gb/s RZ-OOK data in a silicon nanowire via cross-phase modulation: Experiment and theoretical investigation. IEEE Journal of Selected Topics in Quantum Electronics. 2010;**16**(5):1448-1459

[35] Xu QF, Lipson M. Carrier-induced optical bistability in silicon ring resonators. Optics Letters. 2006;**31**(3):341-343

[36] Khan M, Shen H, Xuan Y, Zhao L, Xiao S, Leaird DE, et al. Ultrabroad-bandwidth arbitrary radiofrequency waveform generation with a silicon photonic chip-based spectral shaper. Nature Photonics. 2010;**4**(2):117-122

[37] Zhang Y, Li DP, Zeng C, Huang ZZ, Wang Y, Huang QZ, et al. Silicon optical diode based on cascaded photonic crystal cavities. Optics Letters. 2014;**39**(6):1370-1373

[38] Bulgakov EN, Sadreev AF. All-optical diode based on dipole modes of Kerr microcavity in asymmetric L-shaped photonic crystal waveguide. Optics Letters. 2014;**39**(7):1787-1790

[39] Liu B, Liu YF, Jia C, He XD. All-optical diode structure based on asymmetrical coupling by a micro-cavity and FP cavity at two sides of photonic crystal waveguide. AIP Advances. 2016;**6**:065316

Chapter 10

Liquid Crystals for Responsible Photonic Crystals

Sunnam Kim and Seiji Kurihara

Additional information is available at the end of the chapter

http://dx.doi.org/10.5772/intechopen.72654

Abstract

The photonic crystals (PCs) exhibit photonic band gap (PBG), inhibiting specific wavelength of light decided by structural periodicity. PCs are a new class of periodic dielectric media that can provide novel ways to manipulate and control light. Researchers have recently devoted extensive efforts to fabricating PCs with controlled symmetry, size, and defects on a large scale and tuning of PBG. Liquid crystalline (LC) materials exhibiting self-organization, phase transition, and molecular orientation behaviors in response to external stimuli are attracting significant attention for the bottom-up nanofabrication and tuning of advanced photonic materials and devices. Here, we will introduce self-organization of PCs from LCs and photoswitching mechanism of PBG based on phase transition and anisotropic orientation of LCs.

Keywords: self-organization, molecular orientation, liquid crystal, azobenzene, light response, photonic band gap

1. Introduction

The structural coloration is basically caused by interaction between light and periodic nanostructured-materials, and light absorption with dyes or pigments is not required for the structural coloration. One can find structural coloration of various materials in nature, and the structural colors are iridescent and very durable, for example, opal, peacock wings, and so on. Recently, there is an increasing interest in developing bioinspired nanostructures with controlled symmetry, size, and defects on a large scale for photonic crystals (PCs) [1–7]. Generally, the fabrication processes are divided into bottom-up and top-down methods. The top-down methods are mostly based on lithography-related techniques, which are expensive and requiring high technology [8–11]. The bottom-up methods are primarily about the self-assembly of component units such as emulsions, colloidal particles, and block copolymers [12–16].

INTECH
open science | open minds

© 2018 The Author(s). Licensee InTech. This chapter is distributed under the terms of the Creative Commons Attribution License (http://creativecommons.org/licenses/by/3.0), which permits unrestricted use, distribution, and reproduction in any medium, provided the original work is properly cited. (cc) BY

Liquid crystalline (LC) materials are attracting significant interest on the bottom-up nanofabrication of advanced photonic materials and devices with unique configuration and novel tuning mechanism. LCs undoubtedly represent such a marvelous dynamic system based on molecular reorientation behavior affecting macroscopic properties [17–19]. On the other hand, the responsiveness of LCs to various stimuli such as temperature, light, mechanical force, and electric and magnetic fields as well as chemical and electrochemical reactions can attribute to fascinating tunability of photonic band gap (PBG) in the periodic nanostructure, which provides numerous opportunities in all-optical integrated circuits and next-generation communication system. Here, several examples of self-organized LC nanostructures are introduced, and fabrication of three-dimensional (3D) PCs containing azobenzene LC molecules and photoswitching of PBG are discussed.

2. Self-organized PCs and PBG tuning based on LCs

2.1. Self-organized LC nanostructures

We introduce several examples for the fabrication methods of precisely constructing of PCs using self-organized LC nanostructures of cholesteric LC (ChLC), LC blue phase, and holographic polymer-dispersed LC.

2.1.1. Cholesteric LC

Dissolving a chiral compound in a nematic LC causes helix formation, resulting in the selective reflection of circular light corresponding to the helical pitch and helical sense. The pitch, the distance over which the director field rotates by 2π radians, is sensitive to external stimuli such as temperature, light, electric field, and mechanical stress [20–25]. Kurihara et al. have reported on the photochemical change in transparency as well as the selective reflection of a cholesteric LC (ChLC) resulting from the photoisomerization of a chiral azobenzene compound doped in a host nematic LC [26–31]. It is based on the photochemical change in helical twisting power (HTP) of the chiral azobenzene compounds under irradiation of UV and visible light. A combination of a rigid chiral diol core and azobenzene group led to larger photochemical changes in HTP, resulting in a reversible change of selective reflection over whole range of visible region with faster photochemical switching as shown in **Figure 1**. HTPs of the chiral azobenzene compounds increased or decreased by UV light, and the color change of reflection light was seen even with naked eyes.

On the other hand, Muševič et al. developed ChLC microdroplets, in which the arrangement of ChLC molecules was with parallel anchoring of the LC molecules at the surface. The helical structure of the LC originates from the center of the droplet and gives rise to concentric shells of constant refractive index. The introduction of chirality in ChLCs in the spherical shape resulted in the formation of a multilayered spherical Bragg resonator with an alternation periodic refractive index [32]. This dielectric structure is optically equivalent to the well-known Bragg-onion optical microcavity. In addition, Li et al. reported photoresponsive ChLC droplets with large helical twisting power [33].

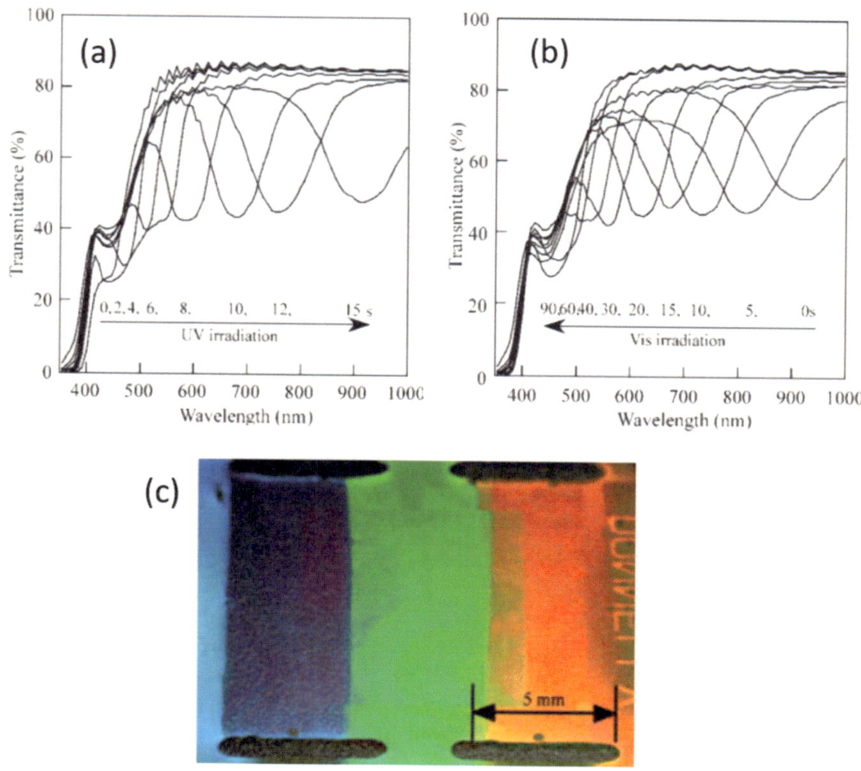

Figure 1. Changes in the selective reflection of the ChLC after UV irradiation (a) and visible light (b) irradiation at 25°C. And the photograph (c) of color change of the ChLC by varying UV irradiation time 0 s (left), 4 s (middle), and 10 s (right). [31] Copyright 2005, Wiley-VCH.

2.1.2. LC blue phase

Blue phases (BPs) are among the most fascinating photonic nanostructures in the area of LCs [34]. They are known to self-organization into the 3D frustrated nanostructures, which originate from the competition between the packing topology and chiral forces. The periodic nanostructures are self-organized from the so-called double-twisted cylinders that are usually stabilized by the formation of defects or disclinations. As the 3D PCs, BPs feature remarkable bandgap tunability over an extensive spectrum in response to various external stimuli such as temperature [35, 36], mechanical strength [37], electric field [38], and light [39–41]. The spectral band gap position of selective reflection can be readily predetermined by varying the chiral fraction of BP mixtures. However, BPs have received increasing attentions especially in recent years, but the reflection wavelength tuning is usually quite narrow. The ability to dynamically tune the photonic bandgap in cubic BPs across a wide wavelength range is highly desirable. Effort to improve the instability and irreversibility of BPs has been made with polymer-stabilized structure [18] and incorporation of unusual bent-core mesogen [42].

2.1.3. Holographic polymer-dispersed LC

Holographic polymer-dispersed liquid crystal (H-PDLC) materials belong to a phase separation material system where the LCs can form droplets, of controllable sizes, that are phase-separated from the polymer-rich regions during the photopolymerization process. Interestingly, it has enabled the in situ fast one-step fabrication of self-organized PCs. The control of the bandwidth of the reflection resonance, related to the light intensity and spatial porosity distributions, was investigated to optimize the optical performance.

For example, Cartwright et al. reported a graded reflection grating based on the H-PDLC material system [43, 44]. An obvious rainbow-colored reflection was observed from the same viewing angle by a modification to this standard configuration where the triangular prism is replaced with a cylindrical lens. Modified optical setup is based on a cylindrical lens to generate a graded holographic interference pattern along the x-axis from the incident and reflected beams. It is expected to compact graded PBG structures for linear variable optical filters.

The marriage of holography and self-organized LC materials is highly expected to produce a diverse range of novel 1D, 2D, and 3D nanostructures for various applications in electro-optical filters, wavelength division multiplexer, free-space optical switches, information display devices, and so on [45–49].

2.2. PBG tuning of 3D photonic crystals

Three-dimensional PCs such as opal and inverse opal structure can be easily prepared by colloidal self-assembly approach. The fabrication of inverse opal structure infiltrated by azobenzene LC molecules for 3D PCs and the mechanism of photoswitching of PBG are described.

For 3D PCs, the reflection wavelength (λ) under normal incident condition is expected by equation of Bragg diffraction as follows [50]

$$\lambda = 2\sqrt{\frac{2}{3}}\, d \sqrt{{n_{sphere}}^2 f + {n_{void}}^2 (1 - f)} \tag{1}$$

where d is the diameter of the sphere, n_{sphere} and n_{void} are the refractive indices of the sphere and the medium in the voids of opal, respectively, and f is the volume fraction of the sphere. In this way, the peak wavelength can be roughly tuned by two approaches: either by control of the diameter of the sphere or by control of the refractive index. If one can control photochemically phase transition or birefringence of LCs in 3D PCs, they will be candidates for applications to optical devices such as display and memory.

On the other hand, light among various promising stimuli shows distinct and significant advantages due to its spatial, remote, and temporal controllability. It is also noteworthy that the light sources with different wavelength, intensity, and polarization are readily accessible. Photocontrol of chemical and physical functions of various molecular systems has been studied vigorously using photochemical configurational change of azobenzene derivatives [51–54]. The photochromism of azobenzene and its derivatives due to cis-trans isomerization has been widely investigated. For PC containing azobenzene molecules, ‘photochemical phase transition' between the alignment phase and isotropic phase, or 'photoinduced optical anisotropy' based on photoselective reorientation behavior is very useful parameter for self-organization and optical photonic crystal switching. Here, fabrication of three-dimensional (3D) PC containing azobenzene molecules and photoswitching of PBG are discussed.

2.2.1. PBG tuning by phase transition change

Three-dimensional PCs such as opal and inverse opal structure can be infiltrated with stimuli-responsive materials, thus enabling PBG tuning. Kubo et al. reported on the drastic change in optical properties for the inverse opal structure containing low molar mass LCs, by taking advantage of the nematic-isotropic phase transition [50]. It is based on change in the light scattering intensity between LC phase and isotropic phase.

SiO_2 inverse opal template could be fabricated by infiltration of SiO_2 spheres into the voids of polystyrene (PS) opal film and calcination of PS as shown in **Figure 2**. When SiO_2 inverse opal film infiltrated with 4-cyano-4′-pentylbiphenyl (5CB), the reflection spectra of the LC-infiltrated inverse opal film were changed with increase in the temperature as shown in **Figure 3**. In the initial state at 25°C, 5CB is in the nematic phase and the film is white due to light scattering. The reflection has two weak peaks. With the gradual temperature rise, the positions of the peaks shifted slowly and a distinct peak at around 610 nm, which is derived

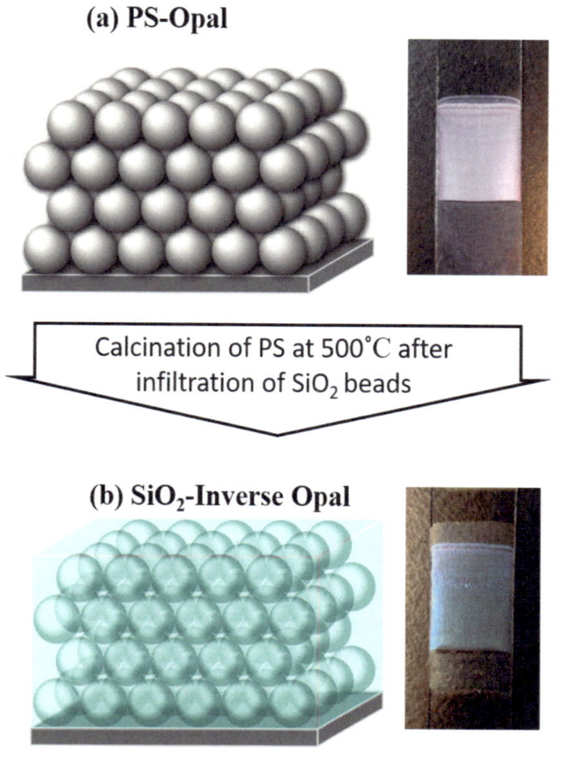

Figure 2. Schematic of PS opal (a) and SiO_2 inverse opal (b) structures and their photographs.

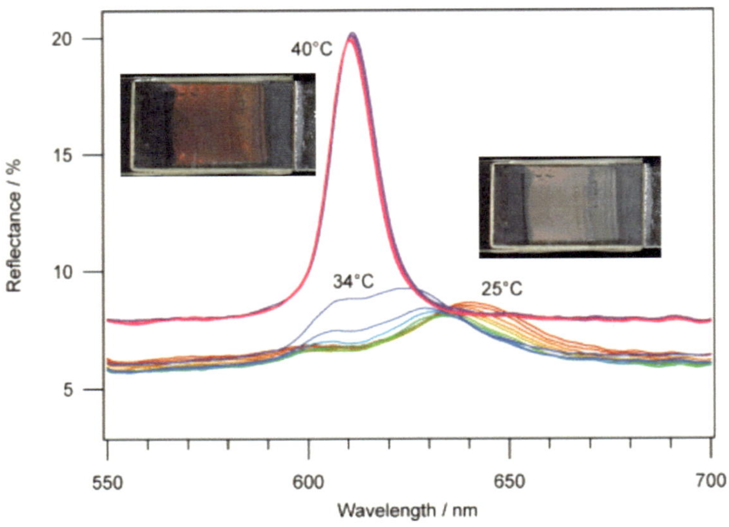

Figure 3. Reflection spectra of SiO_2 inverse opal film infiltrated with 5CB at the temperature from 25°C to 40°C [50]. Reproduced. Copyright 2004, American Chemical Society.

from PBG, appeared rapidly at the phase transition temperature into the isotropic phase at 40°C. The color of film becomes red. This result shows that switching of the optical properties can be realized due to a phase transition of the LCs, leading to thermos-tunable PCs.

Kubo et al. also reported the photoswitching of reflection for SiO_2 inverse opal film infiltrated with photo-response azobenzene containing LC mixture [55]. By UV light, a reflection peak increased rapidly and then saturated, indicating appearance of PBG. The trigger for this drastic change was the trans-cis photoisomerization of the azobenzene derivatives in the films. The excitation of the π-π^* transition of azobenzene molecules at around 360 nm results in a transformation from trans to cis isomers. The photoinduced cis form of azobenzene group has a bent shape and hence tends to disorganize the phase structure of nematic LCs.

The observed change could be reversed to the original state, decrease of reflectance, by irradiating with visible light. The recovery was induced by cis-trans photoisomerization, followed by a phase transition from the isotropic to the nematic phase under visible light (436 nm). Such a switching effect could also be observed in transmission spectra. It is supposed due to the fact that the LC molecules in the nematic phase in the spheroidal voids are aligned parallel to the void surfaces, but the orientation is completely random among spheres. It is similar to the phenomena observed for polymer-dispersed liquid crystal (PDLC). On the other hand, when the LC was transformed into an isotropic phase by UV light, the anisotropy in the dielectric constant disappeared. That is, the dielectric constant of the LC in all of the spheroidal voids became the same, and hence, a reflection peak due to Bragg diffraction appears [56–58]. On the basis of this scenario, dynamic change in the optical properties was observed repeatedly by alternating irradiation with UV and visible light. It was completely reversible, exhibiting good stability and reproducibility during the light irradiation cycle as shown in **Figure 4**.

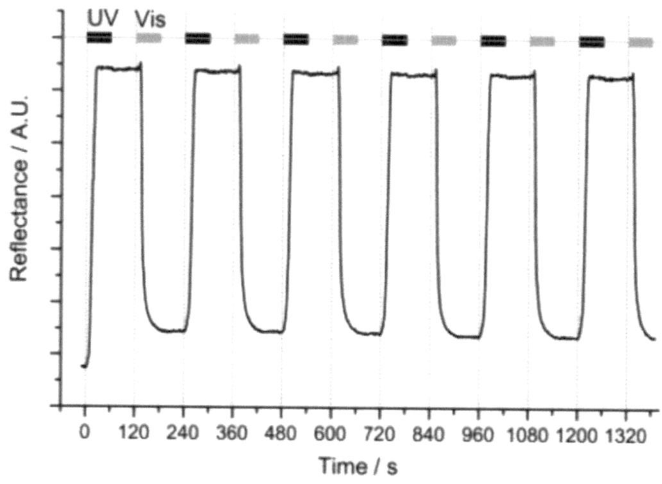

Figure 4. Time-dependent change in the reflectance of 5CB/azobenzene LC mixture-infiltrated inverse opal film at 603 nm by UV and visible light irradiation. The black bars show the periods for the irradiation of UV light (<400 nm), and the gray bars show the periods for irradiation of visible light (436 nm). For other periods, the sample was kept in the dark [58]. Copyright 2002, American Chemical Society.

2.2.2. *PBG tuning by molecular reorientation*

For low-molar-mass LC-infiltrated inverse opal film as mentioned in the previous section, the stability of the switched state was not enough although the photoswitching was reversible. It is due to the thermal back reaction of the azobenzene compound and fluidity of the low-molar-mass LC. In contrast with low-molar-mass LCs, polymeric LCs are well known to show excellent storage stability. Kurihara et al. reported the stable and reversible shift of the Bragg diffraction band for SiO_2 inverse opal structure infiltrated with 4,4′-methoxy hexyloxy azobenzene polymethacrylate, PAz (schematic molecular structure is shown in **Figure 5**), and reflection change were investigated when irradiation of linearly polarized light (LPL) and circularly polarize light (CPL) of Ar + later ($\lambda = 488$ nm) [59].

Photoisomerization behavior is demonstrated by absorption band change as shown in **Figure 6**. By UV light, trans-cis isomerization is induced, the absorption peak at 360 nm decreases, and the peak at 450 nm increases (**Figure 5(a)** and **(c)**). By irradiation of visible light, cis-trans photoisomerization is induced, and reverse changes in absorption spectra are induced (**Figure 6(b)** and **(d)**).

Anisotropic state of azobenzene compounds based on their molecular orientation can be controlled via reversible isomerization behavior by light and thermal stimuli [60–62]. Namely, alignment state of trans-form is perpendicular to the direction of polarized light (known as Weigert effect [20, 63–64]), and the alignment is destroyed by formation of cis-form.

The orientation state is evaluated by polarized absorption spectra and by birefringence, Δn, measurement. From angle dependence of absorbance at the peak of 360 nm to the polarized axis, orientation direction perpendicular to the polarized direction of induced LPL was demonstrated. When Δn is estimated by change in transmitted light of the azobenzene film

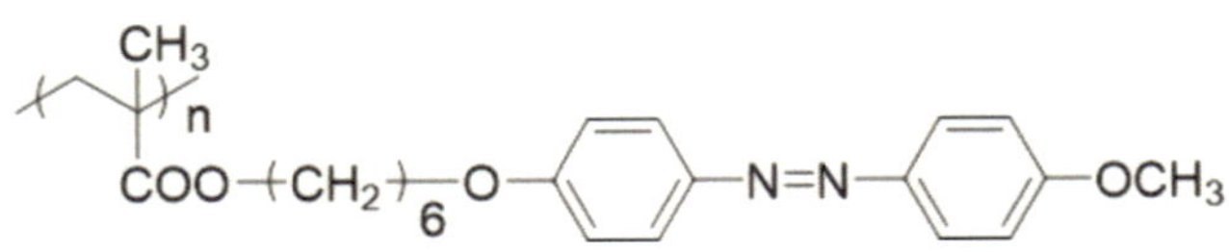

Figure 5. Schematic molecular structure of PAz.

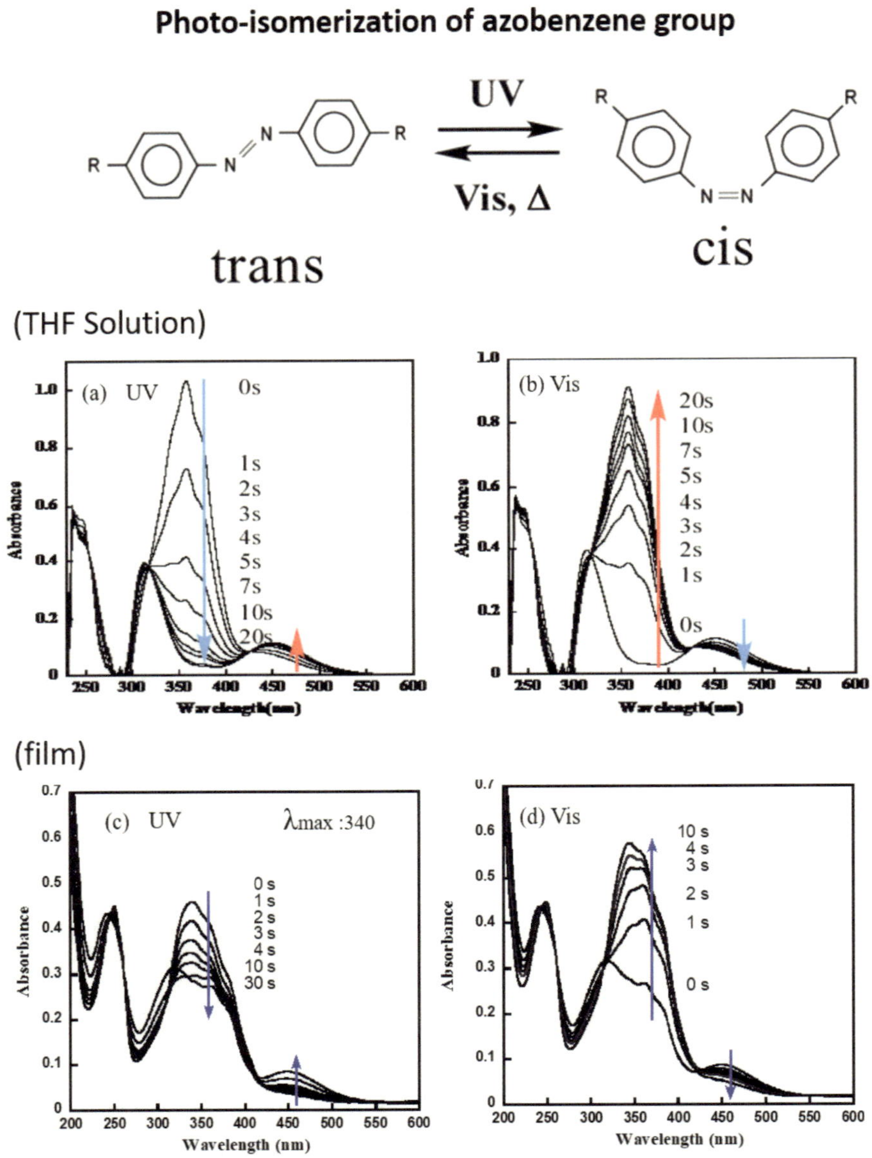

Figure 6. Changes in absorption spectra of PAz in THF after UV light (365 nm, 8 mW/cm^2) (a) and visible light (436 nm, 60 mW/cm^2) (b) irradiation. Changes in absorption spectra of PAz film irradiated UV light (365 nm, 10 mW/cm^2) (c) and following visible light (436 nm, 58 mW/cm^2) (d) at room temperature.

placed between two crossed polarizers, Δn of PAz was estimated to be 0.128, as shown in **Figure 7** [65].

Azobenzene polymer can be infiltrated into the SiO_2 inverse opal film by heating to melt. The pore diameter and volume fraction of SiO_2 inverse opal, d and f, were estimated by comparing wavelength of the reflection peaks of the SiO_2 inverse opal films infiltrated with various solvents with different refractive indices. For SiO_2 inverse opal film, d = 278 nm and f = 0.118, reflection spectra before and after injection of PAz are shown in **Figure 8**. The injection of PAz caused a shift of the reflection peak from 480 nm to 710 nm, because of an increase in the refractive index from 1.0 of air to 1.6 of PAz.

By LPL irradiation of light, reflection measured with polarizer parallel and perpendicular to the polarization direction of LPL is plotted in **Figure 9**. The reflection peaks depend on the irradiation time and polarization direction. By irradiation of a linearly polarized light at 488 nm (Ar + laser), azobenzene groups were aligned in the direction perpendicular to the polarization direction of the laser light.

After irradiation of LPL, reflection band shifted to longer wavelength with little shoulder at shorter wavelength. By observation with parallel and perpendicular polarizers, different two bands were respectively observed. The maximum wavelengths of the reflection bands at shorter and longer wavelength regions were 706 nm and 725 nm, respectively. Therefore, the refractive indices of PAz in the pores after the LPL irradiation, ordinary refractive index (n_o) and extraordinary refractive index (n_e), were estimated to be, respectively, 1.57 and 1.62 by using Bragg diffraction (Eq. (1)) with the parameters d (278 nm), f (0.118), and $nSiO_2$ (1.45). Namely, the photoinduced birefringence of PAz was 0.05 in the pores of SiO_2 inverse opal. It is less optical anisotropy comparing to solid film of PAz (0.13), indicating insufficient alignment

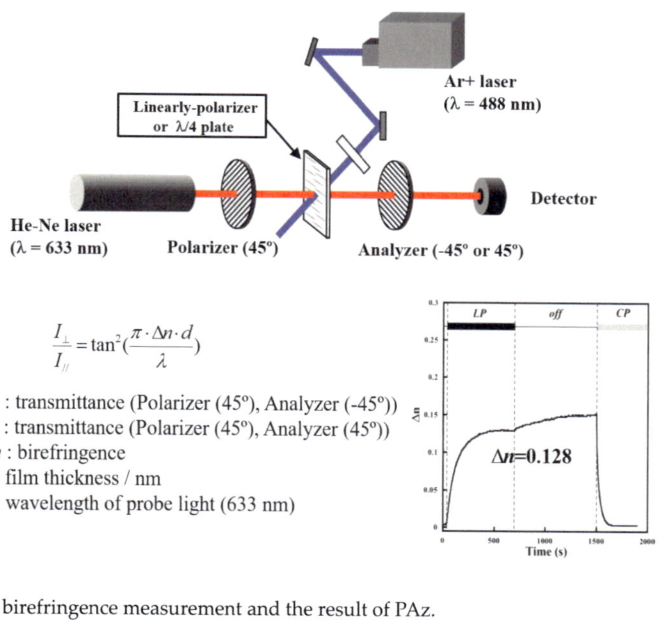

Figure 7. Schematic for birefringence measurement and the result of PAz.

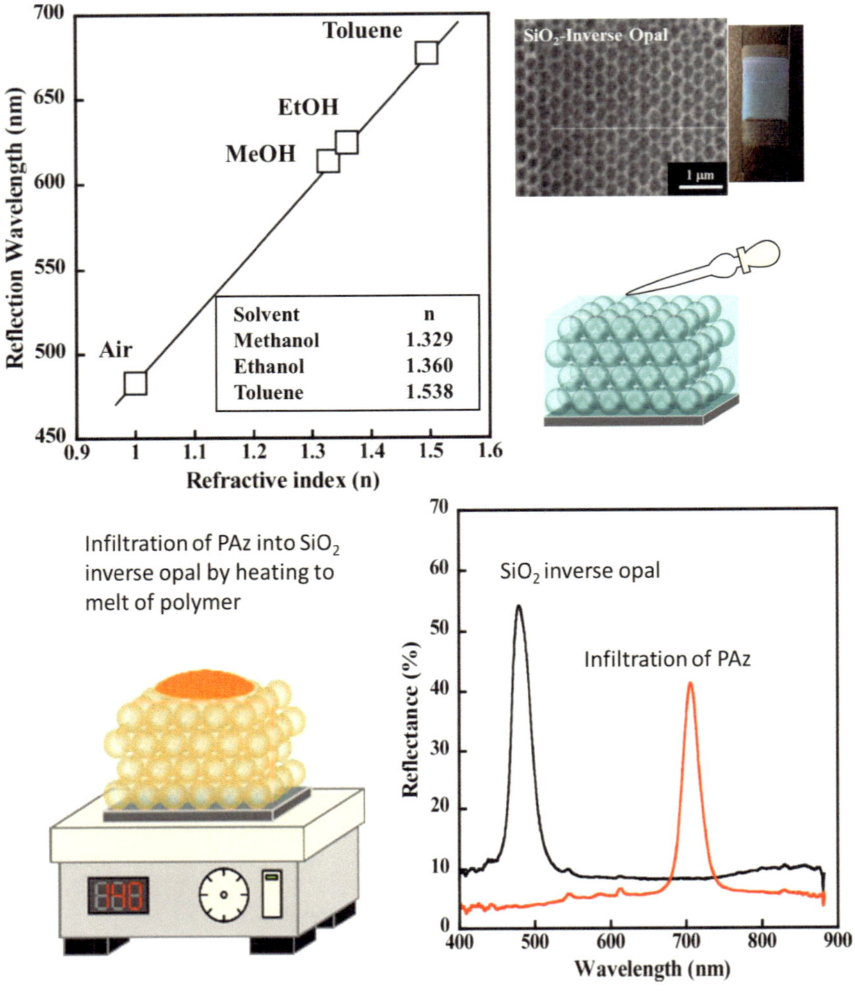

Figure 8. Plot of reflection wavelength change of SiO_2 inverse opal structure depending on solvents with various refractive indices (a) and reflection spectra before and after infiltration of PAz (b).

due to huge surface energy in the pores of SiO_2. In addition, reflection intensity decreased and increased by small and larger refractive index differences between PAz (1.57 and 1.62) and SiO_2 (1.45).

3. Conclusion

Optical properties of LCs represent the responsibility of external stimuli such as temperature, light, mechanical force, and electric and magnetic fields as well as chemical and electrochemical reactions, which are a driving force leading to dynamic change of materials from molecular level to macroscopic level. Dynamic system of LCs makes it a quite advantageous

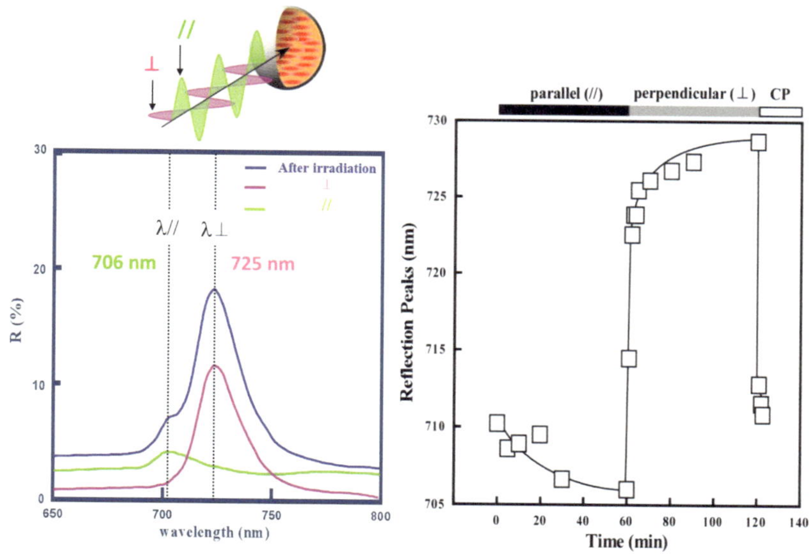

Figure 9. Reflection spectra of Paz-infiltrated SiO_2 film with parallel and perpendicular LPL irradiation of Ar + laser and plot of the wavelength of reflection peak [59].

approach toward bottom-up nanofabrication of photonic materials with novel tuning mechanism. In addition, the responsiveness to various stimuli can attribute to fascinating tunability of PBG, which provides numerous opportunities in all-optical integrated circuits and next-generation communication system.

Author details

Sunnam Kim* and Seiji Kurihara

*Address all correspondence to: sn-kim@kumamoto-u.ac.jp

Department of Applied Chemistry and Biochemistry, Kumamoto University, Kumamoto, Japan

References

[1] Zan G, Wu Q. Biomimetic and bioinspired synthesis of nanomaterials/nanostructures. Advanced Materials (Deerfield Beach, Fla.). 2016;**28**:2099-2147

[2] Pokroy B, Epstein AK, Persson-Gulda MCM, Aizenberg J. Fabrication of bioinspired actuated nanostructures with arbitrary geometry and stiffness. Advanced Materials. 2009;**21**:463-469

[3] Pris AD, Utturkar Y, Surman C, Morris WG, Vert A, Zalyubovskiy S, et al. Towards high-speed imaging of infrared photons with bio-inspired nanoarchitectures. Nature Photonics. 2012;**6**:195-200. DOI: 10.1038/nphoton.2011.355

[4] Bai L, Xie Z, Wang W, Yuan C, Zhao Y, Mu Z, et al. Bio-inspired vapor-responsive colloidal photonic crystal patterns by inkjet printing. ACS Nano. 2014;**8**:11094-11100

[5] Ye B, Rong F, Gu H, Xie Z, Cheng Y, Zhao Y, et al. Bioinspired angle-independent photonic crystal colorimetric sensing. Chemical Communications. 2013;**49**:5331-5333

[6] Hou J, Zhang H, Yang Q, Li M, Song Y, Jiang L. Bio-inspired photonic-crystal microchip for fluorescent ultratrace detection. Angewandte Chemie International Edition. 2014;**53**:5791-5795

[7] Biró LP, Vigneron JP. Photonic nanoarchitectures in butterflies and beetles: Valuable sources for bioinspiration. Laser & Photonics Reviews. 2011;**5**:27-51

[8] Campbell M, Sharp DN, Harrison MT, Denning RG, Turberfield AJ. Fabrication of photonic crystals for the visible spectrum by holographic lithography. Nature. 2000;**404**:53-56

[9] Blanco A, Chomski E, Grabtchak S, Ibisate M, John S, Leonard SW, et al. Large-scale synthesis of a silicon photonic crystal with a complete three-dimensional bandgap near 1.5 micrometres. Nature. 2000;**405**:437-440

[10] Deubel M, Freymann G von, Wegener M, Pereira S, Busch K, Soukoulis CM. Direct laser writing of three-dimensional photonic-crystal templates for telecommunications. Nature Materials 2004;**3**:444-447

[11] Qi M, Lidorikis E, Rakich PT, Johnson SG, Joannopoulos JD, Ippen EP, et al. A three-dimensional optical photonic crystal with designed point defects. Nature. 2004;**429**:538-542

[12] Hynninen A-P, Thijssen JHJ, Vermolen ECM, Dijkstra M, van Blaaderen A Self-assembly route for photonic crystals with a bandgap in the visible region. Nature Materials 2007; **6**:202-205

[13] Jenekhe S, Chen X. Self-assembly of ordered microporous materials from rod-coil block copolymers. Science. 1999;**283**:372-375

[14] Vlasov YA, Bo X-Z, Sturm JC, Norris DJ. On-chip natural assembly of silicon photonic bandgap crystals. Nature. 2001;**414**:289-293

[15] Winfree E, Liu F, Wenzler LA, Seeman NC. Design and self-assembly of two-dimensional DNA crystals. Nature. 1998;**394**:539-544

[16] Joannopoulos JD. Photonics: Self-assembly lights up. Nature. 2001;**414**:257-258

[17] Gibbons WM, Shannon PJ, Sun S-T, Swetlin BJ. Surface-mediated alignment of nematic liquid crystals with polarized laser light. Nature. 1991;**351**:49-50

[18] Kikuchi H, Yokota M, Hisakado Y, Yang H, Kajiyama T. Polymer-stabilized liquid crystal blue phases. Nature Materials. 2002;**1**:64-68

[19] Shannon PJ, Gibbons WM, Sun ST. Patterned optical properties in photopolymerized surface-aligned liquid-crystal films. Nature. 1994;**368**:532-533

[20] Ikeda T, Sasaki T, Ichimura K. Photochemical switching of polarization in ferroelectric liquid-crystal films. Nature. 1993;**361**:428-430

[21] Kurihara S, Nomiyama S, Nonaka T. Photochemical control of the macrostructure of cholesteric liquid crystals by means of photoisomerization of chiral azobenzene molecules. Chemistry of Materials. 2001;**13**:1992-1997

[22] Huang Y, Zhou Y, Doyle C, Wu S-T. Tuning the photonic band gap in cholesteric liquid crystals by temperature-dependent dopant solubility. Optics Express. 2006;**14**:1236-1242

[23] Moreira M et al. Cholesteric liquid-crystal laser as an optic fiber-based temperature sensor. Applied Physics Letters. 2004;**85**:2691-2693

[24] Baessler H, Laronge TM, Labes MM. Electric field effects on the optical rotatory power of a compensated cholesteric liquid crystal. The Journal of Chemical Physics. 1969;**51**:3213-3219

[25] Finkelmann H, Kim ST, Munoz A, Palffy-Muhoray P, Taheri B. Tunable mirrorless lasing in cholesteric liquid crystalline elastomers. Advanced Materials. 2001;**13**:1069-1072

[26] Kurihara S et al. Photo-tuning of lasing from a dye-doped cholesteric liquid crystals by photoisomerization of a sugar derivative having plural azobenzene groups. Applied Physics Letters. 2006;**88**:103121

[27] Hatae Y, Moritsugu M, Ogata T, Kurihara S, Nonaka T. Photochemical modulation of the wavelength of lasing from a dye-doped cholesteric liquid crystal. Molecular Crystals and Liquid Crystals. 2005;**443**:87-94

[28] Yoshioka T, Alam MZ, Ogata T, Nonaka T, Kurihara S. Photochemical tuning of the helical structure of cholesteric liquid crystals by photoisomerization of chiral azobenzenes, and their structural effects. Liquid Crystals. 2004;**31**:1285-1291

[29] Yoshioka T, Ogata T, Zahangir AM, Nonaka T. Discontinuous change in the helical pitch of cholesteric liquid crystals by photoisomerization of a chiral azobenzene molecule. Liquid Crystals. 2004;**31**:15-19

[30] Kurihara S, Yoshioka T, Zahangir A, Ogata T, Nonaka T. Preparation of composites of polymer liquid crystal/cholesteric liquid crystal and their photochemical switching and memory properties by photoisomerization of a chiral azobenzene molecule. Journal of Applied Polymer Science. 2004;**92**:2577-2580

[31] Yoshioka T et al. Reversible-photon-mode full-color display by means of photochemical modulation of a helically cholesteric structure. Advanced Materials. 2005;**17**:1226-1229

[32] Humar M, Muševič I. 3D microlasers from self-assembled cholesteric liquid-crystal microdroplets. Optics Express. 2010;**18**:26995-27003

[33] Chen L et al. Photoresponsive monodisperse cholesteric liquid crystalline microshells for tunable omnidirectional lasing enabled by a visible light-driven chiral molecular switch. Advanced Optical Materials. 2014;**2**:845-848

[34] Cao W, Muñoz A, Palffy-Muhoray P, Taheri B. Lasing in a three-dimensional photonic crystal of the liquid crystal blue phase II. Nature Materials. 2002;**1**:111-113

[35] Hur S-T et al. Liquid-crystalline blue phase laser with widely tunable wavelength. Advanced Materials. 2013;**25**:3002-3006

[36] Coles H, Pivnenko M. Liquid crystal 'blue phases' with a wide temperature range. Nature. 2005;**436**:997-1000

[37] Castles F et al. Stretchable liquid crystal blue phase gels. Nature Materials. 2014;**13**:817-821

[38] Lu S, Chien L. Electrically switched color with polymer-stabilized blue-phase liquid crystals. Optics Letters. 2010;**35**:562-564

[39] Lin T-H et al. Red, green and blue reflections enabled in an optically tunable self-organized 3D cubic nanostructured thin film. Advanced Materials. 2013;**25**:5050-5054

[40] Wang L et al. NIR light-directing self-organized 3D photonic superstructures loaded with anisotropic plasmonic hybrid nanorods. Chemical Communications. 2015;**51**:15039-15042

[41] Chen X et al. Light-controllable reflection wavelength of blue phase liquid crystals doped with azobenzene-dimers. Chemical Communications. 2013;**49**:10097-10099

[42] Lee M et al. Liquid crystalline blue phase I observed for a bent-core molecule and its electro-optical performance. Journal of Materials Chemistry. 2010;**20**:5813-5816

[43] Liu K, Xu H, Hu H, Gan Q, Cartwright AN. One-step fabrication of graded rainbow-colored holographic photopolymer reflection gratings. Advanced Materials. 2012;**24**:1604-1609

[44] Moein T et al. Holographic photopolymer linear variable filter with enhanced blue reflection. ACS Applied Materials & Interfaces. 2014;**6**:3081-3087

[45] Coles H, Morris S. Liquid-crystal lasers. Nature Photonics. 2010;**4**:676-685

[46] Jakubiak R, Bunning TJ, Vaia RA, Natarajan LV, Tondiglia VP. Electrically switchable, one-dimensional polymeric resonators from holographic photopolymerization: A new approach for active photonic bandgap materials. Advanced Materials. 2003; **15**:241-244

[47] Escuti MJ, Qi J, Crawford GP. Tunable face-centered-cubic photonic crystal formed in holographic polymer dispersed liquid crystals. Optics Letters. 2003;**28**:522-524

[48] He GS et al. Tunable two-photon pumped lasing using a holographic polymer-dispersed liquid-crystal grating as a distributed feedback element. Applied Physics Letters. 2003;**83**:2733-2735

[49] Wu S, Fuh A. Lasing in photonic crystals based on dye-doped holographic polymer-dispersed liquid crystal reflection gratings. Japanese Journal of Applied Physics. 2005;**44**:977

[50] Kubo S et al. Tunable photonic band gap crystals based on a liquid crystal-infiltrated inverse opal structure. Journal of the American Chemical Society. 2004;**126**:8314-8319

[51] Shinkai S, Nakaji T, Nishida Y, Ogawa T, Manabe O. Photoresponsive crown ethers. 1. Cis-trans isomerism of azobenzene as a tool to enforce conformational changes of crown ethers and polymers. Journal of the American Chemical Society. 1980;**102**:5860-5865

[52] Ogawa M. Preparation of a cationic azobenzene derivative-montmorillonite intercalation compound and the photochemical behavior. Chemistry of Materials. 1996;**8**:1347-1349

[53] Seki T et al. Photochemical alignment regulation of a nematic liquid crystal by Langmuir-Blodgett layers of azobenzene polymers as "command surfaces". Macromolecules. 1989;**22**:3505-3506

[54] Ikeda T, Tsutsumi O. Optical switching and image storage by means of azobenzene liquid-crystal films. Science. 1995;**268**:1873-1875

[55] Kubo S et al. Control of the optical properties of liquid crystal-infiltrated inverse opal structures using photo irradiation and/or an electric field. Chemistry of Materials. 2005;**17**:2298-2309

[56] Tarhan İİ, Watson GH. Photonic band structure of fcc colloidal crystals. Physical Review Letters. 1996;**76**:315-318

[57] Park SH, Xia Y. Assembly of mesoscale particles over large areas and its application in fabricating tunable optical filters. Langmuir. 1999;**15**:266-273

[58] Kubo S, Gu Z, Takahashi K, Ohko Y, Sato O, Fujishima A. Control of the optical band structure of liquid crystal infiltrated inverse opal by a photoinduced nematic–isotropic phase transition. Journal of the American Chemical Society. 2002;**124**:10950-10951

[59] Kurihara S et al. Photoswitching properties of photonic band gap material containing azo-polymer liquid crystal. European Polymer Journal. 2007;**43**:4951-4960

[60] Natansohn A, Rochon P. Photoinduced motions in azo-containing polymers. Chemical Reviews. 2002;**102**:4139-4176

[61] Alam Z, Ohmachi T, Ogata T, Nonaka T. Kurihara S, Photoisomerization behavior and photoinduced surface relief gratings on azopolymer film by a monochromatic light irradiation. Optical Materials. 2006;**29**:365-370

[62] Kim S, Ogata T, Kurihara S. Azobenzene-containing polymers for photonic crystal materials. Polymer Journal. 2017;**49**:407-412

[63] Shishido A. Rewritable holograms based on azobenzene-containing liquid-crystalline polymers. Polymer Journal. 2010;**42**:525-533

[64] Ichimura K, Suzuki Y, Seki T, Hosoki A, Aoki K. Reversible change in alignment mode of nematic liquid crystals regulated photochemically by command surfaces modified with an azobenzene monolayer. Langmuir. 1988;**4**:1214-1216

[65] Ishiguro M, Sato D, Shishido A, Ikeda T. Bragg-type polarization gratings formed in thick polymer films containing azobenzene and tolane moieties. Langmuir. 2007;**23**:332-338